T0171789

Biocontrol
of
Plant Diseases

Volume I

Editors

K. G. Mukerji, Ph.D.
Professor, Department of Botany
University of Delhi
Delhi, India

K. L. Garg
Mycology Laboratory
Department of Botany
University of Delhi
Delhi, India

CRC Press
Taylor & Francis Group
Boca Raton London New York

CRC Press is an imprint of the
Taylor & Francis Group, an **informa** business

CRC Press
Taylor & Francis Group
6000 Broken Sound Parkway NW, Suite 300
Boca Raton, FL 33487-2742

Reissued 2019 by CRC Press

© 1988 by Taylor & Francis Group, LLC
CRC Press is an imprint of Taylor & Francis Group, an Informa business

No claim to original U.S. Government works

This book contains information obtained from authentic and highly regarded sources. Reasonable efforts have been made to publish reliable data and information, but the author and publisher cannot assume responsibility for the validity of all materials or the consequences of their use. The authors and publishers have attempted to trace the copyright holders of all material reproduced in this publication and apologize to copyright holders if permission to publish in this form has not been obtained. If any copyright material has not been acknowledged please write and let us know so we may rectify in any future reprint.

Except as permitted under U.S. Copyright Law, no part of this book may be reprinted, reproduced, transmitted, or utilized in any form by any electronic, mechanical, or other means, now known or hereafter invented, including photocopying, microfilming, and recording, or in any information storage or retrieval system, without written permission from the publishers.

For permission to photocopy or use material electronically from this work, please access www. copyright.com (http://www.copyright.com/) or contact the Copyright Clearance Center, Inc. (CCC), 222 Rosewood Drive, Danvers, MA 01923, 978-750-8400. CCC is a not-for-profit organization that provides licenses and registration for a variety of users. For organizations that have been granted a photocopy license by the CCC, a separate system of payment has been arranged.

Trademark Notice: Product or corporate names may be trademarks or registered trademarks, and are used only for identification and explanation without intent to infringe.

A Library of Congress record exists under LC control number:

Publisher's Note
The publisher has gone to great lengths to ensure the quality of this reprint but points out that some imperfections in the original copies may be apparent.

Disclaimer
The publisher has made every effort to trace copyright holders and welcomes correspondence from those they have been unable to contact.

ISBN 13: 978-0-367-26273-0 (hbk)
ISBN 13: 978-0-367-26278-5 (pbk)
ISBN 13: 978-0-429-29233-0 (ebk)

Visit the Taylor & Francis Web site at http://www.taylorandfrancis.com and the
CRC Press Web site at http://www.crcpress.com

PREFACE

In the last few years research interest in biological control of plant diseases has become extremely active. One could easily notice this trend by following the number of publications in several reputable journals in plant pathology, mycology, and microbial ecology, where at least three to four hundred papers appear on this subject every year. One of the main reasons for this burst of research activity is the hazardous impact of various fungicides (pesticides) and other agrochemicals on the ecosystem. Therefore, it has become essential to do more work on the biological control of diseases and to avoid the use of fungicides and other chemicals, considering the ecological damage which may result. We were convinced that there was a need for a book which could summarize the existing work on the biocontrol of plant diseases. It is our good fortune that many of the leading researchers, who are actively studying different aspects of this problem, became enthusiastic and agreed to contribute the reviews which form this book.

Under natural and normal conditions the microbes on the plant surface are in a state of dynamic equilibrium. Interactions between pathogens and nonpathogens, and also amongst themselves, are constantly in process. When a pathogen is established on its host it is evident that it is more or less safe from the influence of a multitude of other organisms. When the spores of a pathogen fall on a fresh host surface they are fully exposed to the influence of many other microbes which are natural inhabitants and already present there. They can live alongside others without causing any noticeable effect on their behavior or they may be suppressed. Plants, in a natural (wild) state, possess a relatively stable biological balance of the microbes on their surface. An alien organism introduced into an area in which it has no natural enemies may increase in number to such an extent that the resident population is unable to redress the balance. This presents a problem which is best solved by introducing a second organism chosen with great care, and taking into account many necessary safeguards which will selectively attack the unwanted species. This is "biological control" (biocontrol). Almost any process, occurring naturally or done artificially, which affects the relationship between organisms in such a way that the natural biological balance is restored, can be regarded as biocontrol.

Pathogens are specific in their attack to a particular organ surface of the host. They may enter through the root from soil, or via air in the aerial parts of the plants, i.e., the leaf, vegetative shoot, bud, or flowers. During the last 15 to 20 years enormous amounts of work have been done on the biological control of root and soil-borne diseases and also on diseases of the aerial parts of plants. Different types of pathogens have been controlled by either antagonistic saprophytes or hyperparasites. This has also been done by artificially inoculating the biocontrol agent, or by increasing its number by amending the soil with nutrients, moisture content, pH, etc., and by crop rotation. Soil solarization is a new approach to soil disinfection and is based on the solar heating of the soil during the hot season, thereby increasing the temperature and killing the pathogens. Crop protection can also be done by inducing resistance in a host by a biocontrol agent. Mutation of the pathogens due to insertions, base pair substitutions, or deletions can be made on isolated genes, and these modified genes can be reintroduced to recipient strains to cause site-specific mutagenesis of these recipients. These strains will be very effective biological control agents.

Each chapter is devoted to a separate aspect of biocontrol of plant diseases, and in general the chapters are arranged in order of increasing technical complexity. Since these chapters have been written by independent authors there is the possibility of a slight overlap or repetition of certain statements, but this is difficult to avoid with this sort of organizational structure. One of the chapters deals with the biocontrol of weeds, which some may feel out of place, but it is important to have it as weeds cause enormous losses to agriculture.

It is our hope that this book will be useful to all students and researchers in plant pathology, microbial ecology, and mycology.

We are indebted to Dr. Nalini Nigam and Dr. Geeta Saxena for help in various ways.

31st July, 1986

K. G. Mukerji
K. L. Garg

EDITORS

K. G. Mukerji, M.Sc., Ph.D., is a Professor and presently the Head of the Department of Botany, University of Delhi, India.

Professor Mukerji did his Masters degree in Botany at Lucknow University, Lucknow, in 1955 and obtained his Ph.D. in Botany at the same University in 1962. He is internationally known for his work on fungal taxonomy and microbial ecology. He has worked, or visited, in most of the important laboratories of the world concerned with Mycology and Plant Pathology.

Professor Mukerji is a member of most of the Societies or Associations concerning Mycology, Plant Pathology, and Microbial Ecology. In some of these societies he has also held important offices.

Professor Mukerji has presented over 20 invited lectures at International Meetings, over 30 invited lectures at National Meetings, and approximately 150 guest lectures at Universities and Institutes. He has published more than 220 research papers. He has coauthored *Taxonomy of Indian Myxomycetes* and *Plant Diseases of India,* and edited *Progress in Microbial Ecology.* Presently he is editing an annual volume, *Frontiers in Applied Microbiology,* two volumes of which are already out. He is on the editorial board of several journals dealing with Mycology and Plant Pathology. His current major interests include microbial ecology, mycorrhizal technology, and biocontrol of plant diseases.

Professor Mukerji is a distinguished mycologist and microbial ecologist and he is well respected for his research contributions all over the world.

K. L. Garg, M.Sc., Ph.D., is presently a Junior Mycologist at the National Research Laboratory for Conservation of Cultural Property, Lucknow, U.P., India. After completing his Ph.D. in Botany from Lucknow University in 1982, he joined Professor K. G. Mukerji at Delhi University as a Post Doctoral Fellow. He has published more than 25 research papers.

CONTRIBUTORS, VOLUME I

Ashok Aggarwal, Ph.D.
Lecturer
Botany Department
Kurukshetra University
Kurukshetra, Haryana, India

K. R. Aneja, Ph.D.
Lecturer
Botany Department
Kurukshetra University
Kurukshetra, Haryana, India

Basil M. Arif, Ph.D.
Research Scientist
Department of Virology
Forest Pest Management Institute
Sault Ste. Marie, Ontario, Canada

Abdul Ghaffar, Ph.D.
Professor
Department of Botany
University of Karachi
Karachi, Pakistan

A. K. Gupta, Ph.D.
Senior Research Fellow
Botany Department
Kurukshetra University
Kurukshetra, Haryana, India

S. Hasan, Ph.D., D.Sc.
Principal Research Scientist
Biological Control Unit
CSIRO
Montpellier, France

Peter Jamieson, M.Sc.
Research Associate
Forest Pest Management Institute
Canadian Forestry Service
Sault Ste. Marie, Ontario, Canada

Guy R. Knudsen, Ph.D.
Assistant Professor
Division of Plant Pathology
Department of Plant, Soil and
  Entomological Sciences
University of Idaho
Moscow, Idaho

R. S. Mehrotra, Ph.D.
Professor and Chairman
Department of Botany
Kurukshetra University
Kurukshetra, Haryana, India

Peter C. Mercer, Ph.D.
Faculty of Agriculture
Queen's University of Belfast
Belfast, Ireland

Nalini Nigam, Ph.D.
Research Associate
Department of Botany
University of Delhi
Delhi, India

Bharat Rai, Ph.D.
Professor
Department of Botany
Banaras Hindu University
Varanasi, India

Geeta Saxena, Ph.D.
Research Associate
Department of Botany
University of Delhi
Delhi, India

H. W. Spurr
Biocontrol Research Unit
U.S. Department of Agriculture
Oxford, North Carolina

Leif Sundheim, Ph.D.
Associate Professor
Department of Plant Pathology
Norwegian Plant Protection Institute
Agricultural University of Norway
AAS-NLH, Norway

Arne Tronsmo
Associate Professor
Department of Microbiology
Agricultural University of Norway
AAS-NLH, Norway

R. S. Upadhyay, Ph.D.
Lecturer
Department of Botany
Banaras Hindu University
Varanasi, India

Homer D. Wells, Ph.D.
Research Plant Pathologist
U.S. Department of Agriculture
Department of Plant Pathology
University of Georgia
Tifton, Georgia

CONTRIBUTORS, VOLUME II

Raul G. Cuero, Ph.D.
Microbiologist
Agricultural Research Station
U.S. Department of Agriculture
Southern Regional Research Center
New Orleans, Louisana

Jane L. Faull, Ph.D.
Lecturer
Department of Botany
Birkbeck College
London, England

D. W. Fulbright, Ph.D.
Associate Professor
Department of Botany and Plant Pathology
Michigan State University
East Lansing, Michigan

C. M. E. Garrett, Ph.D.
Plant Bacteriologist
Department of Plant Pathology
Institute of Horticultural Research
East Malling, Maidstone
Kent, England

Sally Westveer Garrod, M.S.
Department of Botany and Plant Pathology
Michigan State University
East Lansing, Michigan

W. Janisiewicz, Ph.D.
Research Plant Pathologist
Agricultural Research Station
Appalachian Fruit Research Station
U.S. Department of Agriculture
Kearneysville, West Virginia

Charles M. Kenerley, Ph.D.
Assistant Professor
Department of Plant Pathology and
  Microbiology
Texas A&M University
College Station, Texas

Gurdev S. Khush, Ph.D.
Principal Plant Breeder
Plant Breeding Department
International Rice Research Institute
Manila, Phillipines

Eivind B. Lillehoj, Ph.D.
Supervisory Microbiologist
Southern Regional Research Center
U.S. Department of Agriculture
New Orleans, Louisana

Alison Murray
Biosciences and Biotechnology Department
University of Strathclyde
Glasgow, Scotland

Cynthia P. Paul, M.S.
Graduate Assistant
Department of Botany and Plant Pathology
Michigan State University
East Lansing, Michigan

R. E. Pettit, Ph.D.
Associate Professor
Department of Plant Pathology
Texas A&M University
College Station, Texas

James E. Rahe
Department of Biological Sciences
Center for Pest Management
Simon Fraser University
Burnaby, British Columbia, Canada

K. V. Sankaran, Ph.D.
Scientist
Division of Forest Pathology
Kerala Forest Research Institute
Peechi, Kerala, India

Rudy J. Scheffer, Ph.D.
Phytopathologist
Willie Commelin Scholten
Phytopathological Laboratory
Baarn, Netherlands

Jyoti K. Sharma, Ph.D.
Scientist-in-Charge
Division of Forest Pathology
Kerala Forest Research Institute
Peechi, Kerala, India

Jagjit Singh, Ph.D.
Research Officer
Department of Biology
Birkbeck College
London, England

John E. Smith, Ph.D.
Professor
Biosciences and Biotechnology Department
University of Strathclyde
Glasgow, Scotland

James P. Stack, Ph.D.
Assistant Professor
Department of Plant Pathology and
  Microbiology
Texas A&M University
College Station, Texas

Gary A. Strobel, Ph.D.
Professor
Department of Plant Pathology
Montana State University
Bozeman, Montana

John Tuite, Ph.D.
Professor
Department of Botany and Plant Pathology
Purdue University
West Lafayette, Indiana

S. S. Virmani, Ph.D.
Plant Breeding Department
International Rice Research Institute
Manila, Phillipines

David M. Wilson, Ph.D.
Professor
Department of Plant Pathology
Coastal Plain Station
University of Georgia
Tifton, Georgia

Marcus Zuber, Ph.D.
Professor Emeritus
Department of Agronomy
University of Missouri
Columbia, Missouri

TABLE OF CONTENTS, VOLUME I

TABLE OF CONTENTS, VOLUME II

Chapter 1

BIOLOGICAL CONTROL — CONCEPTS AND PRACTICES

Nalini Nigam and K. G. Mukerji

TABLE OF CONTENTS

I. INTRODUCTION

Exploratory and developmental research on biological control for crop diseases is being undertaken by both governmental agencies and private industrial research laboratories on an unprecedented scale. The reasons for a major change in outlook are several and varied. One significant factor has been a general widening of the concept of biological control. The concept of biological control today embodies not only introduction of antagonists into cropping systems, but also manipulation of the environment designed to favor resident beneficial organisms via crop rotation, residue management, and a wide range of other cultural practices.

Naturally occurring biological control is an effective factor leading to ecological balance between pathogens and their antagonists.[127] Microorganisms in the phylloplane can influence the incidence of leaf pathogens by direct antagonism or by induction of resistance in a host plant. Increased resistance can be activated by pre-infection of the host as well as by metabolites produced by saprophytic, nonpathogenic microorganisms.

Biological control is the direct or indirect manipulation by man of living natural control agents to increase their attack on pest species. The biological relationships between control agents and pest species are rather specific, thus the control method must be worked out for each pest. Emphasis on the research and practice of biological control has steadily increased in recent years and the development of an integrated management approach can sufficiently amplify the economic effectiveness of biological control in certain areas. Cook[32] reviewed the recent progress made towards biological control of plant pathogens, with special reference to the take-all disease of wheat. Webster[128] has discussed another aspect of biological control involving nematodes. Currently, with increasing awareness of possible deleterious effects of fungicides on the ecosystem, growing interest in pesticide-free agricultural products, and time consuming breeding programs, the biological control of plant pathogenic fungi have received considerable attention.[88] A major challenge in plant pathology is to introduce or develop new disease strategies, as the more traditional controls become obsolete, and to do so without greater use of chemicals. Biological control offers many advantages to the growers and to society in general and must be pursued on all fronts. The ultimate pay-off of a more stable, sustainable, and safer food supply often produced at lower cost is worth the effect.

Biological control of plant pathogens is the use of one or more biological processes to lower inoculum density of the pathogen or reduce its disease producing activities. Plant pathogens include fungi, several kinds of prokaryotic microorganisms (bacteria, actinomycetes, mycoplasma), nematodes, seed plants, viruses, and viroids. Baker and Cook[11] defined biological control as the "reduction of inoculum density or disease producing activities of a pathogen or parasite in its active or dormant state by one or more organisms accomplished naturally or through manipulation of the environment, host, or antagonists." Lupton[82] stated that accelerating or diverting evolutionary processes in order to obtain genotypes adapted to man's need are the most important examples of application of biological control to agricultural and horticultural crops.

Cook[32] grouped different facets of biological control of plant pathogens under three broad headings:

1. The reduction of the pathogen population through use of antagonistic microorganisms that destroy the pathogen inoculum and reduce the vigor or aggressiveness of the inoculum
2. The protection of the plant surface with microorganisms established in wounds, on leaves, or in the rhizosphere, where they serve as a biological barrier through their competitive, antibiotic, or parasitic action inhibitory to the pathogens
3. The establishment of nonpathogenic microorganisms or agents within the plant or infected area to stimulate greater resistance of the plant to the pathogen or to occupy the infected site and starve the pathogen or displace it in the lesion

The latter mechanisms may not save the plant or leaf already infected, but help to reduce inoculum for spread to plants not yet infected. Antagonists are microorganisms with potential to interfere with the growth or survival of plant pathogens and thereby contribute to biological control. Antagonists may be resident or introduced. Resident antagonists are natural inhabitants of soil, the rhizosphere, phylloplane, or other sites occupied by pathogens. Introduced antagonists are those cultured under special conditions and applied to sites where needed, e.g., the soil, seed, or sprayed on leaves or any other plant organ.

II. RESIDENT ANTAGONISTS AS BIOLOGICAL AGENTS

Generally, biological control of plant pathogens is achieved with resident antagonists. They are either managed or fostered by cultural practices. Crop rotation controls soil borne and residue inhabiting pathogens. These pathogens present in soil or plant residue vanish (do not exist) in the absence of their host. The rate of biological destruction of some active pathogen's inocula can also be accelerated through tillage. In this process the infected crop residue is fragmented and buried in soil, thereby increasing the vulnerability of the pathogen to biological stresses in soil. Another method is to flood the soil for a few weeks, or 1 to 2 months in warmer climates, as is accomplished by rotating the field of paddy rice.

Biological control is adjusting cultural practices, such as terminating a niche, encouraging competitors of the pathogens to fill the niche, or providing the crop with a better means to resist, tolerate, or escape the pathogen.

III. INTRODUCED ANTAGONISTS AS BIOLOGICAL AGENTS

Enormous work has been done in this field during the last 10 years. Control of *Agrobacterium radiobacter* var. *tumefaciens* (causal organism of crown gall of roses, fruit trees, and other plants) using the nonpathogenic strain K-84 of *A. radiobacter* var. *radiobacter* is one example of biological control achieved with introduced antagonists. The antagonist produces antibiotic-like substances which are effective against the pathogens. Control of *Heterobasidion annosum*, cause of pine-root rot, by inoculation of freshly cut stumps with *Peniophora gigantia,* is another example. There are several examples where an introduced antagonist has been used as a biological agent for disease control.

There are many factors and various substances both organic and inorganic which play an important role in the process of biological control.

IV. BIOCONTROL OF VIRAL DISEASES

Selective inhibition and decreasing virus infectivity by different species of fungi and their growth products have been observed by several workers.[17,18,27,53,54] These fungi, e.g., *Trichothecium roseum,* produce certain antiviral substances which act indirectly via the host, causing alterations in the latter's susceptibility to viruses.[49] The above work mostly referred to plant viruses. Substances possessing host directed virus inhibitory properties are called "antiviral agents". Antiviral agents, with similar modes of action (prohost), from *Penicillium stoloniferum* and *P. funiculosum* were subsequently discovered to be effective against a number of animal viruses.[52]

Aqueous extracts for certain nonhost plants are known to induce systemic resistance against viral attack in several susceptible plants.[119] Subjecting susceptible plants to successive treatments with extracts from leaves of *Clerodendrum* spp., *Pseuderanthemum* spp., and *Mirabilis jalapa,* and root extracts of *Boerhaavia diffusa* enabled them to resist attack with commonly prevalent viruses and thus these plants escaped infection. Several diseases caused by viruses in crops such as potato, tomato, urd, mung, sunhemp, cucurbits, and *Nicotiana* spp. are checked by these extracts.[120,121,122] Since no direct control measures are available for plant virus infections, prophylactic control offered by plant extracts appears interesting. Plant extracts are not directly virucidal but they exert their antiviral effect by stimulating the repressed resistance mechanism in otherwise susceptible plants.

If vegetatively propagated plants are systemically infected once with a viral disease, the pathogen readily passes from one generation to the next. The entire population of a clonal variety may become infected with the same pathogen.[124] In addition to virus transmission by asexual plant propagation, viral diseases are also spread by infected seeds.[63] A large number of plants have been successfully freed from viral infection through tissue-culture technique.[63] In addition to disease elimination, tissue-culture techniques are also important in the development of disease resistant plants.[40,58]

V. ROLE OF CHEMICALS PRESENT ON THE LEAF SURFACE

Various chemicals present on the leaf surface, in the form of exudate or leachates, play an important role in the process of infection and its pathogenicity. The presence of anti-microbial substances or chemicals on the surface of aerial plant parts are active against microorganisms. Blakeman and Atkinson[14] categorized most of these compounds into two major groups, i.e., terpenoids and phenolic compounds. Amongst terpenoids the monoterpenes (essential oils) are mainly volatile compounds and are formed usually in ducts or glands. Phenolic compounds have been more frequently reported from the leaf surfaces than terpenoids. These antimicrobial substances may act directly on pathogens to limit disease development, or indirectly by reducing the saprophytic microflora leading to alterations in competitive relationships with pathogens. The surface of plants may provide a useful source of novel chemicals for disease control. Dunn et al.[39] found that leaf washings of five different plants inhibited the fungicidal activity of ethylene thiurum disulphide (ETD) in vitro to *Alternaria brassicicola*. Here D-glucose reduced the efficacy of the fungicide deposited on leaves. Several others have reported the mycostatic activity of the host leaf exudates.[57,104,108,118] Brillova[19] showed that germination of the conidia of *Cercospora beticola* was inhibited by 3-hydrotriamine present in the sugar beet leaf washings and suggested that the resident strain of the host produced more of the inhibitors than the susceptible strain. Egawa et al.[41] identified antifungal substances in a number of leaves and observed that these could be leached from intact leaves. Kumar and Nene[72] showed strong antifungal activity of leaf extract from *Cleome isocandra* on several fungi.

VI. ROLE OF PHYTOALEXINS IN BIOLOGICAL CONTROL

Müller[90] defined phytoalexins as "antibiotics which are the result of an interaction of two different metabolic systems, the host and parasite, and which inhibit the growth of microorganisms pathogenic to plants." In other words phytoalexins are the postinfectional metabolites which are formed by depression or activation of a latent enzyme system.[65] In the simplest way they can be defined as antifungal stress metabolites formed in plants as a result of physical, chemical, or biological stress, which resist or suppress the activity of the invader. Hence, these active principles play an important role in disease resistance and in the control of plant diseases. Since phytoalexins are natural products, their use in controlling plant disease is more reliable, harmless, and has no chance of residual effects which are common in the case of fungicides and pesticides. Sometimes plants acquire immunity against a particular disease. According to Chester,[28] plants react to bacteria, fungi, and viruses in the same way as animals react to microorganisms or viral infections — by producing specific antibodies. Later he pointed out that since plants have no circulatory system like animals there is no possibility of antibody formation in plants.

Activation of phytoalexin synthesis by application of biotic and abiotic elicitors may enhance field resistance of plants to some extent. Treatment of seeds with phytoalexin elicitors was also found to be effective in some cases.[110] The direct and indirect role of phytoalexins in plant protection has been conclusively demonstrated in some instances, e.g., capsidiol. Therefore it seems reasonable to speculate that more potential phytoalexins will be available in the future to control various plant diseases.

VII. BIOLOGICAL CONTROL BY GENETIC MANIPULATION OF AGENTS

Upgrading plant health by genetic modification, either by classical or nonconventional means, may in theory be used to improve the biological performance of microbial biocontrol agents.[91] Intergeneric protoplast fusion, incorporation of prokaryotic DNA into eucaryotes, will help in evolving the desired disease resistant lines in a highly controlled manner.[42]

The "life" of a resistant variety is extremely important because the longer the life the greater the net returns.[36] Tactics and feasibility of genetic engineering of biocontrol agents can help in many ways including:

1. Expanding the range of target species
2. Expanding the agent's environmental range beyond its congenial habitat
3. Restricting the range of nontarget organisms.
4. Restricting the agent's potential for becoming established in previously uncongenial environments
5. Augmenting or decreasing its ecological persistence or survival from one season to the next
6. Improving its durability and thus reducing the probability of biocontrol failure after prolonged use

Genetic manipulation of biocontrol agents is now used for improving strains for industrial uses through mutation and other changes in the genetic composition of the agents. This is also widely practiced for amplifying productivity.[83] Genetic make-up of some fungi are so poorly understood that it is difficult to suggest genetic improvements in fungi which may act as biological control agents.[129] However, in bacteria it is possible to increase the efficiency of biological control function by genetic manipulation. This can be done by both genotypic as well as phenotypic changes.[30,73,107,125,126] To make a microbe an efficient biocontrol agent, genes which code for this function should be added and genes which code for pathogenicity should be deleted from the genome of a potential biocontrol agent. Pathogenicity-minus mutants of several *Pseudomonas syringae* pathovars, *Erwinia amylovora*, *E. carotovora*, *E. chrysanthemum*, and *Xanthomonas campestris* pv *vesicatoria*, are now available and can be tested for their potential as biocontrol agents.[76] Cook and Baker[33] suggested that there can be transfer of genes from an antagonistic microorganism into a plant genome so that a resistant cultivar is produced. They indicated that in the case of *Agrobacterium radiobacter* strain K-84, if the bacteriocin producing genes are introduced into the host genome, they develop resistance against crown gall. This modification is the result of recombinant DNA technology. Kerr[67], Ellis et al.,[43] and Cooksey and Moore[35] showed the same results by using a bacteriocin-84 mutant of *A. tumefaciens* to protect peaches against crown gall pathogens. The biological control of crown gall diseases of various plants (due to *A. tumefaciens*) by the prior spray of a saprophytic "agrocin" (a toxin) producing strain of *A. radiobacter* has regenerated interest in saprophytic bacteria inhabiting plants and their role in disease development. Furthermore, the T-DNA portion of Ti-plasmid will integrate with host DNA and is the cause of tumor induction; and thus can be used as a vehicle for introducing procaryotic DNA segments (specifically antibiotic/toxin producing genes) into the host genome. In this way it has been demonstrated that genetic engineering can be used in the management of bacterial disease of plants.[102]

Exploitation of genetic engineering in the service of biological control is being done by various workers.[4] Papavizas and Lewis[97] have given an interesting report in this connection. They induced mutation by UV-radiation in *Trichoderma harzianum*, a biological control agent which controlled the damping off and white rot of onion caused by *Rhizoctonia*.

The insecticide was created by transferring a gene with insecticidal properties from the bacterium *Bacillus thuringiensis* into the cells of *Pseudomonas fluorescens,* which lives on vegetable leaves. When insects feed on the genetically engineered leaf bacteria, toxins are released. Because the toxins are stored in cells they will not reproduce and the bacteria are considered environmentally safe by the Environmental Protection Agency. The root eating insects of corn plants are poisoned by these toxin-bearing bacteria when they colonize the corn roots. Scientists continue to debate the safety of using these organisms in agriculture. *P. syringae,* a genetically engineered strain of bacterium, produces a protein causing ice formation on plant leaves. Lindow[77] produced a mutant strain of bacteria that does not cause ice formation. The "ice-minus" strain has been tested on more than 50 greenhouse species with no harmful effects on plant development.

Jones Anderson[130] says, "I believe we should manipulate biological process for the good of mankind, and I have no hangups about creating modified life forms." Fulbright,[46] while working on the fungus *Endothia parasitica,* which causes diseases of the chestnut, found that some diseased trees have managed to recover. He found not just one strain, but several, that they categorized into two groups: (1) virulent strains, which could infect trees and cause cankers to form under the bark, girdling branches or trunks, and (2) hypovirulent strains, weaker versions of the same fungus, that could check the growth of cankers by merging with virulent forms. The apparent difference between the virulent form and hypovirulent form of *E. parasitica* is the presence of a virus-like molecule in the hypovirulent strain. This molecule contains double stranded RNA similar to DNA, of which almost all genes are made. The fungus reverts to the virulent form after removal of this molecule and turns hypovirulent when this molecule is inserted. In this way genetic manipulation can help in controlling various diseases of important plants.

VIII. ROLE OF MISCELLANEOUS FACTORS IN BIOLOGICAL CONTROL

A. Soil

Suppressive soil is now used by various workers[62] for controlling various diseases of soil borne pathogens. The suppressive soil is that which is not favorable for the development of certain diseases. According to Cook and Baker[33] these soils reduce saprophytic growth and reduce the activity and survival of the pathogen. The germination of chlamydospores or conidia is more suppressed in these soils than in conducive soils.[64,70,111,112] This suppression is directly linked with the microbial activity, and hence can be involved in biocontrol. This type of soil is rich in montmorillonile clay, which favors the development of antagonistic microorganisms, which in turn suppresses the pathogens' activity.[113] The mechanism of suppression in soil is not well known; however, it includes antibiosis, competition, parasitism, and predation.[62] Some important plant pathogens which are suppressed by these soils are the *Fusarium* sp.,[10,47,87] *Rhizoctonia solani,*[69,73] the *Pythium* sp.,[15,16,55,56] and *Phytophthora cinnamoni.*[20,21,22]

B. Soil Fungistasis and Biological Control

The possibility of biological control by soil fungistasis has been reviewed by several workers.[11,48,80,98,126] Fungistasis may be elevated to a level beyond favoring the germination of propagules even in the presence of a susceptible host.[98] It is also possible to manipulate the soil environment to such an extent that propagules remain inactivated even in the presence of root exudates, by increasing the toxicity level. In nature the large amount of organic matter and various sources of energy nullify fungistasis.[79]

C. Organic Amendments

Singh and Singh[109] extensively reviewed the significance of organic amendment in biological control of soil-borne plant pathogens. The organic amendment is more advantageous than the use of any other chemical. The effect of the organic compound will last longer, it may have a multiple pathogen suppression effect, and the use involves a lower cost and fewer hazards than other chemicals produce. This also enhances soil fertility.[109] Green manuring is known to control the potato scab disease.[101] Heavy application of organic matter controlled infection of *Phymatotrichum omnivorum* in cotton.[29,68] Many review papers have appeared on the biological control of soil-borne diseases.[11,31,33,34,48,75,95,96,99] The process of fungistasis in soil is also enhanced by the organic amendments with alfalfa,[1] corn stover,[44] oat straw,[2] soyabean, cotton, linseed meal,[60,61] cellulose,[23] and chitin[51] and thereby plays an important role in biological control of plant diseases.

D. Solarization

A recent innovation from Israel is solar heating of the soil to control diseases. Death of the pathogen is caused by a combination of the predisposing physical environment and the action of soil microorganisms. Solar heating is now in commercial practice to control *Verticillium* wilt of pistachio trees in Southern California.[6,7] Katan[66] has discussed the mechanism of biocontrol by solarization. Solarization increases the activity of microbes towards pathogens and also increases vulnerability of pathogens to soil microorganisms. Solar heating effects the survival ability of the pathogen for a very long period.

E. Drought and Flood

Ayres and Wollacott[9] pointed out that barley grown in dry soil developed greater adult plant resistance to powdery mildew than barley grown in wet soil. Conidial germination and the formation of appressoria and elongating secondary hyphae on upper leaves were inhibited when adult plants were grown in dry soil. They concluded that dry soil conditions in the field will reduce the risk of severe mildew infection. Similarly, flooding of soil has been used as a tool for biological control of many diseases. The resting propagules of pathogens are eliminated by excess of water.[33] Stover[114] found that flooding of soil reduced the population of *Fusarium oxysporum* F. sp. *cubense*.

F. Climate

The microclimate at the plant surface and seasonal variations change the microhabitat of the plant surface. The environment at the leaf surface fluctuates much more rapidly than at the root surface.[13] Naturally occurring plant surface microbes distinctly affect the incidence of disease on the aerial surfaces of plants. If the environment of the plant surface can be manipulated to enhance the activity of antagonists it will be indirectly effective in disease control.[33,100]

It is difficult to change the microclimate of the plant surface; however, the architecture of the plant surface and its nutrition status can be changed. The nutrients available on the plant surfaces influence the microbial population and it can be easily manipulated by application of appropriate nutrients.[12]

IX. ROLE OF DIFFERENT MICROORGANISMS IN BIOLOGICAL CONTROL

A. Amoebae

Very little importance has been placed on soil amoebae in comparison to bacteria, fungi, and actinomycetes. Now some workers have started exploring the potential of soil amoebae in biological control.[92-94] There are many examples where these amoebae feed on the conidia, chlamydospores, sporangia, teleutospores, basidiospores, and hyphae of various pathogens.[94] The fungal propagules and mycelia are ingested by soil amoebae,[44,92,94] e.g., *Alternaria, Helminthosporium, Curvularia,* and *Fusarium.*[44] In *Thielaviopsis basicola*[5], *Fusarium oxysporum, F. solani, F. graminearum,* and *Verticillum dahliae,*[3] amoebae cause perforation and lysis of spores and hyphae. The important parasitic amoebae are *Arachnula impatiens* and *Thecamoeba granifera* sub sp *minor.*[3]

B. Mycoviruses

Recently, a few workers have suggested the possibility of using mycoviruses in the biological control of plant pathogens.[31,33,98] In case of *Rhizoctonia solani* (causal organism of damping off of sugar beet) it was found that some virus-like particles or double stranded RNA mycoviruses[26,74] were able to reduce the infection.[24,25,26] Another report revealed that hypovirulent strains of *Gaeumannomyces graminis* var. *tritici* lost their virulence when associated with mycoviruses.[45]

C. Mycorrhizal Fungi

Marx[84,85] suggested that ectomycorrhizal fungi increased resistance in the plants indirectly by using available nutrients on the root surface. The fungi also behaved as a mechanical barrier, thus checking penetration of root cells by the pathogens. It also helped in the production of volatile and nonvolatile inhibitors by root cortical cells which inhibit pathogens that infect the host. The fungi also supported good growth of antagonists in the rhizosphere. There are several reports[89] of soil borne plant pathogens that are prevented from causing root disease by vesicular arbuscular mycorrhiza (VAM). *Thielaviopsis basicola* causes less damage to the roots of cotton with VAM fungi, than the roots without VAM.[106] Similarly, many workers have reported control of several important plant diseases, e.g., citrus root rot caused by *Phytophthora parasitica*[37], *Verticillium* wilt of cotton,[38] *Brassica* against *Olpidium brassicae*,[105] and many others in the presence of the VAM fungi. In California the soil inoculum of *Glomus deserticola* is commercially used by cultivators for production of a healthy stand.[33]

D. Mycoparasites

The use of mycoparasites for biological control was first demonstrated by Tribe.[115] Enormous amounts of literature have accumulated on this aspect of biological control since then.[11,59,71,81,98] A large number of mycoparasites used successfully for biological control have been listed.[8,116]

Although many mycoparasites have been isolated and are known as biocontrol agents in the laboratory only a few of them have been successfully exploited for biocontrol of diseases in the field. Of these, the *Trichoderma* spp. are parasitic on a large number of fungi and they have been reported to control several soil-borne diseases.[50] *Coniothyrium minitans* grows as a hyperparasite of *Sclerotinia sclerotiorum* under natural conditions.[86,103,115] In recent years, besides the isolation of these parasites, much emphasis has been placed on the ecological relationship with other microorganisms. Efforts have also been made to develop procedures for culturing and application to biocontrol practices.[117]

X. CONCLUSION

It is well known that the application of biological agents to control important diseases needs more time and effort, but once it is successful, it will be a fruitful method in agricultural practices of any developing country. Biological control, no doubt, can overcome various problems arising due to the use of pesticides and other poisonous substances. It will, for example, certainly reduce the pollution risk.

REFERENCES

1. **Abd-El Moity, T. H. and Shatla, M. N.,** Biological control of white rot disease of onion (*Sclerotium cepivorum*) by *Trichoderma harzianum, Phytopathology,* 2(100), 29, 1981.
2. **Adams, P. B.,** Effect of soil temperature and soil amendments on *Thielaviopsis* root rot of sesame, *Phytopathology,* 61, 93, 1971.
3. **Alabouvette, C., Rouxel, F., and Louvet, J.,** Characteristics of fusarium wilt suppressive soils and prospects for their utilization in biological control, in *Soil Borne Plant Pathogens,* Schippers, B. and Gams, W., Eds., Academic Press, London, 1979, 165.
4. **van Alfen, N. K.,** Can pathogen's genes be engineered to control disease?, *Trends Biotechnol.,* 3(10), 250, 1985.
5. **Anderson, T. R. and Patrick, Z. A.,** Soil vampyrellid amoebae that cause small perforations in conidia of *Cochliobolus sativus, Soil. Biol. Biochem.,* 12, 159, 1980.
6. **Ashworth, L. J., Jr.,** Polyethylene trapping of soil in a *Pistachio* nut grove for control of *Verticillium dahleae, Phytopathology,* 69, 913, 1979.
7. **Ashworth, L. J., Jr., Morgan, D. P., Gaona, S. A., and McCain, A. H.,** Polyethylene trapping controls verticillium wilt in Pistachio, *Calif. Agric.,* 36(5—6), 17, 1982.
8. **Ayers, W. A. and Adams, P. B.,** Mycoparasitism and its application to biological control of plant disease, in *Biological Control In Crop Production,* Papavizas, G. C., Ed., Allanheld, Totowa, N.J., 1981, 91.
9. **Ayers, P. G. and Woolacott, B.,** Effect of water level on the development of adult plant resistance to powdery mildew in barley, *Ann. Appl. Biol.,* 94, 255, 1980.
10. **Baker, K. F.,** Microbial antagonism: the potential for biological control, 2nd Int. Symp. Microbial Ecology, Coventry, U.K., 1980, 327.
11. **Baker, K. F. and Cook, R. J.,** *Biological Control of Plant Pathogens,* W. H. Freeman, San Francisco, 1974, 433.
12. **Bakshi, E. and Fokkema, N. J.,** Environmental factors limiting growth of *Sporobolomyces roseus* an antagonist of *Cochliobolus sativus* on wheat leaves, *Trans. Br. Mycol. Soc.,* 68, 17, 1977.
13. **Blakeman, J. P.,** Ecological succession of leaf surface microorganisms in relation to biological control, in *Biological Control on the Phylloplane,* Windels, C. E. and Lindow, S. E., Eds., American Phytopathological Society, St. Paul, Minn., 1985, 6.
14. **Blakeman, J. P. and Atkinson, P.,** Antimicrobial substances associated with the aerial surface of plants, in *Microbial Ecology of the Phylloplane,* Blakeman, J. P., Ed., Academic Press, London, 1981, 245.
15. **Bouhot, D.,** Estimation of inoculum density and inoculum potential: techniques and their value for disease prediction, in *Soil Borne Plant Pathogens,* Schippers, B. and Gams, W., Eds., Academic Press, London, 1979, 21.
16. **Bouhot, D. and Joannes, H.,** Ecologie des champignous parasites dans le soil-IX. Measures du potentiel infectieux des sols naturellement infestes de *Pythium* sp., *Soil. Biol. Biochem.,* 11, 417, 1979.
17. **Bowden, F. C.,** Inhibitors and plant viruses, *Adv. Virus Res.,* 2, 31, 1954.
18. **Bowden, F. C. and Freeman, G. G.,** The nature and behaviour of inhibitors of plant viruses produced by *Trichothecium roseum* Link, *J. Gen. Microbiol.,* 7, 155, 1952.
19. **Brillova, D.,** Influence of phenolic substances on the germination of conidia of *Cercospora beticola,* Sacc., *Biologia (Bratislava),* 26, 717, 1971.
20. **Broadbent, P. and Baker, K. F.,** Behaviour of *Phytophthora cinnamonii* in soils suppressive and conducive to root rot, *Aust. J. Agric. Res.,* 25, 121, 1974.
21. **Broadbent, P. and Baker, K. F.,** Association of bacteria with sporangium formation and breakdown of sporangia in *Phytophthora* spp., *Aust. J. Agric. Res.,* 25, 139, 1974.
22. **Broadbent, P. and Baker, K. F.,** *Biology and Control of Soil-Borne Plant Pathogens,* Bruehl, G. W., Ed., American Phytopathological Society, St. Paul, Minn., 1975, 152.
23. **Broadbent, P., Baker, K. F., and Waterworth, Y.,** Bacteria and actinomycetes antagonistic to fungal root pathogens in Australian soils, *Aust. J. Soil. Sci.,* 24, 925, 1971.
24. **Castanho, B. and Butler, E. E.,** *Rhizoctonia* decline: a degenerative disease of *Rhizoctonia solani, Phytopathology,* 68, 1505, 1978.
25. **Castanho, B. and Butler, E. E.,** *Rhizoctonia* decline: studies on hypovirulence and potential use in biological control, *Phytopathology,* 68, 1511, 1978.

26. **Castanho, B., Butler, E. E., and Sheperd, R. J.,** The association of double-stranded RNA with *Rhizoctonia* decline, *Phytopathology*, 68, 1515, 1978.

27. **Chessin, M.,** Is there a plant interferon?, *Bot. Rev.*, 49(1), 1, 1983.

28. **Chester, K. S.,** The problem of acquired physiological immunity in plants, *Q. Rev. Biol.*, 8, 129, 1933.

29. **Clark, F. E.,** Experiment toward the control of the takeall disease of wheat and the *Phymatotrichum* root rot of cotton, *U.S. Dept. Agric. Tech. Bull.*, 835, 27, 1942.

30. **Clark, F. E. and Warren, C. J.,** Conjugal transmission of plasmids, *Annu. Rev. Genet.*, 13, 99, 1979.

31. **Cook, R. J.,** Management of associated microbiota, in *Plant Disease*, Vol. 1, Horsefall, J. G. and Cowling, E. B., Eds., Academic Press, New York, 1977, 145.

32. **Cook, R. J.,** Progress toward biological control of plant pathogens, with special reference to take-all of wheat, *Agric. For. Bull.*, 5(3), 22, 1982.

33. **Cook, R. J. and Baker, K. F.,** *The Nature and Practice of Biological Control of Plant Pathogens*, American Phytopathological Society, St. Paul, Minn., 1983, 539.

34. **Cook, R. J., Boosalis, M. G., and Dounik, B.,** Crop residue management systems, in *Agron. Soc. Am. Spec. Publ. No. 31*, Madison, Wis., 1978, 147.

35. **Cooksey, D. A. and Moore, L. W.,** Biological control of crown gall with an Agrocin mutant of *Agrobacterium radiobacter*, *Phytopathology*, 72, 919, 1982.

36. **Curlson, G. A. and Main, C. E.,** Economics of disease-loss management, *Annu. Rev. Phytopathol.*, 14, 381, 1976.

37. **Davis, R. M. and Menge, J. A.,** Influence of *Glomus fasciculatus* and soil phosphorus on *Phytophthora* root rot of citrus, *Phytopathology*, 70, 447, 1980.

38. **Davis, R. M., Menge, J. A., and Ervin, D. C.,** Influence of *Glomus fasciculatus* and soil phosphorus on Verticillium wilt of cotton, *Phytopathology*, 69, 453, 1979.

39. **Dunn, C. L., Benyon, K. I., Brown, K. F., and Mantagne, J. T. W.,** The effect of glucose in leaf exudates upon biological activity of some fungicides, in *Ecology of Leaf Surface Microorganisms*, Preece, T. F. and Dickinson, C. H., Eds., Academic Press, London, 1971, 491.

40. **Earle, E. D.,** Application of plant tissue culture in the study of host pathogen interactions, in *Application of Plant Cell and Tissue Culture to Agriculture and Industry*, Tomes, D. T., Ellis, B. E., Horney, P. M., Kasha, K. J., and Peterson, R. L., Eds., University of Guelph Press, Ontario, 1982, 45.

41. **Egawa, H., Kadzumori, T., Shigeyasu, A,. Chung, Y. H., Jiro, F., and Kin, I. K.,** Leaching of antifungal substances onto the leaves and the total number of microorganisms on the leaves of several plants containing antifungal substances, *Bull. Fac. Agric. Shimane Univ.*, 7, 23, 1973.

42. **Ellingboe, A. H.,** Changing concepts in host-pathogen genetics, *Annu. Rev. Phytopathol.*, 9, 125, 1981.

43. **Ellis, J. G., Kerr, A., vanMontague, M., and Schell, J.,** *Agrobacterium:* genetic studies on agrocin 84 production and the biological control of crown gall, *Physiol. Plant Pathol.*, 15, 311, 1979.

44. **Esser, R. P., Ridings, W. H., and Sobers, E. K.,** Ingestion of fungus spores by protozoa, *Proc. Soil Crop Sci. Soc. Florida*, 34, 206, 1975.

45. **Férault, A. C., Tiudi, B., Lamaire, J. M., and Spire, D.,** Etude de l'evolution comparée du niveau d'aggresivité du contenu en partieules de type viral d'une such de *Gaeumannomyces graminis* (Sacc) Arx et Oliver (*Ophiobolous graminis* Sacc), *Ann. Phytophthol.*, 11, 185, 1979.

46. **Fulbright, D.,** Fighting blight with biotechnology, *Futures*, 4(1), 15, 1985.

47. **Furuya, H. and Ui, T.,** The significance of soil microorganisms on the inhibition of the macroconidial germination of *Fusarium solani f.* sp. *phaseoli* in a soil suppressive to common bean root rot, *Ann. Phytopathol. Soc. Japan*, 47, 42, 1981.

48. **Garrett, S. D.,** *Pathogenic Root Infecting Fungi*, Cambridge University Press, London, 1970.

49. **Gendron, Y. and Kassanis, B.,** The importance of the host species in determining the action of virus inhibitors, *Ann. Appl. Biol.*, 41, 183, 1954.

50. **Gindrant, D.,** Biocontrol of plant diseases by inoculation of fresh wounds, seeds, and soil with antagonists, in *Soil Borne Plant Pathogens*, Schippers, B. and Gams, W., Eds., Academic Press, London, 1979, 537.

51. **Gindrant, D., Vander Hoeven, E., and Moody, A. R.,** Control of *Phomopsis sclerotioides* with *Gliocladium roseum* or *Trichoderma*, *Neth. J. Plant Pathol.*, 83(1), 429, 1977.

52. **Gupta, B. M.,** Developments in antiviral chemotherapeutic research, *Indian Phytopathol.*, 38(3), 401, 1985.

53. **Gupta, B. M. and Price, W. C.,** Production of plant virus inhibitors by fungi, *Phytopathology*, 40, 642, 1950.

54. **Gupta, B. M. and Price, W. C.,** Mechanism of inhibition of plant virus infection by fungal growth products, *Phytopathology*, 42, 45, 1952.

55. **Hancock, J. G.,** Factors affecting soil populations of *Pythium ultimum* in the San Joaquin valley of California, *Hilgardia*, 45, 107, 1977.

56. **Hancock, J. G.,** Occurrence of soils suppressive to *Pythium ultimum*, in *Soil Borne Plant Pathogens*, Schippers, B. and Gams, W., Eds., Academic Press, London, 1979, 183.

57. **Hartman, J. R., Arthur, K., and Upper, C. D.,** Differential inhibitory activity of a corn extract to *Erwinia* spp. causing soft rot, *Phytopathology,* 65, 1082, 1975.
58. **Helgeson, J. P. and Haberlach, G. T.,** Disease resistance studies with tissue culture, in *Tissue Culture Methods for Plant Pathologists,* Ingrain, D. S. and Helgeson, J. P., Eds., Blackwell Scientific, Oxford, 1980, 179.
59. **Henis, Y. and Chet, I.,** Microbiological control of plant pathogens, in *Advances in Applied Microbiology,* Vol. 19, Perlman, D., Ed., Academic Press, London, 1975, 85.
60. **Hornby, D.,** Microbial antagonisms in the rhizosphere, *Ann. Appl. Biol.,* 89, 97, 1978.
61. **Hornby, D.,** Take-all decline: a theorist's paradise, in *Soil Borne Plant Pathogens,* Schippers, B. and Games, W., Eds., Academic Press, London, 1979, 133.
62. **Hornby, D.,** Suppressive soils, *Annu. Rev. Phytopathol.,* 21, 65, 1983.
63. **Hu, C. Y. and Wang, P. J.,** Meristem, shoot tip, and bud culture, in *Handbook of Plant Cell Culture. 1. Techniques for Propagation and Breeding,* Evans, D. A., Sharp, W. R., Ammirato, P. V., and Yamanda, Y., Eds., Macmillan, New York, 1983, 177.
64. **Hwang, S. F., Cook, R. J., and Haglund, W. A.,** Mechanisms of suppression of chlamydospores germination for *Fusarium oxysporum* F. sp. *pisi* in Soils, *Phytopathology,* 72, 948, 1982.
65. **Ingham, J. L.,** Disease resistance in higher plants: the concept of pre-infectional and post-infectional resistance, *Phytopathol. Z.,* 78, 314, 1973.
66. **Katan, J.,** Solar heating (solarization) of soil for control of soil borne pests, *Annu. Rev. Phytopathol.,* 19, 211, 1981.
67. **Kerr, A.,** Soil microbiological studies on *Agrobasidium radiobacter* and biological control of crown gall, *Soil Sci.,* 118, 168, 1974.
68. **King, C. J., Hope, C., and Eaton, E. D.,** Some microbiological activities affected in manurial control of cotton root rot, *J. Agric. Res.,* 42, 827, 1934.
69. **Ko, W. H. and Ho, W. C.,** Screening soils for suppressiveness to *Rhizoctonia solani* and *Pythium splendens,* *Ann. Phytopathol. Soc. Japan,* 49, 1, 1983.
70. **Komada, H.,** *Proc. 1st Int. Congr. Int. Assoc. Microbiological Societies,* Vol. 2, Developmental Microbial Ecology Science Council of Japan, 1975, 451.
71. **Kuhlman, E. G.,** *Plant Disease: An Advanced Treatise,* Vol. 5, Horsfall, J. G. and Cowling, E. B., Eds., Academic Press, New York, 1980, 363.
72. **Kumar, K. and Nene, Y. L.,** Antifungal properties of *Cleome isocandra* L. extracts, *Indian Phytopathol.,* 21, 445, 1968.
73. **Lacy, G. H. and Patil, S. S.,** Why genetics?, in *Phytopathogenic Prokaryotes,* Vol. 2, Mount, M. S. and Lacy, G. H., Eds., Academic Press, New York, 1982, 221.
74. **Lemke, P. A.,** *Belstville Symposium in Agricultural Research. 1. Virology in Agriculture,* Romberger, J. A., Ed., Allenheld, Totowa, N. J., 1977, 159.
75. **Lewis, J. A. and Papavizas, G. C.,** *Biology and Control of Soil Borne Plant Pathogens,* Bruchl, G. W., Ed., Academic Press, London, 1975, 84.
76. **Lindemann, J.,** Genetic manipulation of microorganisms for biological control, in *Biological Control on the Phylloplane,* Windels, C. E. and Lindow, S. E., Eds., American Phytopathological Society, St. Paul, Minn., 1985, 116.
77. **Lindow, S. E.,** Methods of preventing frost injury caused by epiphytic ice-nucleation active bacteria, *Plant Dis.,* 67, 327, 1983.
78. **Liu, S. and Baker, R.,** Mechanism of biological control in soil suppressive to *Rhizoctonia solani,* *Phytopathology,* 70, 404, 1980.
79. **Lockwood, J. L.,** Fungistasis in soils, *Biol. Rev.,* 52, 1, 1977.
80. **Lockwood, J. L.,** Soil mycostasis: concluding remarks, in *Soil-borne Plant Pathogens,* Schippers, B. and Gams, W., Eds., Academic Press, London, 1979, 121.
81. **Lumsden, R. D.,** *The Fungal Community and its Role in the Ecosystem,* Wicklow, D. T. and Carroll, G. W., Eds., Marcel Dekker, New York, 1981, 295.
82. **Lupton, F. G. H.,** Biological control: the plant breeding objectives, *Ann. Appl. Biol.,* 104, 1, 1984.
83. **Martin, P. A. W. and Dean, D. H.,** *Microbial Control of Pests and Plant Diseases, 1970—1980,* Burges, H. D., Ed., Academic Press, New York, 1981, 299.
84. **Marx, D. H.,** Ectomycorrhizae as biological deterrents 3558 to pathogenic root infections, *Annu. Rev. Phytopathol.,* 10, 429, 1972.
85. **Marx, D. H.,** *Biology and Control of Soil Borne Plant Pathogens,* Bruehl, G. W., Ed., Academic Press, London, 1975.
86. **Merriman, P. R., Pywell, M., Harrison, G., and Nancarrow, J.,** Survival of sclerotia of *Sclerotinia sclerotiorum* and effect of cultivation practices on disease, *Soil Biol. Biochem.,* 11, 567, 1979.
87. **Mitchell, R. and Hurwitz, E.,** Suppression of *Pythium debaryanum* by lytic rhizosphere bacteria, *Phytopathology,* 55, 156, 1965.

88. **Mukerji, K. G.,** Biocontrol of plant diseases, in *Recent Advances in Plant Pathology,* Husain, A., et al., Eds., Lucknow Print House, Lucknow, India, 1983, 1.

89. **Mukerji, K. G., Sabharwal, A., Kochar, B., and Ardey, J.,** Vesicular arbuscular mycorrhiza: concepts and advances, in *Progress in Microbial Ecology,* Mukerji, K. G., Agnihotri, V. P., and Singh, R. P., Eds., Lucknow Print House, Lucknow, India, 1984, 489.

90. **Müller, K. O.,** Einige einfache Versuche zum Nachweis Von Phytoalexinen, *Phytopathol. Z.,* 27, 237, 1956.

91. **Nicholas, J. and Panopoulos, J.,** Tactis and feasibility of genetic engineering of biocontrol agents, in *Microbiology of the Phyllosphere,* Fokkema, N. J. and Van den Heuvel, J., Eds., Cambridge University Press, Cambridge, 1986, 312.

92. **Old, K. M.,** Giant soil amoebae cause perforation of conidia of *Cochliobolus sativus, Trans. Br. Mycol. Soc.,* 68, 277, 1977.

93. **Old, K. M.,** Perforation of conidia of *Cochliobolus sativus* by soil amoebae, *Acta Phytopathol. Acad. Sci. Hung.,* 12(1—2), 113, 1977.

94. **Old, K. M. and Patrick, Z. A.,** Perforation and lysis of spores of *Cochliobolus sativus* and *Thielaviopsis basicola* in natural soils, *Can. J. Bot.,* 54, 2798, 1976.

95. **Papavizas, G. C.,** in *Pest control by chemical, biological, genetic, and physical means — a symposium,* Agricultural Research Service 33-110, U.S. Government Printing Office, Washington, D.C., 1966, 214.

96. **Papavizas, G. C.,** Status of biological control of soil borne plant pathogens, *Soil Biol. Biochem.,* 5, 709, 1973.

97. **Papavizas, G. C. and Lewis, J. A.,** Induction of new biotypes of *Trichoderma harzianum* resistant to benomyl and other fungicides, *Phytopathology,* 71, 247, 1981.

98. **Papavizas, G. C. and Lumsden, R. D.,** Biological control of soil borne fungal pathogens, *Annu. Rev. Phytopathol.,* 18, 389, 1980.

99. **Patrick, Z. A. and Toussoun, T. A.,** Plant residues and organic amendments in relation to biological control, in *Ecology of Soil Borne Plant Pathogens,* Baker, K.F. and Snyder, W. C., Eds., University of California Press, Berkeley, 1965, 440.

100. **Rodger, G. and Blakeman, J. P.,** Microbial colonization and uptake of ^{14}C label on leaves of sycamore, *Trans. Br. Mycol. Soc.,* 82, 45, 1984.

101. **Sanford, G. B.,** Some factors affecting the pathogenicity of *Actinomyces scabies, Phytopathology,* 16, 525, 1926.

102. **Schmidt, R. M., Lippincott, B. B., and Lippincott, J. A.,** Growth requirements and infectivity of auxotrophic adenine-dependent mutants of *Agrobacterium tumefaciens, Phytopathology,* 59, 1451, 1969.

103. **Schmidt, H. H.,** Untersuchungen über die Lebensdaur der Sklerotien von *Sclerotinia sclerotiorum* (Lib) de Bary im Boden unter dem Einfluss verchiedener Pfangenarten und nach Infektion mit *Coniothyrium minitans* Campb., *Arch. Planzenschutz,* 6, 321, 1970.

104. **Schneider, R. W. and Sinclair, J. B.,** Inhibition of conidial germination and germ tube growth of *Cereospora canescens* by cowpea leaf diffusates, *Phytopathology,* 65, 63, 1975.

105. **Schonbeck, F.,** Endomycorrhiza in relation to plant diseases, in *Soil Borne Plant Pathogens,* Schippers, B. and Gams, W., Eds., Academic Press, London, 1979, 271.

106. **Schonbeck, F. and Dehne, H. W.,** Damage to mycorrhizal and non-mycorrhizal cotton seedlings by *Thielaviopsis basicola, Plant Dis. Rep.,* 61, 266, 1977.

107. **Schroth, M. N., Loper, J. E. and Hildebrand, D. C.,** Bacteria as biocontrol agents of plant disease, in *Current Perspectives in Microbial Ecology,* Klug, M. J. and Reddy, C. A., Eds., American Society for Microbiology, Washington, D.C., 1984, 362.

108. **Sharma, J. K. and Sinha, S.,** Effect of leaf exudates of *Sorghum* varieties varying in susceptibility and maturity on the germination of conidia of *Colletrotrichum graminicola* Ces. Wilson, in *Ecology of Leaf Surface Microorganism,* Preece, T. F. and Dickinson, C. H., Eds., Academic Press, London, 1971, 597.

109. **Singh, N. and Singh, R. S.,** Significance to organic amendment of soil in biological control of soil-borne plant pathogens, in *Progress in Microbial Ecology,* Mukerji, K. G., et al., Eds., Lucknow Print House, Lucknow, India, 1984, 303.

110. **Sinha, A. K. and Hait, G. N.,** Host sensitization as a factor in induction of resistance in rice against *Dechslera* by seed treatment with phytoalexin inducers, *Trans. Br. Mycol. Soc.,* 79, 212, 1982.

111. **Smith, S. N.,** Comparison of germination of pathogenic *Fusarium oxysporum* chlamydospores in host rhizosphere soils conducive and suppressive to wilts, *Phytopathology,* 67, 502, 1977.

112. **Smith, S. N. and Snyder, W. C.,** Germination of *Fusarium oxysporum* chlamydospores in soils favourable and unfavourable to wilt establishment, *Phytopathology,* 62, 273, 1972.

113. **Stotzky, G.,** Techniques to study interaction between microorganisms and clay minerals in vivo and in vitro, *Bull. Ecol. Res. Comm. Stockhom,* 17, 17, 1973.

114. **Stover, R. H.,** The use of organic amendment and green manure in the control of soil-borne phytopathogens, *Recent. Prog. Microbiol.,* 8, 267, 1962.

115. **Tribe, H. T.,** On the parasitism of *Sclerotinia trifoliorum* by *Coniothyrium minitans, Trans. Br. Mycol. Soc.,* 40, 489, 1957.
116. **Upadhyay, R. S. and Rai, B.,** Mycoparasitism with reference to biological control of plant disease, in *Recent Advances in Plant Pathology,* Husain, A. et al., Eds., Lucknow Print House, Lucknow, India, 1983, 48.
117. **Upadhyay, R. S. and Rai, B.,** Biological control of soil borne plant pathogens, in *Frontiers in Applied Microbiology,* Vol. 1, Mukerji, K. G. et al., Eds., Lucknow Print House, Lucknow, India, 1985, 261.
118. **Velson, R. J.,** A study of the Pathogenicity of *Helminthosporium* sp., B.S. thesis, University of Adelaide, Adelaide, Australia, 1957.
119. **Verma, H. N.,** Nature and behaviour of plant virus inhibitors present in certain plants, in *Perspectives in Plant Virology,* Gupta, B. M., Singh, B. P., Verma, H. N., and Srivastava, K. M., Eds., Lucknow Print House, Lucknow, India, 1985.
120. **Verma, H. N. and Dwivedi, S. D.,** Prevention of plant virus diseases in some economically important plants by *Bougainvillea* leaf extract, *Indian J. Plant Pathol.,* 1, 97, 1983.
121. **Verma, H. N. and Khan, Abid Ali, M. M.,** Management of plant virus diseases by *Pseudernthemum bicolor* leaf extract, *Z. Pflanzenkr. Pflanzenschutz,* 91, 266, 1984.
122. **Verma, H. N. and Kumar, V.,** Prevention of potato plants from viruses and insect vectors, *J. Indian Potato Assoc.,* 6, 157, 1979.
123. **Vidaver, A. K.,** Biological control of plant pathogens with prokaryotes, in *Phytopathogenic Prokaryotes,* Vol. 2, Mount, M. S. and Lacy, G. H., Eds., Academic Press, New York, 1982, 387.
124. **Wang, P. J. and Hu, C. Y.,** Regeneration of virus-free plants through in vitro culture, *Adv. Biochem. Eng.,* 18, 61, 1980.
125. **Warren, G. J., Corollo, L. V., and Green, R. L.,** Conjugal transmission of integrating plasmids into *Pseudomonas syringae* and *Pseudomonas fluorescens,* 2nd Int. Symp. Mol. Gen. Bacteria Plant Interaction, Cornell University, at Ithaca, N.Y., (Abstr.), 1984.
126. **Watson, A. G. and Ford, E. J.,** Soil fungistasis — A reappraisal, *Annu. Rev. Phytopathol.,* 10, 327, 1972.
127. **Webber, J. F. and Heger, J. N.,** Comparison of interactions between *Ceratocystis ulmi* and Elm bark saprobes in vitro and in vivo, *Trans. Br. Mycol. Soc.,* 86(1), 93, 1986.
128. **Webster, J. M.,** Nematodes and biological control, in *Economic Nematology,* Webster, J. M., Ed., Academic Press, London, 1972, 469.
129. **Yoder, O. C.,** Recombinant DNA technology for fungal pathogens, *Phytopathology,* 73(Abstr.), 783, 1983.
130. **Anderson, J.,** *Futures,* Vol. 4, Michigan State University Agricultural Experimental Station, East Lansing, Mich., 1985, 7.

Chapter 2

BIOCONTROL AGENTS OF PLANT PATHOGENS: THEIR USE AND PRACTICAL CONSTRAINTS

R. S. Upadhyay and Bharat Rai

TABLE OF CONTENTS

I. INTRODUCTION

Biological control of plant pathogens has been considered as a potential control strategy in recent years. However, finding a practical approach to the biological control of plant pathogens has been considerably slow due to scientific reasons. Since 1963 a number of international symposia have been held on this subject and the subject has been reviewed extensively.[14,16-19,24,36-38,43,77,80,86,108,110,112,113,116,117,128-130,136,150] However, because of practical constraints there are only a few examples of the commercial use of antagonists as biocontrol agents.[38] Bowers[22] remarked that "a transitional period of several years will probably pass before microbial pest control agents are fully accepted for agricultural use."

Much has already been said and done about the chemical approach to disease control, and we have realized the problems associated with the use of hazardous chemicals for pest control. In recent years there has been a new awareness of the problems posed by the synthesis, production, and application of hazardous chemicals. The "Bhopal tragedy" in India has caught attention worldwide and is an example of the magnitude of the dangers involved in chemical pesticides. The Bhopal gas tragedy occurred on December 2, 1984 when methyl isocyanate leaked from a Union Carbide factory that was engaged in the manufacture and commercial production of a pesticide (Sevin). More than 2000 people were killed and over 50,000 were treated for injuries. This incident is considered the worst industrial accident in history. It is time to opt for nonchemical control of plant diseases. When plant pathologists started work on biological control the question was *why* biological control.[38] At the time they had no answer. However, with patience and a tremendous amount of effort, plant pathologists have now answered that question, and now they are asking *how* to use biological control.

This chapter will be devoted mainly to a review of this question, and to the practical constraints associated with the discovery, formulation, and use of biocontrol agents for successful plant disease control.

II. MASS PRODUCTION OF BIOCONTROL AGENTS

The mass production of biocontrol agents has been discussed by Kenny and Couch[73] and Churchill.[35] It has frequently been pointed out that if the biological control of plant pathogens is to be accomplished on a field scale it will be necessary to produce (*en masse*) biocontrol agents in the form of spores or other propagules. The viable inocula must be produced in an inexpensive medium and the cost of production for treatment of large areas must be competitive with that of the chemical pesticides. The stock of the biocontrol agent is generally made in soil cultures, lypholized cultures, agar slants, or spore suspensions, and is maintained at low temperatures.[35] This is done to make the working stocks uniform over an extended period of time. Since microorganisms lose their pathogenic potential when they are repeatedly subcultured in the laboratory, it is necessary to periodically determine the biocontrol potential of the antagonists. And, if the potential is found to be low, a new culture may be isolated from the biocontrol agent's original habitat. Churchill[35] has pointed out that in order to retain the viability of biocontrol agents stored for prolonged periods it is necessary to store the stock cultures in a medium that has the same osmotic concentration and nutrients as the growth medium.

The economics of the mass production of biocontrol agents is an important aspect and should be kept in mind because the success of any biocontrol technology will be judged by the economic feasibility of its implementation.[119] The media on which the microorganisms are mass produced are generally inexpensive and readily available. Churchill[35] suggested that the economic mass production of microorganisms can be achieved by using readily available crude agricultural products. In such crude products the level of carbon, nitrogen,

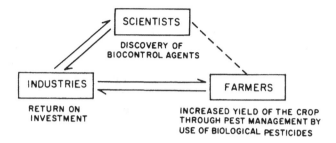

FIGURE 1. A model showing the interaction amongst scientists, industries, and farmers indicating their interests. (The dotted line indicates that there has been little interaction between the two components).

and minerals can be artificially balanced to give maximum spore yield. If the monetary returns of biocontrol technology are not equal or greater than those obtained from other control practices, farmers will hesitate to use these biocontrol methods. This is one of the greatest challenges faced by people advocating the use of biocontrol technology in agricultural practices. Maintenance of sterility in the mass inocula of biocontrol agents is another important problem. Bacteria frequently contaminate the microbial pesticides,[35] thus reducing the efficacy of the inocula. This can be avoided by doing all transfers of inocula through sterilized conditions to minimize the exposure of cultures to air-borne contaminants.

III. COMMERCIALIZATION OF MICROBIAL BIOCONTROL AGENTS

Commercialization of microorganisms for their use in biological control is now increasing and this aspect has been reviewed by Corke and Rishbeth[43] and Bowers.[22] However, microbial pesticides constitute less than 1% of the total pesticide sales in the world.[22] This indicates the slow development of the commercialization process of biocontrol agents. There are now approximately a dozen microbial pesticides registered by the U.S. Environmental Protection Agency. These include five bacteria, three viruses, one protozoan, and one fungus. To substantiate the need for more efforts in this area of research, Bowers[22] has pointed out that when this list is compared with that of approximately 1500 registered chemical pesticides it is easy to see that biological control agents have not been a major component of the modern agricultural pest management program.

In view of the increasing need for biological controls as an alternative means of pest control, industries should take a positive initiative towards commercializing biocontrol agents as microbial pesticides. So far as this aspect of the problem is concerned, both scientists and industry have their own constraints. For example, industries do not have expert personnel or the incentive to discover and study biocontrol agents and industries always look for a profitable return from their investment. For them this has not been guaranteed in the case of biological control agents. On the other hand, scientists are generally involved in the discovery of biocontrol agents during small scale experiments in the laboratory as well as in the field. They publish their results in scientific journals and do not care about commercialization. Most of these discoveries are a result of a research curriculum which they do for academic achievement and satisfaction. Thus, a collaboration of industry with scientists working in different universities, government agencies, and research institutions is strongly needed. With an integrated approach these agencies can develop a suitable technology for the application of microbes for the biological control of plant diseases (Figure 1). It has been suggested that in order for microbial pesticides to be accepted by the agricultural community they must be of high quality, efficacious, safe, and economical. For industry to develop and market the new biological control agents, patents must be obtained. Bowers[22] has stated that a patent may often be the key in an industry's involvement with development and marketing of a specific microbial biological control agent. The patenting procedures of the biological pesticides has not been well defined in most countries, particularly in third world countries. Prior to 1980 the U.S. Patent and Trademark office had maintained that micro-organisms were not patentable. However, since 1980, patenting microorganisms has been made possible.[52]

The following steps should be followed in sequence for the production of biocontrol agents:[22]

1. Discovery and identification of biocontrol agents (so far this has been the responsibility of the scientists only)
2. Efficacy test
3. Safety test (user's safety, environmental safety, and safety of nontarget organisms)
4. Test the genetic stability of the biocontrol agents (during their use biocontrol agents should not lose their virulence)
5. Study the potential for mass production
6. Formulation of biocontrol agents in a container which may enhance their efficacy
7. Test stability and shelf life
8. Test market potential
9. Evaluation of product cost
10. Return on investment analysis
11. Field trials
12. Patent the biocontrol agents
13. Commercialization and delivery of biopesticide products to the users

IV. MASS INTRODUCTION OF BIOCONTROL AGENTS INTO SOIL

Information on the mass introduction of biocontrol agents into soil is fragmented and unclear.[113] However, in recent years there has been a great emphasis on the methods of growth and delivery of antagonists into the natural system (Table 1). This has been reviewed briefly by Papavizas and Lewis[113] and Churchill.[35]

The mass production of bacterial biocontrol agents in nutrient broth has been used frequently.[23,86,94,95] The bacterial cultures obtained are then concentrated by centrifugation. Sun and Huang[139] used the *Arthobacter* sp. on a mixture of sugarcane, bagasse, and urea, and delivered this mixture into soil before planting the host.

The growth media used for the production of an inoculum of antagonistic fungi used in field studies have been summarized by Papavizas and Lewis.[113] *Trichoderma* spp. have frequently been grown on grain bran,[156] celatom and molasses,[15] bark pellet,[140] wheat straw,[6] composted hardwood bark,[66,101,102] chopped straw moistened by an acid-mineral solution,[44] wheat bran,[61,64] wheat bran and peat,[132] cereal meal and sand,[86] wheat bran and saw dust,[84] sand and corn meal,[84] colonized peat and sand mixtures or bark,[58] barley grains,[1] and pellet formulations.[12] Backman and Rodriguez-Kabana[15] have developed a method for the mass production and delivery of *Trichoderma* in diatomaceous earth impregnated with molasses as a food base. Commercialized products of *Trichoderma*, by the use of this and other methods, are now available.[113,120]

Potato dextrose has been used for the growth of *Chaetomium globosum*[78] and *Penicillium oxalicum*.[79] For biological control, *Coniothyrium minitans* was grown on milled rice[5] or barley, rye, and sunflower seed in a ratio of 1:1:1.[68] *Corticium* sp. was grown on corn-leaf meal[65] and sand cornmeal,[84] and *Gliocladium roseum* was grown on peat or soil.[96] These are effective media for the delivery of biocontrol agents into soil. The mode of introduction,

Table 1

SOME POTENTIAL BIOCONTROL AGENTS: THEIR MASS PRODUCTION AND USE IN THE BIOLOGICAL CONTROL OF PLANT PATHOGENS

Biocontrol agents	Habitat	Pathogens/diseases controlled	Mechanisms	Mass Production	Application	Ref.
Agrobacterium radiobacter K84	Soil, rhizosphere	Crown gall caused by *Agrobacterium radiobacter* pv. *tumefaciens* on a wide range of woody and herbaceous angiosperms	Pathogenic strains sensitive to strain K84 are prevented from transferring DNA from their tumor inducing (Ti) plasmid (TDNA) to the wounded host by the antagonist which either kills (by production of agrocin 84) or prevents attachment of the pathogen to the host receptor site	On solid nutrient agar, aqueous cell suspension, or in peat	Applied as cell suspension to the seeds, bare roots, graft unions, or cuttings. Transplants are inoculated before replanting with a suspension of cells of the biocontrol agent to give it an inoculum density greater than the pathogen	40, 41, 74, 97, 98
Bacillus subtilis	Soil	*Sclerotium cepivorum, Fusarium roseum* 'Graminearum' (seedling blight), *F. roseum* 'Avenaceum', *Pythium ultimum, Rhizoctonia solani* (damping-off), *Phytophthora cinamomi,* and *Nectria galligena* (leaf scars infection)	Antibiosis, inhibition of germination of sclerotia of *S. cepivorum* by the antagonist	Easily grown in mass culture	Inoculation on seeds. Not tried on mass scale	7, 23, 29, 106, 107, 141, 151
B. cereus	Soil	*Gaeumannomyces graminis* var. *tritici* (take-all of wheat)	Lysis of the hyphae	Not tried	Suspension of *B. cereus,* isolated from take-all decline soil, were introduced into soil	27

Table 1 (continued)
SOME POTENTIAL BIOCONTROL AGENTS: THEIR MASS PRODUCTION AND USE IN THE BIOLOGICAL CONTROL OF PLANT PATHOGENS

Biocontrol agents	Habitat	Pathogens/diseases controlled	Mechanisms	Mass Production	Application	Ref.
Chaetomium globosum	Decaying organic matter and soil	Seedling blight of corn by *F. roseum* 'Graminearum', *Fusarium nivale* on oat	Production of the antibiotic cochliodinol (?)	Grown on agar culture and on cellulosic material	Inoculation on seeds	29, 161
Mycophagous amoebae *Arachnula impatiens, Biomyxa vagans, Theratomyxa weberi, Vampyrella vorax V. lateritia*	Soil, water films around soil, water filled pores	*Cochliobolus sativus, Thielaviopsis basicola*	Penetration of spores' cell wall by the amoebae	Not tried	Field trials rare, have great potential in control of soil-borne plant pathogens	9, 10, 105
Coniothyrium minitans	Soil	*Sclerotinia sclerotiorum* on sunflower and beans, *S. trifoliorum* on clover, *Sclerotium cepivorum* on onion	A potential microparasite of many sclerotia forming fungi, *C. minitans* colonizes and penetrates the sclerotia and thereby reduces the inoculum of the pathogen in soil	Antagonist is grown on suitable media, Pycnidia and hyphal fragments are scraped from the surface of the culture and put in distilled water to make a suspension. Pycnidia are also mass produced on autoclaved milled rice in plastic bags at 20°C. Milled rice containing pycnidia is air dried for 1 week, ground, and sieved through a 600 μm sieve. The antagonist is also grown on a mixture of autoclaved barley, rye, and sunflower seeds in polyethylene bags	Sprayed on soil surface, seed coating of onion seeds in methyl-cellulose sticker	5, 68, 69, 143, 144, 145, 148, 149

Organism	Occurrence	Disease	Mechanism	Mass production	Application	Ref.
Gliocladium roseum	Soil	Black root of cucumber caused by *Phomopsis sclerotioides*	Biocontrol agent parasitizes the pathogen and kills it by enzymatic or toxic effect	Grown in sterile mixture of peat, soil, leaf compost, sand, conifer bark, and fresh barley flour	Applied in soil, used at a rate of 50g/1500 mℓ of soil	38, 96
G. virens	Soil	Damping-off of cotton seedlings caused by *Pythium ultimum* and *R. solani*, *Rhizoctonia* root rot of white bean on sclerotia of *Sclerotinia sclerotiorum*	Direct parasitism, produces antibiotics which are active against the pathogens	Not tried	Inoculation of soil through seeds	67, 146, 147
Glomus tulasne	Mycorrhizal association with large number of plants	*Phytophthora cinnamomi* on *Chamaecyparis lawsoniana*	The mycorrhiza enhances uptake of phosphorus and other nutrients thereby increasing host plant resistance against pathogens. Other mechanisms involved are production of antibiotics, stimulation of rhizosphere microflora, and induction of biochemical changes in host tissues unfavorable for the pathogens	Mass production is difficult because the biocontrol agent is an obligate parasite. Certain plants which stimulate chlamydospore formation may serve as a medium for the mass production of mycorrhizae. Soil may also serve as a medium for mass production. Studied inadequately	The medium containing the inoculum is mixed in soil. Naturally mixed population of VA mycorrhizae applied to *C. lawsoniana* 6 months before inoculation with *P. cinnamomi* gave 93.5% control of the root rot	20, 115, 127
G. fasciculatum	Same as *G. tulasne*	*Phytophthora parasitica* root-rot of citrus	Same as *G. tulasne*	Not tried	Not tried	45
Laetisaria arvalis (*Corticium* sp.)	Soil	Damping-off of beans, soybeans, and sugar beets caused by *R. solani*, fruit rot of cucumber caused by *R. solani*	The biocontrol agent prevents the saprophytic ability of *R. solani*, it is also parasitic on *R. solani*	The fungus may be grown commercially on diatomaceous earth granules for field inoculation; can also be produced on pellets of sugar-beet pulp or corn leaf meal	Added to soil as mycelium and sclerotia at 704 kg/ha. Also used as seed coating	25, 65, 84, 107

Table 1 (continued)
SOME POTENTIAL BIOCONTROL AGENTS: THEIR MASS PRODUCTION AND USE IN THE BIOLOGICAL
CONTROL OF PLANT PATHOGENS

Biocontrol agents	Habitat	Pathogens/diseases controlled	Mechanisms	Mass Production	Application	Ref.
Leucopaxillus cerealis var. *piceina*	Ectomycorrhizal fungus found on a variety of trees	Protects roots of *Pinus* against infection by *Phytophthora cinamomi*	Protection by providing physical barriers; also secretes an antibiotic diatretyne nitrile	Not tried	Not tried (Natural control)	90, 93
Panicillium oxalicum	Soil	Pre-emergence damping-off of peas by a complex of *Aphanomyces, Fusarium, Pythium,* and *Rhizoctonia* spp.	Forms a protective covering on seeds against the diseases; also uses the seed exudates to suppress the colonization of seeds by the pathogens	Grown on potato dextrose agar in Petri dishes and after 1 to 2 weeks the dry spores are harvested	The spores are inoculated on seeds. The treated seeds are stored at 5°C before sowing	79, 159, 160
Peniophora gigantea	Wood and bark in temperate areas	*Heterobasidion annosum* causing stump rot on conifers, Norway spruce, and Douglas fir	Hyphal interferrence, replacement of the pathogen from the stumps through competition	Grown on malt agar; oidia are harvested with a sucrose solution; talc and cellophos B600 are added in the suspension, poured into molds, and dessicated to form tablets	Each tablet is dissolved in 100 mℓ of water to give 1 × 10⁶ viable propagule/mℓ from 100 mℓ of the suspension. 100 stumps are inoculated. This biocontrol agent is now commercially available in England	59, 70, 123, 124, 125

Organism	Source	Target/disease	Mode of action	Mass production	Application	Ref.
Phialophora graminicola	Soil	Take-all of wheat caused by *Gaeumannomyces graminis* var. *tritici*	The antagonist occupies the infection court by occupying the cortical tissue thus reducing the pathogen's invasion of the roots	Mass production on oat grains in glass jars at 25°C. The infested grains are air dried and stored at 1.4°C	The infested grains are applied directly in soil	38, 46
Pisolithus tinctorius	Ectomycorrhizal association with a large number of forest trees	*Phytophthora cinnamomi* on pines	Produces hydroxyamate siderophore iron chelators which increase the uptake of iron from the soil, forms mechanical barrier, and produces antibiotics	Mass production of spores and mycelia is obtained by growing susceptible plants in infested soil	Mycelium has been used for inoculation in nurseries	91, 92, 126, 142
Pseudomonas fluorescens	Soil, rhizosphere	Brown blotch of mushroom caused by *P. totaasi*, responsible for natural suppression of *Gaeumannomyces graminis* var. *tritici* in the wheat rhizosphere. Natural suppression of take-all in soil. Also active against many fungi: *Sclerotinia sclerotiorum, Phymatotrichum omnivorum, Phytophthora megasperma, Pythium aphanidermatum, Thielavia basicola, Verticillium dehliae,* and *Alternaria* spp.	Competition for nutrients with *P. totaasi,* produces a siderophore (Pseudobactiri, that deprives the pathogen of iron). Also induces natural suppression of soils against *Gaeumannomyces graminis* var. *tritici* and *F. oxysporum* f. sp. *lini*	Easily grown on culture media in Petri dishes. Mass culture on gamma irradiated peat used in polyethylene bags. Dry powder formula of the agent is obtained by inoculating xanthan gum with the bacterial suspension (10^9 cells/mℓ) and dry talc. Suspension of the bacterial cell is also produced in 1.5% methyl cellulose	The inoculum was mixed with neutralized horticultural grade peat used in casing mushroom beds. Seed inoculation at a rate of 10^7—10^8 cells/seed. The seeds were dried before sowing. Seed inoculation during sowing	39, 99, 133, 134, 135
Pythium oligandrum	Soil, crop residues	Suppresses damping-off caused by *P. ultimum* on many plants	Antagonism is through mycoparasitism	Not tried	Sugar-beet seeds are treated with a suspension of mycelia or oospores before sowing	47, 88, 89, 152

Table 1 (continued)

SOME POTENTIAL BIOCONTROL AGENTS: THEIR MASS PRODUCTION AND USE IN THE BIOLOGICAL CONTROL OF PLANT PATHOGENS

Biocontrol agents	Habitat	Pathogens/diseases controlled	Mechanisms	Mass Production	Application	Ref.
Scytalidium uredinicola	As mycoparasite of rust	Rust of slash and loblolly pine, decay of Douglas-fir caused by *Poria carbonica*	It reduces the amount of aeciospore inoculum available to infect oak (an alternate host), produces a thermostable antibiotic (scytalidin), and suppresses aeciospore germination	Grows readily on agar media	Ricard and Laird[121] inoculated standing conifers with the biocontrol agent by firing metal tipped inoculated birch dowels into the trunk with a rifle. Ricard et al.[122] drilled holes in Douglas-fir poles and pushed softwood dowels impregnated with the antagonist into the holes to prevent decay	82, 83, 121, 122
Sporidesmium sclerotivorum	Soil	Lettuce drop caused by *Sclerotinia minor*. The biocontrol agent is also a potent mycoparasite of the sclerotia of *Sclerotinia sclerotiorum*, *S. trifoliorum*, *Sclerotium cepivorum*, and *Botrytis cinerea*	Parasitism and destruction of the sclerotia of the pathogens	Can be grown on host sclerotia on quartz sand at 24—26°C. Conidia are then harvested	The biocontrol agent is added to field soil	3, 4, 13, 14

Organism	Substrate	Disease/use	Mode of action	Production	Application	Ref.
Trichoderma viride	Soil	Damping-off of seedlings caused by *R. solani*, also effective in decreasing the number of sclerotia of *S. sclerotiorum* in soil	Mycoparasitic on many pathogens; involved in natural competition with other microorganisms; host hyphae are disintegrated by mycoparasitism; also produces antibiotics	Commercially produced as pellets (commercial name BINAB T SEPPIC). Also produced on wheat bran, saw dust, and tap water (3:1:4). Have been produced on a variety of growth media (autoclaved barley, rye, and sunflower seeds)	Applied directly to soil along with the food base	50, 51, 69, 153, 154
T. harzianum	Soil	Damping-off of cotton, radish, and snap bean, responsible for suppressing *R. solani* in soil. Diseases caused by *S. rolfsii* and *R. solani*	Same as *T. viride*	As in *T. viride*, also produced on molasses and enriched clay granules as a food base	Backman and Rodriguez-Kabana[15] applied it at a rate of 140 kg/ha after 70 days of planting	15, 30, 53, 54, 55, 56, 57, 64, 87, 156
T. hamatum	Soil	Damping-off of cotton, responsible for suppressing *R. solani* in soil. Protection of pea and radish seedlings from *Pythium* spp. and *R. solani*	Mycoparasitism, produces (1-3)-glucanase and chitinase which act on hyphae. Also produces cellulase which acts on the cell wall of *Pythium* spp.	Not tried	Not tried	30, 31, 56, 62, 63
Tuberculina maxima	On cankers and galls caused by *Cronartium* spp.	White-pine blister rust caused by *Cronartium ribicola*. Also attacks *C. quercuum* f. sp. *fusiforme*	Attacks the aecial cankers in white pines, grows into the pycnia and aecia, and kills the mycelium	Can be readily grown on agar media	Spore suspension applied on rust cankers	75, 82, 157
Verticillium lecanii	Soil	Rust in carnations caused by *Uromyces dianthi*, rust of bean by *U. appendiculatus*	Attacks the rust sori, reduces the number of rust pustules	Grown in submerged cultures, then transferred to husk millet. The millet is ground in a ball mill for application	Spore suspensions sprayed on plants	8, 137, 138

in most cases, was a broadcast method which has been recommended as a dry weight preparation of 250 to 500 kg/ha. In recent years much emphasis has been placed on developing a growth medium and delivery systems for biocontrol agents. Papavizas and Lewis[113] have pointed out that the composition of a growth media determines the efficacy of the biocontrol agents. For the effective use of biocontrol agents in soil a proper food base is necessary in the soil, otherwise, the fungal propagules may be prevented from germinating and growing well due to soil fungistasis. Wright[162,163] has shown that a proper food base is essential for the production of antibiotics by a particular microorganism in soil. The necessity of an appropriate foodbase for successful biological control has also been advocated by Moody and Gindrat,[96] Ahmed and Tribe,[5] and Lewis and Papavizas.[84] Therefore, much more work is still needed in this direction for developing successful processes for biological control of pathogens.

V. TECHNIQUES FOR APPLICATION OF BIOCONTROL AGENTS

Exploration into techniques for the accurate application of biocontrol agents to specific sites is needed. Available field data show that at least some success has been achieved in this area of research. For example, a commercial inoculum of *Peniophora gigantea* spores is now distributed in sachets by Ecological Laboratories Ltd., c/o John Lawrence (Dover Ltd., England). The storage life of the biocontrol agent is 4 months at 20°C or less. To apply the biocontrol agent the contents of a sachet are mixed well with 5 ℓ of water, and the spore suspension obtained is applied to the stump surface from a spouted polyethylene container and spread with a brush.[43,124] For the biological control of wound pathogens (particularly wounds developed by pruning) modification of pruning shears is done in such a way as to deliver the biocontrol agent directly to the pruned branches. This technique has been developed by Grosclaude et al.[60] and Carter and Mullett.[28] With this device the biocontrol agent may be delivered by a pruning action. The inoculum suspension, enclosed in a container, is forced through a tube leading to the cutting blade where it comes in contact with the cuttings. Other methods for delivering antagonists to plants are (1) forcing impregnated hard wood dowels into timber with a nail gun[122] and (2) firing metal tipped birch dowels into the trunk with a rifle.[121] Corke[42] inoculated the antagonists into trunks using wooden dowels impregnated with *Trichoderma viride*. He pushed the dowels into holes drilled into the trunk and then sealed the holes with a nontoxic dressing.

Lewis and Papavizas[84] grew *Corticium* sp., *Trichoderma hamatum, T. harzianum,* and *Gliocladium roseum* on sterilized sand and corn meal (96% quartz sand + 4% corn meal; water 20% v/w) for 1 month. To control belly-rot caused by *Rhizoctonia solani*, the antagonists were broadcast in the field (700 kg wet weight per hectare) 1 day before planting cucumber seeds. *Corticium* and *Trichoderma* reduced the fruit-rot to a level comparable with Captafol. Based on this experiment, Papavizas and Lewis[113] concluded that the method of inoculum preparation of biocontrol agents, their mode of application, use of different food bases, and an integrated application of other chemicals or cultural practices can improve the effectiveness of the biocontrol agents. The application of biocontrol agents by broadcast spreaders, or by direct application in the seed furrow at the time of sowing, has the greatest potential for use in commercial glasshouse operations or small scale field experiments.[38] For biological control of root pathogens Cook and Baker[38] have suggested the introduction of antagonists as a seed treatment. This has been clearly demonstrated in the experiments of Weller and Cook[155] and Wilkinson et al.[158] This method gives the antagonist an opportunity to possess a great ''competitive advantage'' when introduced on the seed rather than in soil because in the soil the roots must be within convenient reach of the antagonist in order for the antagonist to colonize on the roots before the pathogen.

VI. PROGRAMS FOR IMPROVING THE EFFECTIVENESS OF BIOCONTROL AGENTS

Genetic manipulation of biocontrol agents offers a possible approach to improving their potential for plant disease control. This approach has been very successful in industrial microbiology. Efforts have been made in this direction by the research workers involved in biological control of insect pests.[26] It has been visualized that the next decade will see a wide exploitation of genetic engineering and biotechnology in the service of biological control.[38,111] Papavizas'[111] work is an example of this area of study. He induced a mutation in *Trichoderma harzianum* by UV-irradiation thus amplifying its biocontrol potential and enabling it to control *Rhizoctonia* damping-off of cotton and radishes, white-rot of onion caused by *Sclerotium cepivorum*, damping-off and blight of bean caused by *Sclerotium rolfsii*, and damping-off of peas by *Pythium ultimum*.[111] The mutation also improved the tolerance of *T. harzianum* to Benomyl as well as its survival in soil. Fungicide tolerant biotypes were also developed by exposing the conidia of *Trichoderma* to increasingly higher concentrations of the fungicides Ronilan®,[109] Chlorothalonil, Procymidone, Iprodion, and Vinclozolin[2] in culture media. *T. harzianum* biotype Th-1 (Ron-2M), which was adapted to 2000 μg active ingredient (a.i.) Ronilan® per milliliter of medium, suppressed the white-rot of onion more effectively than the wild type. New Benomyl resistant biotypes of *T. viride*[85] and *T. harzianum*[114] have also been isolated and possess better biocontrol capabilities than their respective wild types. A new biotype of *T. viride,* i.e., T-1-R$_9$ isolated recently by these workers very effectively controlled *Fusarium* wilt of chrysanthemum[85] and *Rhizoctonia* scurf of potato.[21] In a similar study by Howell,[67] he reported a higher production of the antibiotic glioviridin by the mutant strains of *Gliocladium virens* than the wild type strains.

VII. INTEGRATED USE OF BIOCONTROL AGENTS

Although it is well established that certain diseases can be controlled either completely or partially by the use of biocontrol agents it is understood that the best method to control diseases is through integrated pest management wherein a biological control component would be significant. In the past, biocontrol agents have only been used in the biological control of plant pathogens. Their integrated use with other management practices has been inadequately studied and practiced. Our experience with antagonists as the only component in biological control programs has not been very encouraging. The integrated biological control of plant pathogens has been discussed by several workers.[17,34,38,53,66,76] In recent reviews by Papavizas[111] and Chet and Henis,[32] the use of the *Trichoderma* species as a potential biocontrol agent in the integrated biological control of plant pathogens, along with other pest management practices, has been discussed. It has been pointed out that since *Trichoderma* dominates in soil treated with a sublethal dosage of pesticides, proliferates easily to produce antibiotics, competes for nutrients, and acts as a mycoparasite, its role in integrated biological control may be of practical significance.

Pest management practices used jointly with biocontrol agents include the use of pesticides and resistant varieties as well as other cultural methods[76] including solarization.[34,71] Important pathogens have been successfully controlled by an integrated approach using biocontrol agents and pesticides. These include:

1.   *Armellaria mellea* on citrus by *Trichoderma* + methyl bromide[104]
2.   *Rhizoctonia solani* on radish and eggplant by *T. harzianum* + PCNB[51,63,64]
3.   *R. solani* on cucumber by *T. harzianum* + chlorothalonil[84]
4.   *Sclerotium rolfsii* on bean seedlings by *T. harzianum* + PCNB[33]

5.   *Phytophthora capsici* on pepper by *T. harzianum* + Ridonil®[113]
6.   *Pythium ultimum* on pea by *T. harzianum* + metalaxyl[81]
7.   Bulb-borne pathogens on iris by *T. harzianum* + PCNB[34]

From these examples it is evident that *Trichoderma harzianum* has been the most important biocontrol agent used in integrated biological control. Papavizas[111] has genetically modified some of the strains of *Trichoderma* for their improved biocontrol potential. The new biotypes of *Trichoderma* tolerated some fungicides and possessed better disease control ability than the wild type strains. These studies have led to another technological advancement for the use of biocontrol agents in integrated pest management programs and have a great scope for future research.

The integrated use of biological control with resistant cultivars has been discussed by Klassen[76] taking Baker and Cook's[17] example of foot-rot of wheat caused by *Fusarium roseum* 'culmorum' (*F. culmorum*). Baker and Cook[17] found that straw residue left on the soil surface permitted the development of antagonists against *F. culmorum*, and the antagonists preempted the substrates, that could have been saprophytically colonized by the pathogen, and thus suppressing them in the habitat. Various steps used in this system have been listed by Klassen.[76]

Recently Kraft and Papavizas[81] used host resistance as a component, along with the biocontrol agent *Trichoderma* and fungicides, for integrated control of some soil-borne diseases of peas. They recorded a maximum yield of peas under field conditions when the seeds of the cultivar (Dark Skin Perfection) susceptible to *Pythium ultimum* were treated with metalaxyl and *Trichoderma harzianum*.

Some success has also been achieved in the integrated use of biocontrol agents with cultural practices. Lewis and Papavizas[84] used the *Corticium* sp. and *Trichoderma* sp. (isolate WT-6) in integration with mechanical plowing to suppress belly-rot of cucumber caused by *R. solani*. The use of antagonists along with plowing controlled the disease more effectively than that of either fungicides or antagonists. Another example of the integrated use of biocontrol agents with cultural practices is the displacement of *Phellinus weirii* in roots of alder by *Trichoderma*, combined with urea.[100]

Enhanced control of several soil-borne diseases by the integrated use of the *Trichoderma* species and soil solarization has been demonstrated.[34,53,72] These workers have observed an enhanced control of the diseases caused by *Rhizoctonia solani* and *Sclerotium rolfsii* by a combination of *T. harzianum* and solarization. The solarization technique has a great potential in biological control particularly when used in conjunction with the biocontrol tactics.[71,72] In this process the pathogens are weakened or killed by a synergistic effect of the two components. The solarization, which involves solar heating (a physical phenomenon), makes the pathogens more vulnerable to antagonism by the biocontrol agents. Another practical application of this technique is that at times the population level of the resident antagonist is augmented by natural solarization and therefore, by using this technique, we can avoid using alien antagonists which are not generally acceptable to the new habitat.

It is evident from the examples discussed here that the integrated use of biocontrol agents with other disease control practices is a better prospect than either used alone. Because of this, in recent years there has been much emphasis placed on the integrated biological control of plant pathogens using different useful components. This approach may prove more effective where no single component is effective in controlling a particular disease. Klassen has stated that a "high priority should be given to systems research to optimize the joint use of biological control agents with other control methods in integrated pest management for insect pests, plant pathogens, nematodes, and weeds."[76]

VIII. PRACTICAL CONSTRAINTS IN THE USE OF BIOCONTROL AGENTS

The following are constraints involved in the practical use of biological pesticides:

A. Economic Considerations

The success of biological control depends on its feasibility in terms of monetary returns. Reichelderfer stated that "should be greater than or equal to those realized from alternative technologies or no control."[119] Biological control will never be adopted by farmers unless it provides them with more incentive than other conventional control practices. This is one of the most important practical constraints related to the use of biocontrol practices, because most of the controls developed for plant diseases have failed to fulfill this aspiration. Recently, Reichelderfer[118,119] reviewed the economic feasibility of the biological control of crop pests. He pointed out that comparatively little research has been conducted in an agricultural economic framework to compare biological control with chemical control options. Such attempts have been made only with some weeds and insect pests.[11,48,49,118,119,131] Reports on economics of biological control with respect to plant diseases are rare. This might be because biocontrol scientists are still involved in finding suitable and practical biocontrol agents and there have been only a few biocontrol agents which have been used on a commercial scale. Another problem associated with it is that biocontrol practices leave a long-term effect on plant pathogens as compared to a short-term kill (mostly one crop season) by chemical control. The period of time of effectiveness of any biocontrol agent on a target pathogen is not certain and assessing the economics of the biological control is difficult. Reichelderfer has stated that "economists are latecomers to the biocontrol evaluation scene."[119]

In evaluating the economic feasibility of biocontrol practices two parameters are important: (1) the price of the biological control agents and (2) the cost associated with utilizing the biocontrol agent. The latter one involves suitable technology for delivering the biocontrol agent. Apart from these constraints, the relative utility of the other control tactics are always kept in mind for economic advantage or disadvantage over the biological control.[119] A comparison of biocontrol tactics should not be made with chemical control tactics because, as pointed out by Reichelderfer,[119] the efficacy of biological control will always be lower than that of chemical control since biological control does not eliminate the pathogens from the habitat; rather, as a result of interaction, it suppresses the disease outbreak. He pointed out that, unfortunately, such comparisons have always been over-emphasized while advocating the profitable use of biological control. Biological controls have many advantages over the chemical control of diseases including a decrease in environmental risk during their manufacture and use.

Another factor involved in the economics of biological control is the value of the crop for which biocontrol practice is to be adopted. The higher the value of crop the greater the value of the protection of the crop by antagonists. Reichelderfer[119] has listed the following characteristics of biocontrol practices for their acceptability and feasibility

1. The target organism is consistently occurring, and a light or moderately damaging single major pest species of a high value crop
2. The biocontrol agent is highly efficacious, relatively risk free, and its effect on the pest population is not highly variable
3. The price of the biocontrol agent, if it is marketed, is low and the research cost to develop the option are justified by the economic impact of the target pest
4. The biocontrol option compares favorably with available nonbiological control alternatives

B. Geographic Variability in the Potential Use of Biocontrol Agents

The biocontrol agents used at one specific geographic location may not do well in other areas. To overcome this problem large scale field trials should be conducted at different geographic locations in order to evaluate the potential of applicability of the biocontrol agents under different environmental conditions. This "efficacy test" also provides data on the optimum conditions necessary for the use of biological pesticides.

C. Other Constraints

• Biological control agents should compete well with chemical pesticides. So far there has been an inadequate amount of technological development of biocontrol agents that can compete well with chemical pesticides. This process is developing very slowly due to various practical constraints.

• Safety testing is mandatory for the practical application of biocontrol agents. Safety testing involves time consuming bioassays of biopesticides for their hazardous impact on various life processes.[12]

• Assurance that the products delivered to the users have a high level of potency, and purity is needed and should be guaranteed on the products' labels.

• Biological control agents are living entities. Therefore handling, transportation, and storage conditions need to be specifically defined, a procedure which is not necessary for chemical pesticides. Furthermore, biocontrol agents depend more on environmental conditions and are more susceptible to them in comparison to pesticides.

• Biological control agents may have a short-term expiration date. Therefore, the expiration date must be carefully examined and placed on the product's label.

• Some broad spectrum pesticides can control many plant pathogens, whereas the biocontrol agents are specific to only one target pathogen. Because of this farmers prefer pesticides over these new methods of disease control.

The above mentioned points add a negative impact to the practical feasibility of biological control. However, there are many advantages to the use of biocontrol agents and these have been discussed frequently elsewhere.[17,38]

REFERENCES

1. **Abd-El-Moity, T. H. and Shatla, M. N.,** Biological control of white rot disease *Sclerotium cepivorum* of onion by *Trichoderma harzianum, Phytopathol. Z.,* 100, 29, 1981.
2. **Abd-El-Moity, T. H., Papavizas, G. C., and Shatla, M. N.,** Induction of new isolates of *Trichoderma harzianum* tolerant to fungicides and their experimental use for biological control of white rot of onion, *Phytopathology,* 77, 396, 1982.
3. **Adams, P. B. and Ayers, W. A.,** Factors affecting parasitic activity of *Sporidesmium sclerotivorum* on sclerotia of *Sclerotinia minor* in soil, *Phytopathology,* 70, 366, 1980.
4. **Adams, P. G. and Ayers, W. A.,** Biological control of *Sclerotinia* lettuce drop in the field by *Sporidesmium sclerotivorum, Phytopathology,* 72, 485, 1982.
5. **Ahmed, A. H. M. and Tribe, H. T.,** Biological control of white rot of onion (*Sclerotium cepivorum*) by *Coniothyrium minitans, Plant Pathol.,* 26, 75, 1977.
6. **Akhtar, C. M.,** Biological control of some plant diseases lacking genetic resistance of the host crops in Pakistan, *Ann. N.Y. Acad. Sci.,* 287, 45, 1977.
7. **Aldrich, J. and Baker, R.,** Biological control of *Fusarium roseum* f. sp. *dianthi* by *Bacillus subtilis. Plant Dis. Rep.,* 54, 446, 1970.
8. **Allen, D. J.,** *Verticillium lecanii* on the bean rust fungus, *Uromyces appendiculatus, Trans. Br. Mycol. Soc.,* 79, 362, 1982.
9. **Anderson, T. R. and Patrick, Z. A.,** Mycophagous amoeboid organisms from soil that perforate spores of *Thielaviopsis basicola* and *Cochliobolus sativus, Phytopathology,* 68, 1618, 1978.

10. **Anderson, T. R. and Patrick, Z. A.,** Soil vampyrellid amoebae that cause small perforations in conidia of *Cochliobolus sativus, Soil Biol. Biochem.,* 12, 159, 1980.

11. **Andres, L. A.,** The economics of biological control of weeds, *Aquat. Bot.,* 3, 111, 1977.

12. **Anon.,** Guidelines for registering pesticides in the U.S., subpart M, in *Data Requirements for Biorational Pesticides: Preamble and Guidelines,* Office of Pesticide Programs, Hazardous Evaluation Division, Environmental Protection Agency, Washington, D.C., 1980, 433.

13. **Ayers, W. A. and Adams, P. B.,** Mycoparasitism of sclerotia of *Sclerotinia* and *Sclerotium* species by *Sporidesmium sclerotivorum, Can. J. Microbiol.,* 25, 17, 1979.

14. **Ayers, W. A. and Adams, P. B.,** Mycoparasitism and its application to biological control of plant diseases, in *Biological Control in Crop Production,* Papavizas, G. C., Ed., Allenheld, Totowa, N.J., 1981, 91.

15. **Backman, P. A. and Rodriguez-Kabana, R.,** A system for the growth and delivery of biological control agents to the soil, *Phytopathology,* 65, 819, 1975.

16. **Baker, K. F.,** Biological control, in *Fungal Wilt Diseases of Plants,* Mace, M. E. et al., Eds., Academic Press, London, 1981, 523.

17. **Baker, K. F. and Cook, R. J.,** *Biological Control of Plant Pathogens,* W. H. Freeman, San Francisco, 1974, 433.

18. **Baker, K. F. and Snyder, W. C., Eds.,** *Ecology of Soil-Borne Plant Pathogens. Prelude to Biological Control,* University of California Press, Berkeley, Calif., 1965, 571.

19. **Baker, R.,** Mechanisms of biological control of soil-borne pathogens, *Annu. Rev. Phytopathol.,* 6, 263, 1968.

20. **Bartschi, H., Gianinazzi-Pearson, V., and Vegh, I.,** Vesicular arbuscular mycorrhiza formation and root rot disease (*Phytophthora cinamomi*) development in *Chamaecyparis lawsoniana, Phytopathol. Z.,* 102, 213, 1981.

21. **Beagle-Ristaino, J. and Papavizas, G. C.,** Reduction of *Rhizoctonia solani* in soil with fermentor preparations of *Trichoderma* and *Gliocladium, Phytopathology,* 74(Abstr.), 836, 1984.

22. **Bowers, R. C.,** Commercialization of microbial biological control agents, in *Biological Control of Weeds with Plant Pathogens,* Charudattan, R. and Walker, H. L., Eds., John Wiley & Sons, New York, 1982, 157.

23. **Broadbent, P., Baker, K. F., and Waterworth, Y.,** Bacteria and actinomycetes antagonistic to fungal root pathogens in Australian soils, *Aust. J. Biol. Sci.,* 24, 925, 1971.

24. **Bruehl, G. W., Ed.,** *Biology and Control of Soil-Borne Plant Pathogens,* American Phytopathological Society, St. Paul, Minn., 1975, 216.

25. **Burdsall, H. H., Jr., Hoch, H. C., Boosalis, M. G., and Setliff, E. C.,** *Laetisaria arvalis* (Aphyllophorales, Corticiaceae). A possible biological control agent for *Rhizoctonia solani* and *Pythium* species, *Mycologia,* 72, 728, 1980.

26. **Burges, H. D., Ed.,** *Microbial Control of Pests and Plant Diseases 1970—1980,* Academic Press, London, 1981, 949.

27. **Campbell, R. and Faull, J. L.,** Biological control of *Gaeumannomyces graminis*: Field trials and the ultrastructure of the interaction between the fungus and a successful antagonistic bacterium, in *Soil-borne Plant Pathogens,* Schippers, B. and Gams, W., Eds., Academic Press, London, 1979, 603.

28. **Carter, M. V. and Mullett, L. F.,** Biological control of *Eutypa armeniacae*. IV. Design and performance of an applicator for metered delivery of protective aerosols during prunning, *Aust. J. Exp. Agric. Anim. Husb.,* 18, 287, 1978.

29. **Chang, I. P. and Kommedahl, T.,** Biological control of seedling blight of corn by coating kernels with antagonistic microorganisms, *Phytopathology,* 58, 1395, 1968.

30. **Chet, I. and Baker, R.,** Induction of suppressiveness to *Rhizoctonia solani* in soil, *Phytopathology,* 70, 994, 1980.

31. **Chet, I. and Baker, R.,** Isolation and biocontrol potential of *Trichoderma hamatum* from soil naturally suppressive to *Rhizoctonia solani, Phytopathology,* 71, 286, 1981.

32. **Chet, I. and Henis, Y.,** *Trichoderma* as a biocontrol agent against soilborne root pathogens, in *Ecology and Management of Soil-borne Plant Pathogens,* Parker, C. A. et al., Eds., American Phytopathological Society, St. Paul, Minn., 1985, 110.

33. **Chet, I., Hadar, Y., Elad, Y., Katan, J., and Henis, Y.,** Biological control of soil-borne pathogens by *Trichoderma harzianum,* in *Soil-borne Plant Pathogens,* Schippers, B. and Gams, W., Eds., Academic Press, London, 1979, 585.

34. **Chet, I., Elad, A., Kalfon, Y., Hadar, Y., and Katan, J.,** Integrated control of soil-borne and bulbborne pathogens in iris, *Phytoparasitica,* 10, 229, 1982.

35. **Churchill, W. B.,** Mass production of microorganisms for biological control, in *Biological Control of Weeds with Plant Pathogens,* Charudattan, R. and Walker, H. L., Eds., John Wiley & Sons, New York, 1982, 139.

36. **Cook, R. J.,** Management of associated microbiota, in *Plant Disease: An Advanced Treatise,* Vol. 1, Horsfall, J. G. and Cowling, E. B., Eds., Academic Press, New York, 1977, 145.

37. **Cook, R. J.**, Biological control of plant pathogens: overview, in *Biological Control in Crop Production*, Papavizas, G. C., Ed., Allenheld, Totowa, N.J., 1981, 23.
38. **Cook, R. J. and Baker, K. F.**, *The Nature and Practice of Biological Control of Plant Pathogens*, American Phytopathological Society, St. Paul, Minn., 1983, 539.
39. **Cook, R. J. and Rovira, A. D.**, The role of bacteria in the biological control of *Gaeumannomyces graminis* by suppressive soils, *Soil Biol. Biochem.*, 8, 267, 1976.
40. **Cooksey, D. A. and Moore, L. W.**, Biological control of crown gall with an agrocin mutant of *Agrobacterium radiobacter*, *Phytopathology*, 72, 919, 1982.
41. **Cooksey, D. A. and Moore, L. W.**, High frequency spontaneous mutations to agrocin 84 resistance in *Agrobacterium tumefaciens* and *A. rhizogenes*, *Physiol. Plant Path.*, 20, 129, 1982.
42. **Corke, A. T. K.**, Interaction between microorganisms, *Ann. Appl. Biol.*, 89, 89, 1978.
43. **Corke, A. T. K. and Rishbeth, J.**, Use of microorganisms to control plant diseases, in *Microbial Control of Pests and Plant Diseases 1970—1980*, Burges, H. D., Ed., Academic Press, London, 1981, 718.
44. **Davet, P., Artigues, M., and Martin, C.**, Production en conditions non aseptiques d' inoculum de *Trichoderma harzianum* Rifai pour des essais de lette biologique, *Agronomie*, 1, 933, 1981.
45. **Davis, R. M. and Menge, J. A.**, Influence of *Glomus fasiculatum* and soil phosphorus on *Phytophthora* root rot of citrus, *Phytopathology*, 70, 447, 1980.
46. **Deacon, J. W.**, Biological control of the take-all fungus, *Gaeumannomyces graminis*, by *Phialophora radicicola* and similar fungi, *Soil Biol. Biochem.*, 8, 275, 1976.
47. **Deacon, J. W.**, Studies on *Pythium oligandrum*, an aggressive parasite of other fungi, *Trans. Br. Mycol. Soc.*, 66, 383, 1976.
48. **Dean, L. A., Schuster, M. F., Boling, J. C., and Riherd, P. T.**, Complete biological control of *Antonina graminis* in Texas with *Neodusmetia sangwani* (a classical example), *Entomol. Soc. Am. Bull.*, 25, 262, 1979.
49. **DeBach, P.**, The scope of biological control, in *Biological Control of Insects and Pests*, DeBach, P., Ed., Reinhold, New York, 1964, 3.
50. **Denis, C. and Webster, J.**, Antagonistic properties of species-group of *Trichoderma*. I. Production of non-volatile antibiotics, *Trans. Br. Mycol. Soc.*, 57, 25, 1971.
51. **Denis, C. and Webster, J.**, Antagonistic properties of species group of *Trichoderma*. II. Production of volatile antibiotics, *Trans. Br. Mycol. Soc.*, 57, 41, 1971.
52. **Dimond, S.**, Commissioner of Patents and Trademarks v. Chakrabarty, A. M., Supreme Court of the United States, *U.S. Quarterly*, 206, 193, 1980.
53. **Elad, Y., Chet, I., and Katan, J.**, *Trichoderma harzianum*: a biocontrol agent effective against *Sclerotium rolfsii* and *Rhizoctonia solani*, *Phytopathology*, 70, 119, 1980.
54. **Elad, Y., Katan, J., and Chet, I.**, Physical, biological, and chemical control integrated for soil-borne diseases in potatoes, *Phytopathology*, 70, 418, 1980.
55. **Elad, Y., Hadar, Y., Hadar, E., Chet, I., and Henis, Y.**, Biological control of *Rhizoctonia solani* by *Trichoderma harzianum* in carnation, *Plant Dis.*, 65, 675, 1981.
56. **Elad, Y., Hadar, Y., Chet, I., and Henis, Y.**, Prevention with *Trichoderma harzianum* Rifai aggr. of reinfestation by *Sclerotium rolfsii* Sacc. and *Rhizoctonia solani* Kühn of soil fumigated with methylbromide and improvement of disease control in tomatoes and plant, *Crop Prot.*, 1, 199, 1982.
57. **Elad, Y., Kalfon, A., and Chet, I.**, Control of *Rhizoctonia solani* in cotton by seed-coating with *Trichoderma* spp., *Plant Soil*, 66, 279, 1982.
58. **Gindrat, D., Van der Hoeven, E., and Moody, A. R.**, Control of *Phomopsis sclerotioides* with *Gliocladium* or *Trichoderma*, *Neth. J. Plant Pathol.*, 83 (Suppl. 1), 429, 1977.
59. **Greig, B. J. W.**, Biological control of *Fomes annosus* by *Peniophora gigantea*, *Eur. J. For. Pathol.*, 6, 65, 1976.
60. **Grosclaude, C., Ricard, J., and Dubos, B.**, Inoculation of *Trichoderma viride* spores via prunning shears for biological control of *Stereum purpureum* on plum tree wounds, *Plant Dis. Rep.*, 57, 25, 1973.
61. **Hadar, Y., Chet, I., and Henis, Y.**, Biological control of *Rhizoctonia solani* damping-off with wheat bran culture of *Trichoderma harzianum*, *Phytopathology*, 69, 64, 1979.
62. **Harman, G. E., Chet, I., and Baker, R.**, *Trichoderma hamatum* effects on seed and seedling disease induced in radish and pea by *Pythium* spp. or *Rhizoctonia solani*, *Phytopathology*, 70, 1167, 1980.
63. **Harman, G. E., Chet, I., and Baker, R.**, Factors affecting *Trichoderma hamatum* applied to seeds as a biocontrol agent, *Phytopathology*, 71, 569, 1981.
64. **Henis, Y., Ghaffar, A., and Baker, R.**, Integrated control of *Rhizoctonia solani* damping-off of radish: effect of successive plantings, PCNB, and *Trichoderma harzianum* on pathogen and disease, *Phytopathology*, 68, 900, 1978.
65. **Hoch, H. C. and Abawi, G. S.**, Biological control of *Pythium* root rot of table beet with *Corticium* sp., *Phytopathology*, 69, 417, 1979.
66. **Hoitink, H. A. J.**, Composted bark, a light weight growth medium with fungicidal properties, *Plant Dis.*, 64, 142, 1980.

67. **Howell, C. R.,** Effect of *Gliocladium virens* on *Pythium ultimum, Rhizoctonia solani,* and damping-off of cotton seedlings, *Phytopathology,* 72, 496, 1982.
68. **Huang, H. C.,** Biological control in *Sclerotinia* wilt in sunflower, *Ann. Conf. Manitoba Agron. 1976,* 1976, 69.
69. **Huang, H. C.,** Control of *Sclerotinia* wilt of sunflower by hyperparasites, *Can. J. Plant Pathol.,* 2, 26, 1980.
70. **Ikediugwu, F. E. O. and Webster, J.,** Hyphal interference by *Peniophora gigantea* against *Heterobasidion annosum, Trans. Br. Mycol. Soc.,* 54, 307, 1970.
71. **Katan, J.,** Solar heating (solarization) of soil for control of soil-borne pests, *Annu. Rev. Phytopathol.,* 19, 211, 1981.
72. **Katan, J.,** Solar disinfestation of soils, in *Ecology and Management of Soil-borne Plant Pathogens,* Parker C. A., et al., Eds. American Phytopathological Society, St. Paul, Minn., 1985, 274.
73. **Kenny, D. S. and Couch, T. L.,** Mass production of biological control agents for plant disease, weed, and insect control, in *Biological Control in Crop Production,* Papavizas, G. C., Ed., Allenheld, Totowa, N.J., 1981, 143.
74. **Kerr, A.,** Biological control of crown gall through production of agrocin 84, *Plant Dis.,* 64, 25, 1980.
75. **Kimmey, J. W.,** Inactivation of lethal-type blister rust cankers on Western White pine, *J. For.,* 67, 296, 1969.
76. **Klassen, W.,** The role of biological control in integrated pest management systems, in *Biological Control in Crop Production,* Papavizas, G. C., Ed., Allenheld, Totowa, N.J., 1981, 433.
77. **Kommedahl, T.,** *Proc. of Symp. IX Int. Congr. Plant Protection,* 2 Vols., Burgess, Minneapolis, Minn., 1981, 630.
78. **Kommedahl, T. and Mew, C.,** Biocontrol of corn root infection in the field by seed treatment with antagonists, *Phytopathology,* 65, 296, 1975.
79. **Kommedahl, T. and Windels, C. E.,** Evaluation of biological seed treatment for controlling root diseases of pea, *Phytopathology,* 68, 1087, 1978.
80. **Kommedahl, T. and Windels, C. E.,** Introduction of microbial antagonists to specific courts of infection: seeds, seedlings and wounds, in *Biological Control in Crop Production,* Papavizas, G. C., Ed., Allenheld, Totowa, N.J., 1981, 227.
81. **Kraft, J. M. and Papavizas, G. C.,** Use of host resistance, *Trichoderma* and fungicide to control soil-borne diseases and increase seed yield of peas, *Plant Dis.,* 67, 1234, 1983.
82. **Kuhlman, E. G.,** Parasitic interaction with sporulation by *Cronartium quercuum* f. sp. *fusiforme* on loblolly and slash pine, *Phytopathology,* 71, 348, 1981.
83. **Kuhlman, E. G., Carmichael, J. W., and Miller, T.,** *Scytalidium uredinicola,* a new mycoparasite of *Cronartium fusiforme* on *Pinus, Mycologia,* 68, 1188, 1976.
84. **Lewis, J. A. and Papavizas, G. C.,** Integrated control of *Rhizoctonia* fruit rot of cucumber, *Phytopathology,* 70, 85, 1980.
85. **Locke, J. C., Morris, J. J., and Papavizas, G. C.,** Biological control of *Fusarium* wilt of green-house grown chrysanthemum, *Plant Dis.,* 69, 167, 1984.
86. **Mangenot, F. and Diem, H. G.,** Fundamentals of biological control, in *Ecology of Root Pathogens,* Krupa, S. V. and Dommergues, Y. R., Eds., Elsevier, Amsterdam, 1979, 107.
87. **Marshall, D. S.,** Effect of *Trichoderma harzianum* seed treatment and *Rhizoctonia solani* inoculum concentration on damping-off of snap bean in acidic soils, *Plant Dis.,* 66, 788, 1982.
88. **Martin, F. N. and Hancock, J. G.,** Relationship between soil salinity and population density of *Pythium ultimum* in the San Joaquin Valley of California, *Phytopathology,* 71(Abstr.), 893, 1981.
89. **Martin, F. N. and Hancock, J. G.,** The effects of Cl⁻ and *Pythium oligandrum* on the ecology of *Pythium ultimum, Phytopathology,* 72(Abstr.), 996, 1982.
90. **Marx, D. H.,** The influence of ectotrophic mycorrhizal fungi on the resistance of pine rots to pathogenic infections. II. Production, identification, and biological activity of antibiotics produced by *Leucopaxilis cerealis* var. *piceina, Phytopathology,* 59, 411, 1969.
91. **Marx, D. H.,** Ectomycorrhizae as biological deterents to pathogenic root infections, *Annu. Rev. Phytopathol.,* 10, 429, 1972.
92. **Marx, D. H. and Bryan, W. C.,** Growth and ectomycorrhizal development of loblolly pine seedlings in fumigated soil infested with the fungal symbiont *Pisolithus tinctorius, For. Sci.,* 21, 245, 1975.
93. **Marx, D. H. and Davey, C. B.,** The influence of ectotrophic mycorrhizal fungi on the resistance of pine roots to pathogenic infections. III. Resistance of aseptically formed mycorrhizae to infection by *Phytophthora cinamomi, Phytopathology,* 59, 549, 1969.
94. **Merriman, P. R., Price, R. D., Kollmorgen, F., Piggot, T., and Ridge, E. H.,** Effect of seed inoculation with *Bacillus subtilis* and *Streptomyces griseus* on the growth of cereals and carrots, *Aust. J. Agric. Res.,* 25, 219, 1974.
95. **Mitchell, R. and Hurwitz, E.,** Suppression of *Pythium debaryanum* by lytic rhizosphere bacteria, *Phytopathology,* 55, 156, 1965.

96. **Moody, A. R. and Gindrat, D.,** Biological control of cucumber black root rot by *Gliocladium roseum*, *Phytopathology*, 67, 1159, 1977.

97. **Moore, L. W.,** Prevention of crown gall on Prunus roots by bacterial antagonists, *Phytopathology*, 67, 139, 1977.

98. **Moore, L. W. and Cooksey, D. A.,** Biology of *Agrobacterium tumefaciens*: plant interactions, *Int. Rev. Cytol.,* Suppl. 13, 15, 1981.

99. **Nair, N. G. and Fahy, P. C.,** Commercial application of biological control of mushroom bacterial blotch, *Aust. J. Agric. Res.,* 27, 415, 1976.

100. **Nelson, E. E.,** Survival of *Poria weirii* on paired plots in alder and conifer stands, *Microbios,* 12, 155, 1975.

101. **Nelson, E. B. and Hoitink, H. A. J.,** The role of microorganisms in the suppression of *Rhizoctonia solani* in container media amended with composted hardwood bark, *Phytopathology,* 73, 274, 1983.

102. **Nelson, E. B., Kuter, G. A., and Hoitink, H. A. J.,** Effects of fungal antagonists and compost age on suppression of *Rhizoctonia* damping-off in container media amended with composted hardwood bark, *Phytopathology,* 73, 1457, 1983.

103. **Odvody, G. N., Boosalis, M. G., and Kerr, E. D.,** Biological control of *Rhizoctonia solani* with a soil inhabiting Basidiomycetes, *Phytopathology,* 70, 655, 1980.

104. **Ohr, H. D., Munnecke, D. E., Bricker, J. L.,** The interaction of *Armillaria mellea* and *Trichoderma* spp. as modified by methyl bromide, *Phytopathology,* 63, 965, 1973.

105. **Old, K. M. and Patrick, Z. A.,** Perforation and lysis of spores of *Cochliobolus sativus* and *Thielaviopsis basicola* in natural soils, *Can. J. Bot.,* 54, 2798, 1976.

106. **Olsen, C. M.,** Antagonistic Effects of Microorganisms on Rhizoctonia solani in Soil, Ph.D. thesis, University of California, Berkeley, Calif., 1964, 152.

107. **Olsen, C. M. and Baker, K. F.,** Selective heat treatment of soil, and its effect on the inhibition of *Rhizoctonia solani* by *Bacillus subtilis,* *Phytopathology,* 58, 79, 1968.

108. **Papavizas, G. C.,** Status of biological control of soil-borne plant pathogens, *Soil Biol. Biochem.,* 5, 709, 1973.

109. **Papavizas, G. C.,** Induced tolerance of *Trichoderma harzianum* to fungicides, *Phytopathology,* 70(Abstr.), 691, 1980.

110. **Papavizas, G. C., Ed.,** *Biological Control in Crop Production,* Allenheld, Totowa, N.J., 1981, 461.

111. **Papavizas, G. C.,** *Trichoderma* and *Gliocladium*: biology, ecology, and potential for biocontrol, *Annu. Rev. Phytopathol.,* 23, 23, 1985.

112. **Papavizas, G. C. and Lumsden, R. D.,** Biological control of soil-borne fungal propagules, *Annu. Rev. Phytopathol.,* 18, 389, 1980.

113. **Papavizas, G. C. and Lewis, J. A.,** Introduction and augmentation of microbial antagonists for the control of soil-borne plant pathogens, in *Biological Control in Crop Production,* Papavizas, G. C., Ed., Allenheld, Totowa, N.J., 1981, 305.

114. **Papavizas, G. C., Lewis, J. A. and Abd-El-Moity, T. H.,** Evaluation of new biotypes of *Trichoderma harzianum* for tolerance to benomyl and enhanced biocontrol capabilities, *Phytopathology,* 72, 126, 1982.

115. **Parke, J. L. and Linderman, R. G.,** Association of vesicular abuscular mycorrhizal fungi with the moss *Funaria hygrometrica, Can. J. Bot.,* 58, 1898, 1980.

116. **Parker, C. A., Rovira, A. D., Moore, K. J., Wong, P. T. W., and Kollmorgen, J. F., Eds.,** *Ecology and Management of Soil-borne Plant Pathogens,* American Phytopathological Society, St. Paul, Minn., 1985, 358.

117. **Pimental, D., Ed.,** *Handbook of Pest Management in Agriculture* Vol. 2, CRC Press, Boca Raton, Fla., 1981, 528.

118. **Reichelderfer, K. H.,** *Economic Feasibility of Biological Control Technology: Using a Parasitic Warp, Rediobius faveolatus to Manage Mexican Bean Beatle on Soybeans,* Agric. Econ. Rep. No. 430, U.S. Dept. Agric., ESCS, 1979.

119. **Reichelderfer, K. H.,** Economic feasibility of biological control of crop pests, in *Biological Control in Crop Production,* Papavizas, G. C., Ed., Allenheld, Totowa, N.J., 1981, 403.

120. **Ricard, J. L.,** Commercialization of a *Trichoderma* based mycofungicide: some problems and solutions, *Biocontrol News Inf.,* 2(2), 95, 1981.

121. **Ricard, J. L. and Laird, P.,** Current research in the control of *Fomes annosus* with *Scytalidium* sp., an immunizing commensal, in *Proc. 3rd Int. Conf. Fomes Annosus,* Hodges, C. S., Ed., U.S. Department of Agriculture Forest Service, Asheville, N.C., 1968.

122. **Ricard, J. L., Wilson, M. M., and Bollen, W. B.,** Biological control of decay in Douglas-fir poles, *For. Prod. J.,* 19, 41, 1969.

123. **Rishbeth, J.,** Stump protection against *Fomes annosus*. III. Inoculation with *Peniophora gigantea, Ann. Appl. Biol.,* 52, 63, 1963.

124. **Rishbeth, J.**, Stump inoculation: a biological conrol of *Fomes annosus*, in *Biology and Control of Soil-borne Plant Pathogens*, Bruehl, G. W., Ed., American Phytopathological Society, St. Paul, Minn., 1975, 158.

125. **Rishbeth, J.**, Modern aspects of biological control of *Fomes* and *Armillaria*, *Eur. J. For. Pathol.*, 9, 331, 1979.

126. **Ross, E. W. and Marx, D. H.**, Susceptibility of sand pine to *Phytophthora cinamomi*, *Phytopathology*, 62, 1197, 1972.

127. **Schenck, N. C. and Kellam, M. K.**, The influence of vesicular arbuscular mycorrhizae on disease development, *Fl. Agric. Exp. Stn. Bull.*, 798, 1, 1978.

128. **Schippers, B. and Gams, W., Eds.**, *Soil-borne Plant Pathogens*, Academic Press, New York, 1979, 686.

129. **Scott, P. R., Ed.**, Interactions between microorganisms, *Ann. Appl. Biol.*, 89, 89, 1978.

130. **Scroth, M. N. and Hancock, J. G.**, Selected topics in biological control, *Annu. Rev. Phytopathol.*, 35, 453, 1981.

131. **Simonds, F. J.**, The economics of biological control, *J. R. Soc. Arts*, p. 880, 1967.

132. **Sivan, A., Elad, Y., and Chet, I.**, Biological control effects of a new isolate of *Trichoderma harzianum* on *Pythium aphanidermatum*, *Phytopathology*, 74, 498, 1984.

133. **Sivasithamparam, K., Parker, C. A., and Edwards, C. S.**, Bacterial antagonists to the take-all fungus and fluorescent pseudomonads in the rhizosphere of wheat, *Soil Biol. Biochem.*, 11, 161, 1979.

134. **Smiley, R. W.**, Antagonists of *Gaeumannomyces graminis* from the rhizoplane of wheat in soils fertilized with ammonium or nitrate nitrogen, *Soil Biol. Biochem.*, 10, 169, 1978.

135. **Smiley, R. W.**, Colonization of wheat roots by *Gaeumannomyces graminis* inhibited by specific soils, microorganisms and, ammonium nitrogen, *Soil Biol. Biochem.*, 10, 175, 1978.

136. **Snyder, W. C., Wallis, G. W., and Smith, S. N.**, Biological control of plant pathogens, in *Theory and Practice of Biological Control*, Haffaker, C. B. and Messenger, P. S., Eds., Academic Press, New York, 1976, 521.

137. **Spencer, D. M.**, Parasitism of carnation rust (*Uromyces dianthi*) by *Verticillium lecanii*, *Trans. Br. Mycol. Soc.*, 74, 191, 1980.

138. **Spencer, D. M. and Atkey, P. T.**, Parasitism of carnation rust and brown rust of wheat by *Verticillium lecanii*, *Glasshouse Crops Res. Inst. Annu. Rep.*, 1980, p. 134, 1981.

139. **Sun, S. K. and Huang, J. W.**, Ecological study and control trials on *Fusarium* wilt of watermelon, *Proc. 3rd Int. Congr. Plant Pathology*, Munich, August, 1978, 189(Abstr.).

140. **Sundheim, L.**, Attempts at biological control of *Phomopsis sclerotioides* in cucumber, *Neth. J. Plant Pathol.*, 83 (Suppl. 1), 439, 1977.

141. **Swinburne, T. R., Barr, J. G., and Brown, A. E.**, Production of antibiotics by *Bacillus subtilis* and their effect on fungal colonists of apple leaf scars, *Trans. Br. Mycol. Soc.*, 65, 203, 1975.

142. **Trappe, J. M.**, Selection of fungi for ectomycorrhizal inoculations in nurseries, *Annu. Rev. Phytopathol.*, 15, 203, 1977.

143. **Tribe, H. T.**, On the parasitism of *Sclerotinia trifoliorum* by *Coniothyrium minitans*, *Trans. Br. Mycol. Soc.*, 40, 489, 1957.

144. **Trutmann, P., Keane, P. J., and Merriman, P. R.**, Reduction of sclerotial inoculum of *Sclerotinia sclerotiorum* with *Coniothyrium minitans*, *Soil Biol. Biochem.*, 12, 461, 1980.

145. **Trutmann, P., Kean, P. J., and Merriman, P. R.**, Biological control of *Sclerotinia sclerotiorum* on aerial parts of plants by the hyperparasite *Coniothyrium minitans*, *Trans. Br. Mycol. Soc.*, 78, 521, 1982.

146. **Tu, J. G.**, *Gliocladium virens*, a destructive mycoparasite of *Sclerotinia sclerotiorum*, *Phytopathology*, 70, 670, 1980.

147. **Tu, J. G. and Vaartaja, O.**, The effect of the hyperparasite (*Gliocladium virens*) on *Rhizoctonia solani* and on *Rhizoctonia* root rot of white beans, *Can. J. Bot.*, 59, 22, 1981.

148. **Turner, G. J. and Tribe, H. T.**, Preliminary field plot trials on biological control of *Sclerotinia trifoliorum* by *Coniothyrium minitans*, *Plant Pathol.*, 24, 109, 1975.

149. **Turner, G. J. and Tribe, H. T.**, On *Coniothyrium minitans* and its parasitism of *Sclerotinia* species, *Trans. Br. Mycol. Soc.*, 66, 97, 1976.

150. **Upadhyay, R. S. and Rai, B.**, Biological control of soil-borne plant pathogens, in *Frontiers in Applied Microbiology*, Vol. 1, Mukerji, K. G. et al. Eds., Lucknow Print House, Lucknow, India, 1985, 261.

151. **Utkhede, R. S. and Rahe, J. L.**, Biological control of onion white rot, *Soil Biol. Biochem.*, 12, 101, 1980.

152. **Vésely, D.**, Parasitic relationship between *Pythium oligandrum* Drechsler and some other species of Oomycetes class, *Zentralbl. Bakteriol. Parasitenkd. Infektionskr. Hyg. Abt. 2*, 133, 341, 1978.

153. **Weindling, R.**, *Trichoderma lignorum* as a parasite of other fungi, *Phytopathology*, 22, 837, 1932.

154. **Weindling, R. and Fawcett, H. S.**, The isolation of a toxic substance from the culture filtrate of *Trichoderma*, *Hilgardia*, 10, 1, 1936.

155. **Weller, D. M. and Cook, R. J.**, Suppression of take-all of wheat by seed treatment with fluorescent pseudomonads, *Phytopathology*, 73, 463, 1983.

156. **Wells, H. D., Bell, D. K., and Jaworski, C. A.,** Efficacy of *Trichoderma harzianum* as a biocontrol of *Sclerotium rolfsii, Phytopathology,* 62, 442, 1972.

157. **Wicker, E. F.,** Natural control of white pine blister rust by *Tuberculina maxima, Phytopathology,* 71, 997, 1981.

158. **Wilkinson, H. T., Weller, D. M., and Alldredge, J. R.,** Enhanced biological control of wheat take-all when inhibitory pseudomonas strains are introduced on inoculum or seed as opposed to directly into soil, *Phytopathology,* 72, 948, 1982.

159. **Windels, C. E.,** Growth of *Penicillium oxalicum* as a biological seed treatment on pea seed in soil, *Phytopathology,* 71, 929, 1981.

160. **Windels, C. E. and Kommedahl, T.,** Factors affecting *Penicillium oxalicum* as a seed protectant against seedling blight of pea, *Phytopathology,* 68, 1656, 1978.

161. **Wood, R. K. S. and Tveit, M.,** Control of plant diseases by use of antagonistic organisms, *Bot. Rev.,* 21, 441, 1955.

162. **Wright, J. M.,** The production of antibiotics in soil. II. Production of griseofulvin by *Penicillium nigricans, Ann. Appl. Biol.,* 43, 288, 1955.

163. **Wright, J. M.,** Biological control of a soil-borne *Pythium* infection by seed inoculation, *Plant Soil,* 8, 132, 1956.

Chapter 3

FUNGI — AGENTS OF BIOLOGICAL CONTROL

R. S. Mehrotra, K. R. Aneja, A. K. Gupta, and Ashok Aggarwal

TABLE OF CONTENTS

I. INTRODUCTION

Today there is a growing movement in many countries to reduce the amount of chemicals being released into the environment, and their persistence and accumulation over a period of time in the environment (especially in the soil and aquatic ecosystem) and to reduce their inherent hazards to plant and animal life. To increase world food production and feed the ever-increasing population, agricultural production can be augmented with biological control instead of pesticides. Cook,[36] in his Presidential address to members of the American Phytopathological Society, states, "Biological control presents one of the greatest challenges of all times to our discipline and it requires the attention of all facets of our discipline. It offers something for commercial interests as well as the public sector and to the practitioner as well as the scientist conducting basic research." According to Papavizas and Lumsden,[96] the enhanced research activity on the subject of biological control is in line with the increased effort and determination by plant pathologists and soil microbiologists to adapt the conceptual scheme of integrated pest management as an acceptable ecosystem approach to disease control.

There are doubts expressed in certain quarters about the possible use of biological control for plant pathogens. However, looking to the recent reports of the successful use of microbes and other organisms for controlling plant pathogens, the pessimistic approach towards biological control should be done away with. Hornby,[73] in his review on microbial antagonism in the rhizosphere, stated, "whilst much has been learnt it remains debatable that our ability to manipulate antagonism and exercise biological control has increased commensurably."

Garrett[62] has defined biological control in plant disease as "any condition under which, or practice whereby survival or activity of a pathogen is reduced through the agency of any other living organism (except man himself), with the result that there is a reduction in incidence of disease caused by the pathogen." According to Garrett, biological control can be brought about either by the introduction or augmentation in numbers of one or more species of controlling organism, by a change in environmental conditions designed to favor the multiplication and activity of such organisms, or by a combination of both procedures. According to Snyder[117] biological control relies largely upon an interruption of host-parasite relationships through biological means. He takes a broader view of biological control and includes imparting resistance to the host, usually through plant breeding or modifying the culture of the crop so as to avoid or reduce infection. Sewell[109] supported Snyder's view and proposed that "biological control is the induced or natural, direct or indirect limitation of a harmful organism, or its effects by another organism or groups of organisms". Baker and Cook[10] stated that "biological control is the reduction of inoculum density or disease producing activity of a pathogen or parasite in its active or dormant stage, by one or more organisms accomplished naturally or through manipulation of the environment, host, or antagonists, or by mass introduction of one or more antagonists." Host plants were included in the concept of biological control since most of the pathogens exist, through much or all of their life cycle, in intimate association with their host; and, most strategies directed at their control have had to consider the host as an integral part of the control. At the very least, the host provides a battleground where antagonists inhibit or displace pathogens. They called this the "passive role" of the host in biological control. Skolko's[115] concept has been accepted by many workers, i.e., that in biological control the fourth component, namely "other living organisms", is brought into play regardless of how it is initiated. For a plant disease to occur three factors are necessary, viz., a susceptible host, a pathogen, and environmental condition suitable for infection. The recognition of a further component, namely other organisms which affect the course of disease either by their direct influence on the pathogen or indirectly through their effect on the host or environment, provides the basis of our concept of biological control.[115] According to Cook[36], development of a resistant variety by conventional plant breeding is just as much a way to achieve biological control as is induced resistance, or the development of a resistant variety by genetic engineering. And, the use of crop rotation combined with tillage, to permit or accelerate the useful biological destruction of the inoculum of some pathogens in the soil, is just as much a biological control as is the introduction of a hyperparasite to destroy the inoculum.

Biological control of plant pathogens is not new. Baker and Snyder[11] have stated that when plants emerged from warm primeval seas and invaded land in the Devonian period, about 300 million years ago, they were undoubtedly accompanied by fungi and bacteria which had been parasitizing them for millions of years. As roots evolved, they were invaded in turn. Thus, parasitic fungi and bacteria have been a part of the environment of roots for at least as long as the soil itself. It is, therefore, to be expected that the interactions between the parasite, root, and soil, and between parasitic and saprophytic microorganisms, have become extremely complex. Those organisms which did not adjust to this competitive state, by one means or another, did not survive. A state of fluctuating biological balance thus developed for each native habitat, and was self-adjusting for the relatively slow evolutionary and climatic changes.

Biological control must no longer be recognized as a science based mainly on the disciplines of ecology, taxonomy, and soil microbiology. Biological control is also based on the disciplines of plant and microbial genetics, molecular biology, cytology, biochemistry, plant physiology, and many other disciplines. Moreover, biocontrol is not just applicable to a few nematodes and phytopathogenic fungi. Kerr[77] and others have now shown its potential against bacteria, insect pests,[2,4] and weeds,[59,124,125] and virologists are coming up with some of the best biocontrols, using avirulent and modified virus strains to obtain cross protection against virulent plant viruses.[42] Biological control can be accomplished by genetic manipulation of the host, antagonist, or even the pathogen itself, and may be directed at the ecosystem, population, or individual level.

Biological control, broadly defined, may occur remote from the plant, it may occur on the plant, or it may take place inside the plant; and, although it often depends on antagonistic microorganisms in the classic sense, it also depends on the plant and may even use the pathogen against itself. It is the broad concept of biological control that makes it so fascinating as a field of study and so potentially useful as a long-term solution and plant disease management strategy.

When man began cultivating crops about 10,000 years ago he had his first experience with plant disease. He found that when farming the same land the yields of a particular crop decreased year after year, but he soon found that by moving to fresh land he could greatly increase his yield. Biological control of plant pathogens has always been instrumental in keeping plants healthy, and we probably do not realize how widespread this effect is and how greatly dependent upon it we are. Though also operating on aerial plant parts, biological control is particularly active in the soil through the interaction of microorganisms with nutrients, moisture, and other physical factors. Their vast interacting network of population is in a state of dynamic equilibrium. It is biologically buffered against the kind of sudden changes induced by the agricultural activity of man, but constantly undergoes gradual changes as the ecosystem evolves. Man has been effecting biological control for centuries through practices such as crop rotation, fallowing, and irrigation.

There are several reviews, symposia, and articles on various aspects of microbial antagonism and biological control.[2,4,10-12,14,22,36,37,41,45,46,58-60,62,63,78,85,95,96,104,108,115,117,118,127] According to Cook[35], research on the biological control of plant pathogens is now proceeding on at least five broad fronts:

1.   Reduction of inoculum density — This is the more classical approach to biological control, aimed at reducing numbers of the target organism.

2.   Replacement of the pathogen with saprophytes — This approach applies mainly to pathogens in host debris where the primary colonist is a pathogen that is quickly succeeded by saprophytes.[22]
3.   Suppression of germination or growth, or interference with pathogenicity of the target organism — This approach includes the use of competition, antibiosis, bacteriocins, myco-viruses, or other means to suppress or cripple the pathogen during attempted pathogenesis.
4.   Protection of the infection court — This approach includes prior inoculation of the freshly cut surfaces with weak or nonpathogens to prevent colonization by a more virulent pathogen.
5.   Stimulation of host resistance or cross protection — This approach includes the prior inoculation of plants with a weakly virulent or avirulent race or strain of a pathogen which results in host resistance to the otherwise virulent race or strain.

Many nonpathogenic microorganisms are known to produce antifungal substances which may inhibit the growth of fungal parasites.[76,99] The antagonistic effects of a fungus on other microorganisms is well understood. It is generally accepted that one organism influences the activities of others when they are competing for space, air, water, and nutrients.[10,110] Many saprophytes are known to synthesize toxic substances which limit the growth of the parasite.[76,89,99,112]

II. CONTROL OF PLANT PATHOGENS

The principles of biological control have been unknowingly exploited in traditional agricultural notations for centuries. Biological control of apple canker by soil amendment was in use "unknowingly" as early as the seventeenth century.[41] Green manuring to control potato scab caused by *Streptomyces scabies* was considered as early as 1921, and a paper by Sanford[105] in 1926, in which it was suggested that antagonistic bacteria were responsible, was considered by Garrett[62] to contain a concise hypothesis of biological control. Millard and Taylor[87] reported the control of potato scab in potatoes grown in sterilized soil and inoculated with *S. scabies* through simultaneous inoculation of the soil with *Streptomyces praecox*, a vigorous saprophytic species. This was a forerunner of the cross-protection experiments so favored in the 1970s. By 1937 there were practical directions for utilizing the biological control of *Phymatotrichum omnivorum* root-rot of cotton by burying organic manure in deep furrows prior to planting. In the present review we propose to discuss some outstanding examples of biological control by fungal antagonists.

A. Soil Treatments
In 1931 Sanford and Broadfoot[106] provided experimental evidence that the infection of wheat seedling by *Ophiobolus graminis* in sterilized soil could be completely suppressed by the antagonistic action of various individually co-inoculated species of fungi and bacteria. Weindling[134] demonstrated the myco-parasitism of *Rhizoctonia* by *Trichoderma lignorum*, and, together with Emerson, isolated the antifungal antibiotic gliotoxin from the latter.[135] This antagonist, later identified as a species of *Gliocladium* rather than *Trichoderma*,[133] controlled damping-off caused by *Rhizoctonia* sp. in citrus seedlings.[136]

Bliss[15] made some unexpected observations during an investigation of carbon-disulphide fumigation of the soil for the elimination of *Armillariella (Armillaria) mellea* from infected citrus roots. Bliss observed that from those root pieces in which *A. mellea* was not viable, *Trichoderma viride* always emerged. He therefore suggested that the destruction of *A. mellea* was due to *T. viride* and not the fumigant. This hypothesis was later tested experimentally by Garrett.[61]

The antagonistic activity of *Trichoderma harzianum* against several fungi, including *Rhizoctonia solani,* was reported by Dennis and Webster.[51,52,53] Wells et al.[137] were the first to report field control of *Sclerotium rolfsii* by the infestation of soil with *T. harzianum* grown on an autoclaved mixture of ryegrass seeds and soil. Backman and Rodriguez-Kabana[9] used molasses and enriched clay granules both as a food base for growing the same antagonist and as a carrier to facilitate dispersal in the field. They observed a significant decrease in *S. rolfsii* damage to peanuts and an increase in yield during a 3-year test. Hadar et al.[70] reported that *T. harzianum,* when applied in the form of wheat bran culture to *R. solani* infested soil, effectively controlled damping-off of bean, tomato, and eggplant seedlings. They also reported an isolate of *T. harzianum* which directly attacked the mycelium of *R. solani.* When applied to soil artificially infested with *R. solani,* a wheat-bran culture of the antagonist effectively controlled it in the greenhouse. This isolate, however, was not effective against *Sclerotium rolfsii,* in culture, or in the greenhouse.

Chet et al.[32] isolated *T. harzianum* antagonistic to *S. rolfsii* and *R. solani* from soil, and no antibiotic activity of *T. harzianum* towards the pathogens could be detected. When grown on the pathogen's cell walls, the fungus produced extracellular β-(1,3)-glucanase and chitinase. When applied in the form of wheat bran culture to soil infected with *R. solani* or *S. rolfsii* in the greenhouse, *T. harzianum* effectively controlled damping-off disease of bean, peanuts, and eggplants caused by these soil-borne plant pathogens. Field experiments were carried out and a significant reduction in disease incidence was obtained. Application of PCNB at subinhibitory doses improved control of the disease when applied together with the *Trichoderma* preparation.

Elad et al.[54,55] reported that solar heating of the soil (by polyethylene mulching) or the fumigant methyl-bromide, when tested under field conditions, significantly reduced diseases caused by *R. solani* and *Verticillium dahliae.* The biocontrol agent *T. harzianum* reduced *R. solani* disease. Combining solar heating or methyl bromide with *T. harzianum* improved their efficiency and also resulted in the control of *Sclerotium rolfsii.* Both *T. harzianum* and solar heating reduced the inoculum potential of *R. solani* as well as its build-up in the field and under greenhouse conditions. These results were more pronounced when the two treatments were combined. Combining heat treatment with *T. harzianum,* both at sublethal doses and under greenhouse conditions, resulted in 90-100% control of *S. rolfsii* disease in beans.

Lui and Baker[81] noted that the numbers of *Trichoderma* spp. propagules in the soil increased as suppressiveness increased, whereas inoculum density of *R. solani* was inversely proportional to the density of these *Trichoderma* spp. following radish monoculture.

Sivan et al.[114] reported that *T. harzianum,* applied to either soil or rooting mixtures, efficiently controlled damping-off induced by *Pythium aphanidermatum* in peas, cucumbers, tomatoes, peppers, and gypsophila. Up to 85% disease reduction was obtained in tomatoes. *T. harzianum* applied in a seed coating mixture containing 5×10^9 conidia/mℓ was as effective in sandy soil as the broadcast application of wheat/bran/peat preparation. However, the broadcast application was superior to seed coating for protecting tomato seedlings in an infested peat/vermiculite rooting mixture. When germinated at a low temperature (22°C) pea seeds coated with conidia of *T. hamatum* were better protected from *P. aphanidermatum* than seeds coated with *T. harzianum,* but this was not the case at 30°C. Extracellular filtrate from cultures of *T. harzianum,* added to a synthetic medium, inhibited linear growth of *P. aphanidermatum* by 83% compared with 8% inhibition by a culture filtrate of *T. hamatum.* Substances excreted by *P. aphanidermatum* into the growth medium enhanced the linear growth of *T. harzianum* by 34% but not that of *T. hamatum.*

B. Seed Treatments

Tveit and Wood[130] found that seed-borne Fusaria, causing blight of oat seedlings, was controlled by treating grains with certain isolates of the *Chaetomium* spp. There are also examples of spontaneous seed protection by antagonists. Species of *Chaetomium* occurring in seeds protected oats from Victoria blight caused by the seed-borne *Cochliobolus victoriae* in Brazil.[129]

Seed inoculation with the antagonistic microorganisms offers a potential for the control of root diseases, but there are fewer reports of success with this method. One good example of this is the control of *Fusarium roseum* f. sp. *cerealis* on corn by seed inoculation with *Chaetomium globosum*.[29]

The inoculation of seeds with microorganisms before planting in nontreated soils has long been studied in Russia and has apparently resulted in increased crop yields and control of some root pathogens. These results have been partially confirmed by workers at Rosthamsted Experimental Station in England[18-20] and at the Commonwealth Scientific and Industrial Research Organization (CSIRO) in Australia. Seed inoculation with selected soil microorganisms as a method of increasing plant growth has been extensively investigated by soil microbiologists and considerable data are available on the use of selected soil microflora as seed inoculants for the growth stimulation of agricultural crops.[16,21,88] Wu[141] reported that the introduction of *Epicoccum nigrum* and *Trichoderma harzianum* into soil infested with *Drechslera sorokiniana*, *Fusarium culmorum*, and *Rhizoctonia solani* improved emergence and vigor of wheat and oat seedlings. Pelleting seed with a mixture of either *Aspergillus clavatus*, *T. aureoviride*, or *T. harzianum* conidia and binder likewise improved seedling emergence and yield in soil infested with *Drechslera sorokiniana*, *F. culmorum*, and *R. solani*.

Application of *Gliocladium virens* to seed potatoes in storage suppressed *Rhizoctonia solani* by the production of two highly potent antibiotics.[5]

Singh and Chohan[113] studied the antagonism of *T. viride* against some pathogenic fungi on groundnut seeds and found that this was highly effective in protecting groundnut seeds from the seed-rot phase caused by the pathogenic fungi.

Harman et al.[71] reported that *T. hamatum*, applied as a seed treatment, controlled seed-rot of pea or radish in soil infested with *Pythium* spp. or *R. solani*, respectively, if soil temperatures were between 17 and 34°C and seeds had been treated with a suspension of conidia with a concentration equal to or greater than $10^6/m\ell$.

Cooksey and Moore[38] found in field tests that isolates of *Penicillium* and *Aspergillus* reduced the incidence of galling on mazzard, cherry seedlings by some of the pathogens tested.

C. Inoculation of Fresh Pruning and Felling Wounds

One of the most successful examples of biological control of a plant disease, and one which is in commercial use, is the protection of pine stumps from infection by *Fomes annosus* through inoculation with *Peniophora gigantea*. This treatment was developed by Rishbeth[103] in Britain where *F. annosus* is the most important pathogen affecting forests. A commercially produced inoculum now protects 62,000 ha of pines against *Fomes* infection.[132] *F. annosus* causes a root-rot of conifers throughout the temperate regions of the world. Basidiospores colonize stumps produced by felling operations and the fungus then attacks live trees through roots in contact with infected stumps.[64] In forests where *Peniophora gigantea* is an abundant natural colonizer of stumps, it not only prevents the infection of stumps by *F. annosus* spores but checks the advance of, and to some extent replaces, *Fomes* by growing down and colonizing the tissues of the stumps and roots of trees which have initial infection from root grafts.[132] Rishbeth[103] proposed the inoculation of pine stumps with *P. gigantea* as an alternative to protection with chemicals such as sodium nitrite. The inoculum is prepared commercially in Britain in 1 mℓ sachets each containing a minimum of 5×10^6 viable oidia. Each sachet added to 5 ℓ of water will treat at least 100 stumps of 20 cm diameter at a cost of about $0.07/stump including inoculum and labor. The inoculation of stumps with *P. gigantea* should take place immediately after felling. Greig[65] has shown that this can be done automatically during felling by mixing the oidia in the chain-saw oil. As this method also inoculates the base of the log and, as *P. gigantea* can cause staining and decay in pine timber, precautions against this should be taken.[65]

Carter[24] reported that *Fusarium lateritium* is a naturally occurring saprophyte on certain pruning wounds of apricot trees that escape invasion by the pathogen *Eutypa armeniacea*. In vitro *F. lateritium* produces a diffusible mycotoxin strongly active against *E. armeniacea*.[25] In orchard experiments, conidial suspensions of *Fusarium lateritium,* benomyl, or an integrated treatment combining both methods were almost equally effective in protecting pruned apricot's sapwood. Since tolerance of *Fusarium* to benomyl is about ten times as great as that of *E. armeniacea*, a supplementary action of the antagonist and the fungicide was suggested: benomyl accords an immediate protection while the antagonist is an effective protectant when established in the infection court.[25,26]

Grosclaude et al.[68] used modified pruning shears to apply a conidial suspension of *Trichoderma* species to plum trees at the time of pruning. In the preliminary orchard experiments on a small number of trees, biological treatment followed by inoculation with basidiospores of *Stereum purpureum,* 2 days later completely prevented the silver leaf disease caused by the basidiomycetes. Later, they[67] found that *Trichoderma* was highly antagonistic to *S. purpureum* in vitro. On agar, killed conidia of the former still exert an antagonistic action on the pathogen.

Corke and Hunter[40] used several microbial species including *Trichoderma viride* to protect pruning wounds on apple tree shoots from invasion by *Nectria galligena*. Inoculation with microorganisms resulted in 45 to 65% fewer infections from subsequent inoculation with the pathogen.

Pottle et al.[98] showed that inoculation of drill wounds in red maple, with conidia of *T. harzianum* in glycerol, gave complete protection against invasion by decay Hymenomycetes within 21 months after treatment. The isolate of *T. harzianum* was obtained from the discolored wood in a red maple.[97]

Tiwari and Mehrotra[126] screened the isolated microflora from the rhizosphere and rhizoplane soil of *Piper betle* and also from other sources and selected four strong antagonists against *Phytophthora parasitica* var. *piperina*. Among these, the fungal antagonists were *T. viride* and *Aspergillus terreus*. Cuttings, at the time of planting, were dipped in a suspension of the antagonists and good control of the disease was obtained with *T. viride*. Mehrotra and Claudius[86] utilized *T. viride* in the control of lentil wilt caused by *Fusarium oxysporum* f. sp. *lentis*.

Ricard[100] developed the term "immunizing commensals" (IC) for harmless microorganisms capable of growing and becoming established in living or dead substrates and inhibiting or preventing in them the development of harmful microorganisms. He and his coworkers studied the microflora of healthy hosts in disease situations and determined which component of the microflora was responsible for the absence of disease. *T. viride* was used as an IC against *S. purpureum*, the cause of silver leaf disease of plum and other fruit trees. The IC was equally effective when applied as a spray to pruning cuts or when applied by means of the pruning shears.[68] Later this IC was tested by Corke[39] on pear trees infected with a silver leaf pathogen using inoculation darts and agar plugs inserted into holes drilled into the trunk, as well as inoculum applied to wounds. There was a visible improvement in the status of more than half of the severely affected trees following inoculation with *T. viride* by any of these methods.

D. Mycoparasitism

Coniothyrium minitans is also an important naturally occurring biological control agent.[74,128] *C. minitans* was instrumental in the natural decrease in the number of viable sclerotia of

Sclerotinia sclerotiorum during the course of a growing season. It can be grown successfully on media and applied to sclerotia to cause their destruction. A *Corticium* sp. on *Rhizoctonia solani*,[80,93] a sterile basidiomycete on *Macrophomina phaseolina*,[50] and *Pythium oligandrum* on other *Pythium* spp.[131] and on *Gaeumannomyces graminis*[49] have also been studied.

Recently, progress has been made with the newly discovered mycoparasite, *Sporidesmium sclerotivorum*, destructive to the *Sclerotinia* spp. and related fungi.[7,8] Ecological studies have shown that the mycoparasite has exacting nutritional requirements for growth and infection of sclerotia, but is active over a broad range of environmental conditions in agricultural soils.[8] New evidence indicates that *Sporidesmium sclerotivorum* can effectively control lettuce-drop in the field caused by *Sporidesmium minor*.[1]

Several other mycoparasites are presently being evaluated in the control of soil-borne pathogens. Each of these biocontrol agents show promise on the basis of limited field testing. A basidiomycetous fungus, *Laetisaria arvalis (Corticium* sp.), is being tested for the control of seedling pathogens, *Pythium ultimum* and *Rhizoctonia solani*;[72,92] and a pythiaceous fungus, *Pythium oligandrum*, is being considered as a biocontrol agent against pathogenic *Pythium* spp.[131]

Elad et al.[55] found an isolate of *Trichoderma harzianum* capable of lysing mycelia of *Sclerotium rolfsii* and *Rhizoctonia solani*. In naturally infested soils, a wheat bran preparation of *T. harzianum* inoculum significantly decreased diseases caused by *S. rolfsii* or *R. solani* in field experiments with beans, cotton, or tomatoes and they significantly increased the yield of beans.

Chand and Logan[30] screened microorganisms for antagonism against *R. solani* and found that *Penicillium cyclopium*, *Chaetomium olivaceum*, and *Gliocladium roseum* were the most effective fungi in reducing stem canker of potato under controlled conditions, and these fungi also penetrated and killed *R. solani* hyphae in pure culture.

E. Weak Pathogens

The approach used by Kerr to control crown gall is significant not only in practical terms, but also because it points the way to the potential for genetic and physiological manipulation of plant pathogens through use of closely related but nonpathogenic strains.[77,78] It is encouraging to note that an equivalent or similar means of biological control has been discovered in fungi, namely, the transmissible hypovirulence reported for *Gaeumannomyces graminis* var. *tritici*,[79] *Rhizoctonia solani*,[27] and *Endothia parasitica*.[66] The hypovirulence agent in *R. solani* and *E. parasitica* have been identified as dsRNA;[28,47] and in both cases, biological control of healthy isolates of respective pathogens has been obtained under controlled conditions using hypovirulent isolates.

F. Cross-Protection

In cross-protection by a virulent or weakly virulent, closely related strains of soil-borne pathogens, there is no direct effect on the pathogen by the accompanying microorganisms, but an indirect one mediated through the host plant.[6,13,48,49,139] Perhaps the most studied example of cross-protection with soil-borne pathogens is that of the control of the take-all fungus, *Gaeumannomyces graminis*, by the closely related nonpathogenic fungus, *Phialophora radicicola*[48,49,139] and avirulent isolates of *G. graminis*.[6,139] These fungi can protect grass and cereal hosts from infection by the pathogen, and the control is probably widespread in natural plant communities.[48]

Wong and Southwell[140] have shown that wheat inoculated at sowing with *Gaeumannomyces graminis* var. *graminis* and *Phialophora radicicola* var. *radicicola* gave significantly greater grain yields than wheat unprotected against take-all, whereas the one isolate of *Phialophora radicicola* var. *graminicola* tested did not protect it.

G. Antagonists

Marois and Mitchell[84] reported that in greenhouse experiments, the fungal antagonist amendment reduced significantly the mean lesion length and the incidence of *Fusarium* crown-rot disease of tomato caused by *Fusarium oxysporum* f. sp. *radicislycopersici*.

Purkayastha and Bhattacharya[99] screened 26 fungi isolated from the phyllosphere of jute and found *Aspergillus nidulans* and *Penicillium oxalicum* to be highly antagonistic to *Colletotrichum corchori*, the cause of anthracnose, an important foliar disease of jute.

Singh and Khara[111] screened phylloplane fungi of tomato against *Alternaria solani* and found the following fungi antagonistic to it, viz., *Sepedonium ochraceum, Aspergillus candidus, A. nidulans* var. *echinulatus,* and *Penicillium funiculosum*.

Gupta et al.[69] screened soil microorganisms against *Sclerotium oryzae* Catt. and found *Aspergillus fumigatus, A. luchuensis, A. niger, Penicillium oxalicum, P. piceum* and *Penicillium* sp. antagonistic to it. Culture filtrates of these antagonistic microorganisms have also shown positive results in vitro.

III. CONTROL OF INSECT PESTS

Insects occupy virtually every ecological niche with both boon and curse to the biological entity it is affecting. Homeostasis, a natural phenomenon, has been applied to control these insect pests. Biological control is a major element of the natural control which keeps all living creatures in a state of balance. Entomogenous fungi have played a significant role in the history of insect pathology and specifically in microbial control.[119]

There are a number of important monographs and reviews which deal with the use of microorganisms as biological control agents of insects.[2-4,23,56,57,75,82,83,90,91,107,120,121,138] The successful control of insect pests has been obtained with entomogenous fungi (Table 1).

According to Aizawa[4] there are several disadvantages in the utilization of entomogenous fungi:

1. It is not easy to estimate the potency of the fungal preparation by bioassay or by conidial counts.
2. Insects are not attacked by the fungus at higher temperatures, sometimes not even at 30°C.
3. The fungal preparation is not stable.
4. The production of mycotoxins is not well known in some cases and the safety of fungal preparations should be investigated in greater detail.

However, the problems of infection at higher temperatures and the stability of fungal preparation can be solved by the selection and improvement of strains.

IV. CONTROL OF WEEDS

Although chemical herbicides are the most effective immediate solution to most weed problems, their frequent use has posed certain problems including pollution hazards, affects on nontarget organisms, short period of effectiveness, and high cost. The possibility of controlling weeds by nonchemical agents, i.e., microbial plant pathogens called biological control agents or microbial herbicides, is assuming a greater importance.[59,124,125]

Some of the most successful examples of the control of weeds by microbial herbicides are

1. *Puccinia chondrilla* was imported into Australia to control rush skeleton weed, *Chondrilla juncea*.[43] This weed invaded millions of acres of wheat and ranges.

Table 1
ENTOMOGENOUS FUNGI AND THEIR HOSTS

Entomogenous fungi	Host		Country
	Common name	Zoological name	
Aschersonia aleyrodis	Citrus whitefly	*Dialeurodes citri*	Japan
A. placenta	Citrus whitefly	*Dialeurodes citri*	U.S.S.R.
Aschersonia sp.	Greenhouse whitefly	*Trialeurodes vaporariolum*	U.S.S.R.
Beauveria bassiana	Colorado potato beetle	*Leptinotarsa decemlineata*	U.S.S.R.
B. bassiana	Codling moth	*Carpocapsa pomonella*	U.S.S.R.
B. tenella (*B. bronghiartii*)	Soybean beetle	*Anomala rufocuprea*	Japan
B. tenella	Larger striated chafer	*Anomala costata*	Japan
Hirsutella thompsoni	Citrus rust mite	*Phyllocoptruta oleivora*	U.S.
Metarhizium anisopliae	Rhinoceros beetle	*Oryctes rhinoceros*	Tonga
M. anisopliae	Black rice bug	*Scotinophara lurida*	Japan
Nomuraea rileyi	Tobacco budworm	*Heliothis virescens*	U.S.
Paecilomyces farinosus	Citrus ground mealy bug	*Rhizoecus kondonis*	Japan
P. fumosoroseus	Peach fruit moth	*Carposinia niponensis*	Japan
P. lilacinus	Potato tuberworm	*Phthorimaea opeculella*	Japan
Verticillium lecanii	Brown soft scale	*Coccus hesperidum*	Czechoslovakia

2.   Wild blackberries, *Rubus constrictus* and *R. ulmifolius,* were controlled in Chile by blackberry rust, *Phragmidium violaceum,* imported from Europe.[94]

3.   E. E. Trujillo[142] effectively controlled pamakani weed (*Agerarina riparia-eupatorium riparia*) in ranges of pastures in Hawaii by introducing *Cercosporella agertinae* (*C. adenophorum*) from Jamaica.

4.   Northern jointvetch, *Aeschynomene virginica,* was controlled by applying *Colletotrichum gloeosporioides* f. sp. *aeschynomene* to rice fields in Arkansas, Mississippi, and Louisiana.[17,44,116,122,123]

Two other excellent examples of the mycoherbicide approach, using indigenous fungi to control weeds, are approaching commercial use. Milkweed vine or strangler vine (*Morrenia odorata*) is a serious problem in Florida citrus groves.[101,102] Field inoculations with the endemic fungus, *Phytophthora citrophthora,* have reduced the establishment of milkweed vines by 50 to 70%.

The control of water hyacinth by *Cercospora rodmanii* is another example of a bioherbicide that is rapidly approaching commercialization. *C. rodmanii* is host specific and under certain field conditions can control populations of water hyacinth.[33,34] *Uredo eichhorniae* is another fungus that shows potential as a mycoherbicide for the control of water hyacinth.[31]

V. CONCLUDING REMARKS

The nature of the pathogen is probably the principal key factor determining the development of successful biological control. In most biocontrol programs, emphasis is to control the most important plant pathogens. Although these pathogens represent a great challenge, much could be gained by placing more emphasis on the control of minor pathogens that are more susceptible to the antagonism of microorganisms and easily manipulated changes in the environment.

Efforts should be made for a more effective use of antagonists. There are two obvious approaches to a more effective use of antagonists in biological control:

1. Learn to enhance, or at least not to upset, the benefits of the resident antagonists that operate in suppressive soil or as suppressive populations of endophytes and epiphytes on our crop plants.
2. Find or develop superior antagonists for introduction into soil or for plant inoculation.

We need to have a simultaneous effort with the above mentioned approaches. They are inseparably interrelated. The resident antagonists provide the pool from which we can obtain our candidates or prototype candidates for use as introduced antagonists. Our aim should be that the introduced antagonists function as residents.

The effective antagonists will be found in those cases where the target pathogen should cause disease but is not able to do so. The antagonists, by inference, might be acting here. If the antagonist is to live in soil, obviously we should seek our candidates from soil, but those to be used for plant protection should be sought from among the epiphytes or endophytes on and in the plants themselves. The most promising approach at present is the development of antagonists through breeding, induced mutations, or by genetic engineering. Pathogens can also be used as an antagonist to bring about its own biological control. One strain of the pathogen may be used to control another strain. We may ultimately find that our best antagonists will be modified strains of the target pathogen, because they will be best adapted to the niches or sites occupied by the pathogen.

The conventional methods of breeding for resistance to pathogens will continue to be our most productive approach to biological control with the host.

Biological control is not a substitute for chemical control but can certainly act as a valuable adjunct for controlling the diseases of plants. Any effort towards controlling plant diseases is a fight for controlling hunger and famine and it becomes all the more important with the rising world population. We have to fight the plant pathogens on all fronts and with all the weapons at our command.

REFERENCES

1. **Adams. P. B. and Ayers, W. A.,** Biological control of *Sclerotinia* luttuce drop in the field by *Sporidesmium sclerotivorum, Phytopathology,* 72, 485, 1982.
2. **Agarwal, G. P. and Rajak, R. C.,** Entomogenous fungi in biological control of insect-pests, in *Trends in Plant Research,* Govil, C. M. and Kumar, V., Eds., Bishen Singh Mahendra Pal Singh, Dehradan, 1985, 485.
3. **Aizawa, K.,** Microbial insecticides in plant protection in Japan, *Agricultural Asia,* 10, 387, 1976.
4. **Aizawa, K.,** Microbial control of insect pests, in *Advances in Agricultural Microbiology,* Subba Rao, N. S., Ed., Oxford and IBH, New Delhi, 1982, 397.
5. **Aluko, M. O. and Hering, T. F.,** The mechanisms associated with the antagonistic relationship between *Corticium solani* and *Gliocladium virens, Trans. Br. Mycol. Soc.,* 55, 173, 1970.

6. **Asher, M. J. C.,** Interactions between isolates of *Gaeumannomyces graminis* var. *tritici, Trans. Br. Mycol. Soc.,* 71, 367, 1978.
7. **Ayers, W. A. and Adams, P. B.,** Mycoparasitism of sclerotia of *Sclerotinia* and *Sclerotium* species by *Sporidesmium sclerotivorum, Can. J. Microbiol.,* 25, 17, 1979.
8. **Ayers, W. A. and Adams, P. B.,** Factors affecting germination, mycoparasitism, and survival of *Sporidesmium sclerotivorum, Can. J. Microbiol.,* 25, 1021, 1979.
9. **Backman, P. A. and Rodriguez-Kabana, R.,** A system for the growth and delivery of biological control agents to the soil, *Phytopathology,* 65, 819, 1975.
10. **Baker, K. F. and Cook, R. J.,** *Biological Control of Plant Pathogens,* Freeman, San Francisco, 1974, 433.
11. **Baker, K. F. and Snyder, W. C.,** *Ecology of Soil-borne Plant Pathogens — Prelude to Biological Control,* University of California Press, Berkeley, 1965, 571.
12. **Baker, R.,** Mechanisms of biological control of soil-borne pathogens, *Annu. Rev. Phytopathol.,* 6, 263, 1968.
13. **Baker, R., Hanchey, P., and Dottarar, S. D.,** Protection of carnation against *Fusarium* stem rot by fungi, *Phytopathology,* 68, 1495, 1978.
14. **Blakeman, J. P. and Fokkema, N. J.,** Potential for biological control of plant diseases on the phylloplane, *Annu. Rev. Phytopathol.,* 20, 167, 1982.
15. **Bliss, D. E.,** The destruction of *Armillaria mellea* in citrus soils, *Phytopathology,* 41, 665, 1951.
16. **Bowen, G. D. and Rovira, A. D.,** Influence of microorganisms on the growth and metabolism of plant roots, in *Root Growth,* Proc. University of Nottingham Easter School in Agricultural Science, Whittington, W. J., Ed., Butterworths, London, 1969, 199.
17. **Boyette, C. D., Templeton, G. E., and Smith, R. J., Jr.,** Control of winged waterprimrose (*Jussiaea decurrens*) and northern jointvetch (*Aeschynomene virginica*) with fungal pathogens, *Weed Sci.,* 27, 497, 1979.
18. **Brown, M. E.,** Plant growth substances produced by microorganisms of soil and rhizosphere, *J. Appl. Bacteriol.,* 35, 443, 1972.
19. **Brown, M. E.,** Seed and root bacterization, *Annu. Rev. Phytopathol.,* 12, 181, 1974.
20. **Brown, M. E., Burlingham, S. K., and Jackson, R. M.,** Studies on *Azotobacter* species in soil. III. Effects of artificial inoculation on crop yields, *Plant Soil,* 20, 194, 1964.
21. **Brown, M. E., Jackson, R. M., and Burlingham, S. K.,** Growth and effects of bacteria introduced into soil, in *Ecology of Soil Bacteria,* Gray, T. K. G. and Parkinson, D., Eds., Liverpool University Press, England, 1968, 531.
22. **Bruehl, G. W.,** Systems and mechanisms of residue possession by pioneer fungal colonists, in *Biological Control of Soil-Borne Plant Pathogens,* Bruehl, G. W., Ed., American Phytopathological Society, St. Paul, Minn., 1975, 77.
23. **Burges, H. D.,** *Microbial Control of Pests and Plant Diseases. 1970—1980,* Academic Press, New York, 1981, 914.
24. **Carter, M. V.,** Biological control of *Eutypa armeniacae, Aust. J. Exp. Agric. Anim. Husb.,* 11, 687, 1971.
25. **Carter, M. V. and Price, T. C.,** Biological control of *Eutypa armeniacae.* II. Studies of the interaction between *E. armeniacae* and *Fusarium lateritium* and their relative sensitivities to benzimidazole chemicals, *Aust. J. Agric. Res.,* 25, 105, 1974.
26. **Carter, M. V. and Price, T. C.,** *Aust. J. Agric. Res.,* 26, 537, 1975.
27. **Castanho, B. and Butler, E. E.,** *Rhizoctonia* decline: a degenerative disease of *Rhizoctonia solani, Phytopathology,* 68, 1505, 1978.
28. **Castanho, B., Butler, E. E., and Shepherd, R. J.,** The association of double stranded RNA with *Rhizoctonia* decline, *Phytopathology,* 68, 1515, 1978.
29. **Chang, I. P. and Kommedahl, T.,** Biological control of seedling blight of corn by coating kernels with antagonistic microorganisms, *Phytopathology,* 58, 1395, 1968.
30. **Chand, T. and Logan, C.,** Antagonists and parasites of *Rhizoctonia solani* and their efficacy in reducing stem-rot canker of potato under controlled conditions, *Trans. Br. Mycol. Soc.,* 83, 107, 1984.
31. **Charudattan, R., McKinney, D. E., Erdo, H. A., and Silveira-Guideo, A.,** *Uredo eichhorniae,* a potential biocontrol agent for water hyacinth, in *Proc. 4th Int. Symp. Biol. Control Weeds,* Freeman, T. E., Ed., University of Florida, Gainesville, 1976, 210.
32. **Chet, I., Hadar, Y., Elad, Y., Katan, J., and Henis, Y.,** Biological control of soil-borne plant pathogens by *Trichoderma harzianum,* in *Soil-Borne Plant Pathogens,* Schippers, B. and Gams, W., Eds., Academic Press, London, 1979, 585.
33. **Conway, K. E., Cullen, R. E., Freeman, T. E., and Cornell, J. A.,** Misc. Pap. A-79-6, Dept. Plant. Pathol., University of Florida, Gainesville, 1979, 50.
34. **Conway, K. E. and Freeman, T. E.,** Host specificity of *Cercospora rodmanii,* a potential biological control of water hyacinth, *Plant Dis. Rep.,* 61, 262, 1977.

35. **Cook, R. J.,** Antagonism and biological control: concluding remarks, in *Soil-Borne Plant Pathogens,* Schippers, B. and Gams, W., Eds., Academic Press, London, 1979, 653.

36. **Cook, R. J.,** Biological control of plant pathogens: theory to application, *Phytopathology,* 75, 25, 1985.

37. **Cook, R. J. and Baker, K. F.,** *The Nature and Practice of Biological Control of Plant Pathogens,* American Phytopathological Society, St. Paul, Minn., 1983, 539.

38. **Cooksey, D. A. and Moore, L. W.,** Biological control of crown gall with fungal and bacterial antagonists, *Phytopathology,* 70, 506, 1980.

39. **Corke, A. T. K.,** The prospect for biotherapy in trees infected by silver leaf, *J. Hortic. Sci.,* 49, 391, 1974.

40. **Corke, A. T. K. and Hunter, T.,** Biocontrol of *Nectria galligena* infections of pruning wounds on apple shoots, *J. Hortic. Sci.,* 54, 47, 1979.

41. **Corke, A. T. K. and Rishbeth, J.,** Use of microorganisms to control plant diseases, in *Microbial Control Pests and Plant Diseases 1970—1980,* Burges, H. D., Ed., Academic Press, London, 1981.

42. **Costa, A. S. and Müller, G. W.,** Tristeza control by cross protection: a U.S.-Brazil cooperative success, *Plant Dis. Rep.,* 64, 538, 1980.

43. **Cullen, J. M.,** Evaluating the success of the program for the biological control of *Chondrilla juncea,* in *Proc. 4th Int. Symp. Biol. Control Weeds,* Freeman, T. E., Ed., University of Florida, Gainesville, 1976, 117.

44. **Daniel, J. T., Templeton, G. E., Smith, R. J., Jr., and Fox, W. T.,** Biological control of northern jointvetch in rice with an endemic fungal disease, *Weed Sci.,* 21, 303, 1973.

45. **Darpoux, H.,** Biological interference with epidemics, in *Plant Pathology — An Advanced Treatise,* Vol. 3, Horsefall, J. G. and Dimond, A. E., Eds., Academic Press, London, 1960, 521.

46. **Davis, D.,** Cross infection in *Fusarium* wilt disease, *Phytopathology,* 56, 825, 1966.

47. **Day, P. R., Dodds, J. A., Elliston, J. E., Jaynes, R. A., and Anagnostakis, S. L.,** Double strand RNA in *Endothia parasitica, Phytopathology,* 67, 1393, 1977.

48. **Deacon, J. W.,** Control of the take-all fungus by grass leys in intensive cereal cropping, *Plant Pathol.,* 22, 88, 1973.

49. **Deacon, J. W.,** Biological control of the take-all fungus, *Gaeumannomyces graminis,* by *Phialophora radicicola* and similar fungi, *Soil Biol. Biochem.,* 8, 275, 1976.

50. **De La Cruz, R. E. and Hubbell, D. H.,** Biological control of the charcoal root rot fungus *Macrophomina phaseolina* on slash pine seedlings by a hyperparasite, *Soil Biol. Biochem.,* 7, 25, 1975.

51. **Dennis, C. and Webster, J.,** Antagonistic properties of species-groups of *Trichoderma.* I. Production of non-volatile antibiotics, *Trans. Br. Mycol. Soc.,* 57, 25, 1971.

52. **Dennis, C. and Webster, J.,** Antagonistic properties of species-groups of *Trichoderma.* II. Production of volatile antibiotics, *Trans. Br. Mycol. Soc.,* 57, 41, 1971.

53. **Dennis, C. and Webster, J.,** Antagonistic properties of species-groups of *Trichoderma.* III. Hyphal interactions, *Trans. Br. Mycol. Soc.,* 57, 363, 1971.

54. **Elad, Y., Chet, I., and Katan, J.,** *Trichoderma harzianum.* A biocontrol agent effective against *Sclerotium rolfsii* and *Rhizoctonia solani, Phytopathology,* 70, 119, 1980.

55. **Elad, Y., Katan, J., and Chet, I.,** Physical, biological, and chemical control integrated for soil-borne diseases in potatoes, *Phytopathology,* 70, 418, 1980.

56. **Fawcett, H. S.,** Fungus and bacterial diseases of insects as factors in biological control, *Bot. Rev.,* 10, 327, 1944.

57. **Ferron, P.,** Biological control of insect pests by entomogenous fungi, *Annu. Rev. Entomol.,* 23, 409, 1978.

58. **Fokkema, N. J.,** Antagonism between fungal saprophytes and pathogens on aerial plant surfaces, in *Microbiology of Aerial Plant Surfaces,* Dickinson, C. H. and Preece, T. F., Eds., Academic Press, London, 1976, 487.

59. **Freeman, T. E.,** Microbial herbicides, in *Advances in Agricultural Microbiology,* Subba Rao, N. S., Ed., Oxford and IBH, New Delhi, 1982, 419.

60. **Garrett, S. D.,** A century of root disease investigation, *Ann. Appl. Biol.,* 42, 211, 1955.

61. **Garrett, S. D.,** Inoculum potential as a factor limiting lethal action by *Trichoderma viride* Fr. on *Armillaria mellea* Fr. Quel., *Trans. Br. Mycol. Soc.,* 41, 157, 1958.

62. **Garrett, S. D.,** Towards biological control of soil-borne plant pathogens, in *Ecology of Soil-Borne Plant Pathogens,* Baker, K. F. and Snyder, W. C., Eds., University of California Press, Berkeley, 1965, 4.

63. **Garrett, S. D.,** Species of *Trichoderma* and *Gliocladium* as agents for biological control of root disease fungi, in *Recent Advances in the Biology of Micro-organisms,* Bilgrami, K. S. and Vyas, K. M., Eds., Bishen Singh Mahendra Pal Singh, Dehra Dun, India, 1980, 1.

64. **Greig, B. J. W.,** Biological control of *Fomes annosus* by *Peniophora gigantea, Eur. J. For. Pathol.,* 6, 65, 1976.

65. **Greig, B. J. W.,** *Eur. J. For. Pathol.,* 6, 286, 1976.

66. **Grente, J. and Sauret, S.,** L hypovirulence exclusive phenomene original en pathologie vegetale, *C.R. Hebd. Seanc. Acad. Sci. Ser. Paris,* 268, 2347, 1969.

67. **Grosclaude, C., Dubos, B., and Ricard, J. L.,** Antagonism between ungerminated spores of *Trichoderma viride* and *Stereum purpureum, Plant Dis. Rep.,* 58, 71, 1974.

68. **Grosclaude, C., Ricard, J. L., and Dubos, B.,** Inoculation of *Trichoderma viride* spores via pruning shears for biological control of *Stereum purpureum* on plum tree wounds, *Plant Dis. Rep.,* 57, 25, 1973.

69. **Gupta, A. K., Ashok Aggarwal, and Mehrotra, R. S.,** In vitro studies on antagonistic microorganisms against *Sclerotium oryzae* Catt., *Geobios,* 12, 3, 1985.

70. **Hadar, Y., Chet, I., and Henis, Y.,** Biological control of *Rhizoctonia solani* damping-off with wheat bran culture of *Trichoderma harzianum, Phytopathology,* 69, 64, 1979.

71. **Harman, G. E., Chet, I., and Baker, R.,** Factors affecting *Trichoderma hamatum* applied to seeds as a biocontrol agent, *Phytopathology,* 71, 569, 1981.

72. **Hoch, H. C. and Abawi, G. S.,** Mycoparasitism of oospores of *Pythium ultimum* by *Fusarium merismoides, Mycologia,* 71, 621, 1979.

73. **Hornby, D.,** Microbial antagonisms in the rhizosphere, *Ann. Appl. Biol.,* 89, 97, 1978.

74. **Huang, H. C.,** Importance of *Coniothyrium minitans* in survival of sclerotia of *Sclerotinia sclerotiorum* in wilted sunflower, *Can. J. Bot.,* 55, 289, 1977.

75. **Kamat, M. N. and Rao, V. G.,** Entomogenous fungi and their taxonomy, in *Advances in Mycology and Plant Pathology,* Raychaudhuri, S. P., Varma, A., Bhargava, K. S., and Mehrotra, B. S., Eds., published by Prof. R. N. Tandon's Birthday Celebration Committee, Allahabad University, New Delhi, 1975, 57.

76. **Kellock, L. and Dix, N. J.,** Antagonism by *Hypomyces aurantius.* I. Toxins and hyphal interaction, *Trans. Br. Mycol. Soc.,* 82, 327, 1984.

77. **Kerr, A.,** Biological control of crown gall through production of agrocin 84, *Plant Dis. Rep.,* 64, 24, 1980.

78. **Kerr, A.,** Biological control of soil-borne microbial pathogens and nematodes, in *Advances in Agricultural Microbiology,* Subba Rao, N. S., Ed., Oxford and IBH, New Delhi, 1982, 429.

79. **LaPierre, H., Lemaire, J. M., Jouan, B., and Molin, G.,** Mise en evidence de particules virales associees a une perte de pathogenicite chez la Pietinechaudage des cereals, *Ophiobolus graminis* Sacc., *C.R. Hebd. Seanc. Acad. Sci. Ser. Paris,* 271, 1833, 1970.

80. **Lewis, J. A. and Papavizas, G. C.,** Integrated control of *Rhizoctonia* fruit rot of cucumber, *Phytopathology,* 70, 85, 1980.

81. **Lui, S. D. and Baker, R.,** Mechanism of biological control in soil suppressive to *Rhizoctonia solani, Phytopathology,* 70, 404, 1980.

82. **Madelin, M. F.,** Fungal parasite of insects, *Annu. Rev. Entomol.,* 11, 423, 1966.

83. **Madelin, M. F.,** Fungal parasites of invertebrates. I. Entomogenous fungi, in *The Fungi — An Advanced Treatise,* Vol. 3, Ainsworth, G. C. and Sussman, A. S., Eds., Academic Press, New York, 1968, 227.

84. **Marois, J. J. and Mitchell, D. J.,** Effects of fumigation and fungal antagonists on the relationships of inoculum density to infection incidence and disease severity in *Fusarium* crown rot of tomato, *Phytopathology,* 71, 167, 1981.

85. **Mehrotra, R. S., Aneja, K. R., and Goel, M. K.,** Approaches to biological control with antagonistic microorganisms, *Annu. Rev. Plant Sci.,* 2, 81, 1980.

86. **Mehrotra, R. S. and Claudius, G. R.,** Biological control of the root rot and wilt diseases of *Lens culinaris* Medic., *Plant Soil,* 36, 657, 1972.

87. **Millard, W. A. and Taylor, C. B.,** Antagonism of microorganisms as the controlling factor in the inhibition of potato scab by green manuring, *Ann. Appl. Biol.,* 10, 70, 1927.

88. **Mishustin, E. N. and Naumova, A. N.,** Bacterial fertilizers, their effectiveness and mode of action, *Mikrobiologia,* 31, 543, 1962.

89. **Myers, D. F. and Strobel, G. A.,** *Pseudomonas syringae* as a microbial antagonist of *Ceratocystis ulmi* in the apoplast of American Elm, *Trans. Br. Mycol. Soc.,* 80, 389, 1983.

90. **Narasimhan, N. J.,** Entomogenous fungi and possibility of their use for biological control of insect pests in India, *Indian Phytopathol.,* 23, 16, 1970.

91. **Norris, J. R.,** *Symp. Soc. Gen. Microbiol.,* 21, 197, 1971.

92. **Odvody, G. N., Boosalis, M. G., and Kerr, C. D.,** Biological control of *Rhizoctonia solani* with a soil-inhabiting basidiomycete, *Phytopathology,* 70, 655, 1980.

93. **Odvody, G. N., Boosalis, M. G., Lewis, J. A., and Papavizas, G. C.,** Biological control of *Rhizoctonia solani, Proc. Am. Phytopathol. Soc.,* 4, 158, 1977.

94. **Oehrens, E.,** Biological control of the blackberry through the introduction of rust, *Phragmidium violaceum,* in Chile, *FAO Plant Prot. Bull.,* 25, 26, 1977.

95. **Papavizas, G. C.,** Introduction and augmentation of microbial antagonists for the control of soil-borne and foliar plant pathogens, in 5th Beltsville Symp. on Agricultural Research, *Trop. Pest Manage.,* 26, 331, 1980.

96. **Papavizas, G. C. and Lumsden, R. D.,** Biological control of soil-borne fungal propagules, *Annu. Rev. Phytopathol.,* 18, 389, 1980.
97. **Pottle, H. W. and Shigo, A. L.,** Treatment of wounds on *Acer rubrum* with *Trichoderma viride, Eur. J. For. Pathol.,* 5, 274, 1975.
98. **Pottle, H. W., Shigo, A. L., and Blanchard, R. O.,** Biological control of wound hymenomycetes by *Trichoderma harzianum, Plant Dis. Rep.,* 61, 687, 1977.
99. **Purkayastha, R. P. and Bhattacharya, B.,** Antagonism of microorganisms from Jute phyllosphere towards *Colletotrichum corchori, Trans. Br. Mycol. Soc.,* 75, 363, 1982.
100. **Ricard, J. L.,** Biological control of *Fomes annosus* in Norway spruce (*Picea abies*) with immunizing commensals, *Stud. For. Suec.,* 84, 1, 1970.
101. **Ridings, W. H., Mitchell, D. J., Schoulties, C. L., and El-Gholl, N. E.,** Biological control of milkweed vine in Florida citrus groves with a pathotype of *Phytophthora citrophthora,* in *Proc. 4th Int. Symp. Biol. Control Weeds,* Freeman, T. E., Ed., University of Florida, Gainesville, 1976, 224.
102. **Ridings, W. H., Schoulties, C. L., El-Gholl, N. E., and Mitchell, D. J.,** *Proc. Int. Soc. Citricult.,* 3, 877, 1977.
103. **Rishbeth, J.,** Stump protection against *Fomes annosus.* III. Inoculation with *Peniophora gigantea, Ann. Appl. Biol.,* 52, 63, 1963.
104. **Saksena, S. B.,** Root diseases and biological control, *J. Indian Bot. Soc.,* 51, 1, 1972.
105. **Sanford, G. B.,** Some factors effecting the pathogenicity of *Actinomyces scabies, Phytopathology,* 16, 525, 1926.
106. **Sanford, G. B. and Broadfoot, W. C.,** Studies of the effects of other soil inhabiting microorganisms on the virulence of *Ophiobolus graminis* Sacc., *Sci. Agric.,* 11, 512, 1931.
107. **Saxena, B. N., Nigam, S. S., and Agarwal, B. N.,** *Labdev J. Sci. Technol.,* 98, 81, 1971.
108. **Schippers, B. and Gams, W.,** *Soil-Borne Plant Pathogens,* Academic Press, London, 1979, 686.
109. **Sewell, G. W. F.,** The effect of altered physical condition of soil on biological control, in *Ecology of Soil-Borne Plant Pathogens,* Baker, K. F. and Snyder, W. C., Eds., University of California Press, Berkeley, 1965, 479.
110. **Sharma, K. R. and Mukerji, K. G.,** Microbial colonization of aerial parts of plants — a review, *Acta Phytopathol. Acad. Sci. Hung.,* 8, 425, 1973.
111. **Singh, J. and Khara, H. S.,** *In vitro* inhibition of *Alternaria solani* by phylloplane fungi of tomato, *Indian Phytopathol.,* 37, 579, 1984.
112. **Singh, R. S.,** Use of *Epicoccum purpurascens* as an antagonist against *Macrophomina phaseolina* and *Colletotrichum capsici, Indian Phytopathol.,* 38, 258, 1985.
113. **Singh, T. and Chohan, J. S.,** Antagonism of *Trichoderma viride* and *Bacillus* sp. against some pathogenic fungi on groundnut seeds, *Indian J. Mycol. Plant Pathol.,* 4, 80, 1974.
114. **Sivan, A., Elad, Y., and Chet, I.,** Biological control effects of a new isolate of *Trichoderma harzianum* on *Pythium aphanidermatum, Phytopathology,* 74, 498, 1984.
115. **Skolko, A. J.,** *Biological Control of Forest Diseases,* The Canadian Forestry Service, Ottawa, Canada, 21, 1972.
116. **Smith, R. J., Jr., Daniel, J. T., Fox, W. T., and Templeton, G. E.,** Distribution in Arkansas of a fungus disease used for biocontrol of northern jointvetch in rice, *Plant Dis. Rep.,* 57, 695, 1973.
117. **Snyder, W. C.,** Antagonism as a plant disease control principle, in *Biological and Chemical Control of Plant and Animal Pests,* Reitz, L. P., Ed., American Association for the Advancement of Science, Washington, D.C., 1960, 127.
118. **Snyder, W. C., Wallis, G. W., and Smith, S. N.,** Biological control of plant pathogens, in *Theory and Practice of Biological Control,* Huffaker, C. B. and Messenger, P. S., Eds., Academic Press, New York, 1976, 521.
119. **Steinhaus, E. A.,** Microbial control — The emergence of an idea: A brief history of insect pathology through the nineteenth century, *Hilgardia,* 26, 107, 1956.
120. **Steinhaus, E. A.,** *Insect Pathology,* Academic Press, New York, 1963, 689.
121. **Steinhaus, E. A.,** *Principles of Insect Pathology,* Hafner, New York, 1967.
122. **Te Beest, D. O., Templeton, G. E., and Smith, R. J., Jr.,** Histopathology of *Colletotrichum gloeosporioides* f. sp. *aeschynomene* in northern jointvetch, *Phytopathology,* 68, 1271, 1978.
123. **Templeton, G. E. and Smith, R. J., Jr.,** Managing weeds with pathogens, in *Plant Disease: An Advanced Treatise,* Vol. 1, Horsefall, J. G. and Cowling, E. B., Eds., Academic Press, New York, 1977, 167.
124. **Templeton, G. E., Te Beest, D. O., and Smith, R. J., Jr.,** Development of an endemic fungal pathogen as a mycoherbicide for biological control of northern jointvetch in rice, *Int. Symp. Biol. Weeds,* University of Florida, Gainesville, 1976, 214.
125. **Templeton, G. E., Te Beest, D. O., and Smith, R. J., Jr.,** Biological weed control with mycoherbicides, *Annu. Rev. Phytopathol.,* 17, 301, 1979.
126. **Tiwari, D. P. and Mehrotra, R. S.,** Rhizosphere and rhizoplane of *P. betle* with special reference to biological control, *Bull. Indian Phytopathol. Soc.,* 4, 79, 1968.

127. **Toussoun, T. A.,** *Fusarium* suppressive soils, in *Biology and Control of Soil-borne Plant Pathogens,* Bruehl, G. W., Ed., American Phytopathological Soc., St. Paul, Minn., 1975, 145.

128. **Turner, G. J. and Tribe, H. T.,** On *Coniothyrium minitans* and its parasitism of *Sclerotinia* species, *Trans. Brit. Mycol. Soc.,* 66, 97, 1976.

129. **Tveit, M. and Moore, M. B.,** Isolates of *Chaetomium* that protect oats from *Helminthosporium victorae, Phytopathology,* 44, 686, 1954.

130. **Tveit, M. and Wood, R. K. S.,** The control of *Fusarium* blight in oat seedlings with antagonistic species of *Chaetomium, Ann. Appl. Biol.,* 43, 538, 1955.

131. **Vesley, D.,** Biological protection of emerging sugarbeet against damping-off established by mycoparasitism in nonsterilized soil, *Zentralbl. Bakteriol.,* 133, 436, 1978.

132. **Webb, P. J.,** An alternative to chemical stump protection against *Fomes annosus* in state and private forestry, *Scott. For.,* 27, 24, 1973.

133. **Webster, J. and Lomas, N.,** Does *Trichoderma viride* produce gliotoxin and viridin?, *Trans. Brit. Mycol. Soc.,* 47, 535, 1964.

134. **Weindling, R.,** *Trichoderma lignorum* as a parasite of other soil fungi, *Phytopathology,* 22, 837, 1932.

135. **Weindling, R. and Emerson, O. H.,** The isolation of a toxic substance from the culture filtrate of *Trichoderma, Phytopathology,* 26, 1068, 1936.

136. **Weindling, R. and Fawcett, H. S.,** Experiments in the control of *Rhizoctonia* damping-off of citrus seedlings, *Hilgardia,* 10, 1, 1936.

137. **Wells, H. D., Bell, D. K., and Jaworski, C. A.,** Efficacy of *Trichoderma harzianum* as a biocontrol for *Sclerotium rolfsii, Phytopathology,* 62, 442, 1972.

138. **Wolf, F. T.,** The biology of *Entomophthora, Nova Hedwigia,* 35, 553, 1981.

139. **Wong, P. T. W. and Siviour, T. R.,** Control of *Ophiobolus* patch in *Agrostis* turf using avirulent fungi and take-all suppressive soils in pot experiments, *Ann. Appl. Biol.,* 92, 191, 1979.

140. **Wong, P. T. W. and Southwell, R. J.,** Biological control of take-all in the field using *Gaeumannomyces graminis* var. *graminis* and related fungi, in *Soil-borne Plant Pathogens,* Schippers, B. and Gams, W., Eds., Academic Press, London, 1979, 597.

141. **Wu, Weu-Shi,** Biological control of seed and air borne fungi associated with wheat and oats, *Bot. Bull. Acad. Sin.,* 17, 161, 1976.

142. **Trujillo, E. E.,** Biological control of *Hamkera pamakani* with plant pathogens, *Proc. Am. Phytopathol. Soc.,* 3, 298, 1976.

Chapter 4

HYPERPARASITES IN BIOLOGICAL CONTROL

L. Sundheim and A. Tronsmo

TABLE OF CONTENTS

I. INTRODUCTION

Parasitism is an antagonistic symbiosis between organisms. In a restricted sense, symbiosis is a relationship between taxonomically different organisms in which there is a permanent or semipermanent physical union between the individuals. Parasitism among fungi is generally named *mycoparasitism*. This phenomenon is common among all groups of fungi, from simple chytrids to higher basidiomycetes. When the host fungus itself is parasitizing a plant or animal host, the parasite is a *hyperparasite*. Recently, there has been a considerable interest in exploiting hyperparasitic relationships in the biological control of plant disease.

Barnett and Binder[15] separated mycoparasites into two groups based on their mode of parasitism. The *necrotrophic* parasite kills the host cells before, or just after invasion, and utilizes the nutrients that are released. The infection is initiated by hyphal contact or coiling around fungal cells, or by direct penetration and invasion of hyphae or survival structures. The *biotrophic* parasite is able to obtain nutrients from living host cells and causes little or no harm to the host in the early stages of parasitism. Apparently, the loss of the ability to synthesize some specific nutrients has caused the parasite to become dependent on the host for survival. The parasite may obtain its nutrients through absorptive cells or through haustoria formed from appressoria-like hyphal swellings.

Many fungi have been recorded growing on other fungi in nature. But until a true nutritional relationship is established, these fungi should be considered *fungicolous*. Hyperparasitism is readily observed on several aerial plant pathogens. Rust-fungi are often parasitized by *Eudarluca caricis* (Fr.) O. Erikss., and the hyperparasite *Ampelomyces quisqualis* Ces. is rather ubiquitous on powdery-mildews.

Few critical investigations have been made of the parasitic interactions in natural soils. Boosalis[19] estimated the frequency of parasitism of *Rhizoctonia solani* Kühn by *Penicillium vermiculatum* Dang in soil by observing the parasitism of the *R. solani* hyphae *in situ*. Living fungal structures are often parasitized in natural soils.[12,58,107] Even with increasing evidence for natural mycoparasitism in soil, the ecological significance of this phenomenon is uncertain. Griffin[41] maintained that mycoparasitism is of no ecological significance in soil. However, others have suggested that mycoparasitism has a great effect on the populations of soil fungi by reducing the longevity of survival structures.[107]

In many antagonistic relationships between microorganisms the actual mechanism is not known. Such information as the relative importance of parasitism, competition, and antibiosis is of importance in planning the application strategy in practical biological control. Organisms that parasitize and destroy the sclerotia, or other surviving structures of the pathogen, should preferably be applied before planting the crop, whereas an antagonist that competes with the pathogen or inhibits its growth by antibiosis may be applied at the time of planting.

There have been a number of studies on the hyperparasites of plant-pathogenic fungi, but the exploitation of the hyperparasites in actual biological control is still limited. Earlier workers did not get encouraging results using hyperparasites in the biological control of plant diseases.[108] They observed that successful laboratory results could seldom be repeated in the field. However, during the last decade, progress has been made with several hyperparasites. Successful disease control has been achieved following the application of hyperparasites to crops both in the field and under controlled environmental conditions. There are an increasing number of reports about promising results from field trials, and some hyperparasites are being developed for commercial utilization.[18,26,60,69,71,82,94,119,128,134,140]

II. HYPERPARASITES ON OOMYCETES

In most plant diseases caused by the *Aphanomyces* spp., *Phytophthora* spp., and *Pythium* spp. the pathogens survive as oospores between crops. Sneh et al.[107] reported a number of microorganisms parasitizing oospores of the *Phytophthora* spp. Ayers and Lumsden[12] identified *Hyphocytrium catenoides* Kerling as a parasite of oospores in soil. Hsu and Lockwood[53] studied the possibility of using *H. catenoides* in the biological control of *Phytophthora megasperma* Drechsler f. sp. *glycinae* in greenhouse tests. Seedling emergence and plant growth was increased when the parasite was added to inoculated soil. Root disease was reduced to a level close to the uninoculated control. The control produced similar results whether the soil was flooded or not before seeding. Filonow and Lockwood[37] coated soybean seeds with four actinomycetes and *H. catenoides*, and planted them in soil naturally infested with *P. megasperma* f. sp. *glycinea*. The actinomycetes *Actinoplanes missouriensis* Couch, *A. utahensis* Couch, and a *Micromonospora* sp. increased the soybean stand as determined one month after planting, while *H. catenoides* had no effect on the plant stand.

A. *Pythium oligandrum*

Pythium oligandrum Drechsler was first described as a plant pathogen causing pea root rot in the U.S., but it has since been found on many other plants. It is frequently associated with *Pythium debaryanum* Hesse and other *Pythium* spp. as a secondary invader of plant tissue, partly parasitic on the primary invaders. Deacon[30] made the first thorough study of the hyperparasitism of *P. oligandrum* and found it to be confined to the immature hyphae of the host fungi. Young hyphae of most fungi are susceptible to infection by *P. oligandrum*, but as the hyphae mature they become resistant to infection. In some fungi this can be related to pigmentation. Deacon also found *P. oligandrum* to be an aggressive, necrotrophic parasite on the plant pathogenic fungi *Gäumannomyces graminis* (Sacc.) v. Arx and Olivier var. *tritici* Walker and *Phialophora radicicola* Cain. The hyperparasite produces numerous thin haustorial threads and coils around the host hyphae.[142]

Damping-off in sugar-beet is a damaging disease, particularly in relation to sugar-beet production based on the use of monogerm seeds. Important pathogens of the emerging sugar-beets are *Pythium ultimum* Throw., *P. debaryanum*, and *Aphanomyces laevis* De By. Veselý and Hejdánek[143] compared the hyperparasite *P. oligandrum* to fungicides in pot trials. The sugar-beet emergence was higher when the hyperparasite was applied as a seed treatment than when a fungicide was used to control damping-off. Application of a *P. oligandrum* preparation at the rate of 100,000 oospores/seed led to the colonization of sugar-beet root surfaces and protection against the pathogenic *Pythium* spp.[141]

In field experiments Ahrens[8] obtained higher emergence and increased plant weight when beet seeds were inoculated with oospores of *P. oligandrum*, compared to seeds treated with the fungicide mancozeb. Al-Hamdani and Cooke[9] demonstrated that the cellulolytic ability and sclerotium production in *Rhizoctonia solani* are markedly reduced when *P. oligandrum* and *R. solani* are inoculated simultaneously. These effects are caused both by parasitism and competition for nutrients.

III. HYPERPARASITES ON SCLEROTIA FORMING FUNGI

Several soil-borne fungi survive as sclerotia.[25] The genus *Sclerotinia* contains several important plant pathogens. The sclerotia of the *Sclerotinia* spp. have been reported to survive for 4 to 8 years in soil. Coley-Smith[24] found that more than 90% of the sclerotia of the onion white-rot fungus, *Sclerotium cepivorum* Berk., survive in natural soils for more than 4 years. Ayers and Adams[11] suggested that the survival of sclerotia in soil is most significantly affected by the biological components of the soil. A literature survey revealed that more than 30 species of fungi and bacteria have been reported to be parasitic on the sclerotia of the *Sclerotinia* spp. Only four of these, *Coniothyrium minitans* Campbell, *Gliocladium* sp., *Sporidesmium sclerotivorum* Uecker, and certain *Trichoderma* sp., were destructive to the *Sclerotinia* spp. in soil.[1]

A. *Coniothyrium minitans*

The hyperparasite *Coniothyrium minitans* was first isolated and described by Campbell.[21] He obtained *C. minitans* from the sclerotia of *Sclerotinia sclerotiorum* (Lib.) D.By. on diseased guayule roots in California. *C. minitans* invades the sclerotia and produces numerous pycnidia in the surface layers. Infected sclerotia become soft and disintegrate easily. Conidia ooze out from the pycnidia as black liquid masses. Campbell[21] reported that sclerotia became infected and covered with pycnidia within 10 days after inoculation.

C. minitans has been identified as a parasite of the *Sclerotinia* spp. in Australia,[87] Canada,[57] England,[126] and several other countries. Turner and Tribe[138] reported that the host range of *C. minitans* includes *S. sclerotiorum*, *S. trifoliorum* Erikss., *Botrytis cinerea* Pers. ex Fr., *B. fabae* Sardina, *B. narcissicola* Kleb., *Sclerotium cepivorum*, and *Claviceps purpurea* Fr. The *Sclerotinia* spp. and *S. cepivorum* were the most susceptible, while 17 species of green plants were not affected. Turner and Tribe[138] were unable to find a natural habitat for the hyperparasite apart from the infected sclerotia. They postulated that *C. minitans* occurs as a saprophyte on aerial parts of plants, and that it invades the *Sclerotinia* mycelium and attacks the sclerotia as they are formed. Based on inoculation experiments they obtained no evidence of damage to green plants.

In field trials, Turner and Tribe[138] found that up to 65% of the sclerotia of *S. trifoliorum* were destroyed following the application of a pycnidial dust of *C. minitans*. Ahmed and Tribe[7] used *C. minitans* as a soil treatment and as a seed dressing to protect onion seeds sown in *S. cepivorum* infested soil. Both of the hyperparasite treatments gave results comparable to the calomel-seed dressing.

Huang[56] found the hyperparasite to be widely distributed in Canada. *C. minitans* occurred in 27% of 122 sunflower fields sampled, and it was detected more readily in fields with a light to severe incidence of *Sclerotinia* wilt than in field with only trace amounts of the disease.

Huang,[55] in greenhouse experiments with *S. sclerotiorum*, found *C. minitans* to be a more effective mycoparasite than *Gliocladium catenulatum* Gilman and Abbott or *Trichoderma viride* Pers. ex Fr. Field trials during 3 years showed that the introduction of *C. minitans* into *S. sclerotiorum*-infested soil at seeding time decreased *Sclerotinia* wilt of sunflowers and reduced yield losses. The number of primary infection loci decreased but the rate of pathogen spread was not significantly reduced. This suggests that the reduction in *Sclerotinia* wilt is due to control of the primary inoculum.[55]

B. *Sporidesmium sclerotivorum*

The dematiaceous hyphomycete, *Sporidesmium sclerotivorum* Uecker, Ayers and Adams, was first described on sclerotia of *Sclerotinia minor* Jagger from Maryland.[139] *S. sclerotivorum* parasitizes the sclerotia of *S. minor* both in vivo and in vitro.[11] More than 95% of *S. minor* sclerotia became parasitized within 10 weeks. Also, sclerotia of *Sclerotinia trifoliorum*, *S. sclerotiorum*, *Sclerotium cepivorum*, and *Botrytis cinerea* can be parasitized by *S. sclerotivorum*. However, Ayers and Adams[11] were unable to infect sclerotia of *Macrophomina phaseolina* Tassi (Goid), *Sclerotium rolfsii* Sacc., and *Rhizoctonia solani*.

S. sclerotiorum infects sclerotia without forming any specialized penetration structures. The rind cells become invaded intracellularly, and the invading hyphae then grows intracellularly through the sclerotium. The sclerotium is eventually filled with brown compacted hyphae. Infected sclerotia become soft and disintegrate after some time.[5,11] In soil, hyphae of *S. sclerotiorum* can grow from one infected sclerotium to other sclerotia by a thin, branching hyphae. Observations have shown that sclerotia as far as 1 cm away from the infected sclerotium have become infected. Macroconidia develop along the hyphae, and within 6 weeks Ayers and Adams[11] observed that the hyphae of the hyperparasite formed a blackened bulk of soil and infected sclerotia, interwoven by a network of mycelium. Adams and Ayers[2] found that *S. sclerotivorum* is active in soils between 15 and 25°C and over a pH range of 5.5 to 7.5. It grows in soil with soil-water potentials of −8 bars and higher. In the U.S. the hyperparasite is widely distributed in sclerotia on host plants. Adams and Ayers[3] did not find evidence that it exists as a saprophyte in nature.

S. sclerotivorum is able to control lettuce-drop caused by *S. minor* in the field.[4] One application of the hyperparasite controlled the disease on three successive crops during 2 years. Adams et al.[6] studied populations of the hyperparasite and its host in experiments where both the concentration of the hyperparasite and the concentrations of the *S. minor* sclerotia were varied. The results indicate that an inoculum as low as 5 macroconidia of *S. sclerotivorum* per gram soil, or 22 kg/ha, successfully infects and decays the sclerotia. Each infected sclerotium supports the production of approximately 15,000 new macroconidia in the soil. However, the time needed for disease control by *S. sclerotivorum* depends both on the density of the *S. minor* sclerotia and on the environmental conditions. Under practical conditions, Adams and Ayers[4] were able to control lettuce-drop caused by *S. minor* with 10 macroconidia/g soil when the density of *S. minor* was 16 to 24 sclerotia/g soil. Once the hyperparasite is established on the sclerotia it increases its population markedly in soil. Adams et al.[6] found a 14,000-fold increase of spores over a 12-week period in a sandy loam soil in vitro.

C. *Laetisaria arvalis*

Laetisaria arvalis Burdsall, a soil-inhabiting basidiomycete, was described by Burdsall et al.[20] It parasitizes *Rhizoctonia solani* and some other fungi and has potential as a biocontrol agent. The teleomorph stage of the hyperparasite may be produced on *R. solani* sclerotia in culture.

Hoch and Abawi[50] tested a number of parasites of *Pythium ultimum* and found *L. arvalis* to be highly effective in reducing the preemergence and postemergence damping-off caused by *P. ultimum*. Soil amendments with the hyperparasite increased beet emergence and subsequent survival of table-beet seedlings. Repeated plantings of table-beets in the same field induced a soil suppressiveness that was characterized by a reduced number of *P. ultimum* propagules, an increased number of *L. arvalis* sclerotia, and a lower seedling disease incidence.[84] The beet seedlings are highly susceptible during the first 3 weeks after planting. Incorporation into soils of large amounts of *L. arvalis* grown on organic sugar-beet pulp was not considered economically feasible for a single crop season, but establishment of the hyperparasite provides control for several crop seasons. Thus, Martin et al.[84] maintained that both the short-term and the long-term suppression of reproduction of *P. ultimum* by *L. arvalis* indicates the potential of *L. arvalis* in a disease management system. *L. arvalis* also prevents infection by the common seed-borne fungus *Phoma betae* Frank on beets.[83] Soil amendments with *L. arvalis*, or coating seedballs with sclerotia of the hyperparasite, greatly reduced the disease induced by *P. betae* in natural soil.

The host range of *L. arvalis* includes a number of other important plant pathogens. Larsen et al.[78] applied *L. arvalis* to sugar-beet fields naturally infested with *Rhizoctonia solani*. The antagonist population reached its peak 3 months after application, and the *R. solani* population was reduced to less than half compared to the untreated plots. However, neither autumn nor spring treatments reduced sugar-beet losses due to *R. solani*. Lewis and Papavizas[79] obtained a 33% reduction in cucumber fruit-rot caused by *R. solani* when *L. arvalis* was applied to field soils naturally infested with the pathogen.

IV. TRICHODERMA AND GLIOCLADIUM AS HYPERPARASITES

The *Trichoderma* spp. and the *Gliocladium* spp. are common soil inhabitants, especially in organic soils. These fungi can live either saprophytically or parasitically on other fungi.

These two fungi are well known for their antagonistic behavior, and have been much used in biocontrol trials.[27,92,128] The actual mechanisms of antagonism are not always elucidated, but competition for nutrients, production of inhibitors, and hyperparasitism are described. There has been much confusion about the accurate classification of *Trichoderma* and *Gliocladium*. Therefore, several research articles may have described results from misclassified isolates.

A. Hyphal Interaction

Hyphal interaction, seen as the coiling of the *Trichoderma* spp. around fungal hyphae, is frequently observed. Dennis and Webster[31] investigated 80 *Trichoderma* isolates in dual cultures against six test fungi (*Rhizoctonia solani* Kühn, *Heterobasidion annosum* (Fr.) Bref., *Fusarium oxysporum* Schlecht. Ex Fr., *Pyronema domesticum* (Snow. ex Grey) Sacc., *Mucor hiemalis* Wehmer, and *Pythium ultimum* Trow.,) and only ten isolates did not show coiling. Most isolates coiled around the hyphae of all six test fungi, but some did not interact with the hyphae of *F. oxysporum*. Interestingly enough, the *Trichoderma* spp. did not show coiling around plastic threads of similar diameter as fungal hyphae.[31] This indicates that a contact stimulus is necessary for coiling.

Hyphal interactions of the *Trichoderma* spp. were studied at pH 4.0 and 6.5, and no differences were found.[31] This contrasts the findings of Aytoun[13] who observed more intense coiling at pH 3.4 than at pH 5.1 or 7.0. The effect of temperature, in the range of 5 to 20°C, has also been studied, but no effects on the hyphal interaction were observed.[130]

Investigations of the infection process of the *Gliocladium* spp. have showed similar behavior as for *Trichoderma* spp. Walker and Maude[144] showed that when *Gliocladium roseum* (Link) Bainer came in contact with *Botrytis allii* Munn, it produced appressoria, and then developed intracellular hyphae within the *Botrytis* hyphae. The same infection process has been observed on *Rhizoctonia solani* Kühn by *G. virens* Miller Giddens and Foster[137] and on *Eutypa armeniacea* Hansf. and Carter[102] and *R. solani* by *G. virens*.[52]

B. Parasitism and Lysis

Even though hyphal interactions are very frequent among isolates of the *Trichoderma* spp., penetration of the hyphae by the *Trichoderma* spp. is seldom observed through a light microscope.[31] It was, therefore, doubted whether the *Trichoderma* spp. really act as hyperparasites or not. However, studies of the cell-wall degrading enzymes from the *Trichoderma* spp. have shown that they produce enzymes necessary for the degradation of fungal walls composed of chitin, cellulose, glucan-, and xylan-polymers.[44,59,62,90,124] Careful studies with light, scanning, and transmission electron microscopes have also demonstrated that the *Gliocladium* spp.[52,54,136] and the *Trichoderma* spp.[33,34,36,131] may act as hyperparasites. Despite the documentation of in vitro mycoparasitism, its ecological significance is not well understood.[92]

C. Recognition Between Host and Hyperparasite

Dennis and Webster[31] observed that the *Trichoderma* spp. did not show coiling around plastic threads, but they do coil around certain fungi. This indicates that a specific recognition is necessary before a hyperparasitic relation can take place. Lectins, sugar-binding proteins or glycoproteins of a nonimmune origin, are known components of fungal cell walls.[99] The possible role of lectins in the interaction between the *Trichoderma* spp. and *Rhizoctonia solani* and *Sclerotium rolfsii* has been studied by Barak et al. and Elad et al.[14,35] They have found indications that lectins are involved in the direct attachment between the *Trichoderma* spp. and its prey, and a specific recognition between the parasite and its host could also be explained by this mechanism. This recognition may be the first, and a necessary step, in the fungus to fungus interaction of a mycoparasite. Further progress is dependent on the ability of the hyperparasite to dissolve and penetrate the host cell wall.

D. *Gliocladium* spp. as Hyperparasite

Gliocladium roseum and *G. virens* have been shown to be parasitic on a wide variety of plant pathogens including *Botrytis allii*,[144] the *Fusarium* spp.,[54] *Eutypa armeniacea*,[102] *Phomopsis sclerotioides* Kest.,[88] *Rhizoctonia solani*,[52] and *Sclerotinia sclerotiorum*.[54,136] *G. roseum* was found to be an almost constant contaminant of isolates of *Botrytis allii* taken from growing plants and onion bulbs, and it also occurred on *Botrytis* sclerotia, many of which failed to germinate.[144] In-furrow treatment of soil, infested with *R. solani*, with *G. virens* reduced the preemergence damping-off from 55 to 11%.[52] Treatment of the soil with *G. virens* resulted in a 63% reduction in the number of viable *R. solani* sclerotia after 3 weeks of incubation.[52] Tu and Vaartaja[137] obtained promising results with *G. virens* against *R. solani* on white beans in artificially infested greenhouse soil, when the hyperparasite was introduced 2 months before or at the time of planting. *G. roseum* has also been reported to reduce cucumber black-root caused by *Phomopsis sclerotioides*.[88]

E. *Trichoderma* spp. as Hyperparasite

The mode of action of the *Trichoderma* spp. against pathogens is often less well-defined than for the *Gliocladium* spp. However, several successful biocontrol trials have been performed with *Trichoderma*, and hyperparasitism may be the most important antagonistic mechanism. Roth[103] was able to control an unidentified stem-rot disease on sugarcane with a hyperparasitic isolate of *T. viride* Pers. ex Fr. The author indicated that loss from stem-rot disease on sugarcane was due to the absence of *T. viride* in the soil. Another disease on aerial plant parts[128] controlled by *T. viride* is *Stereum purpureum* on pears and plums.[28,42] *T. harzianum* has been reported to control *Botrytis cinerea* Pers. on apples,[132] and the same fungi has been controlled by *T. viride* and *T. harzianum* Rifai on grapes[32,43] and on strawberries.[129] Many successful biocontrol trials have been performed on soil-borne pathogens[27,92] (see also, Wells, Chapter 5, this volume), but it has been difficult to prove that hyperparasitism is the actual antagonistic mechanism in a natural soil system.[92]

F. Integrated Control

Even if one disease is controlled by a hyperparasite, other diseases on the same plant usually have to be controlled by chemicals. In this context, resistance or tolerance in the antagonist against fungicides is needed. Resistant isolates of *Trichoderma* have been obtained by gamma- or UV-irradiation and by selection of mutants from media containing fungicides.[128]

Many of these mutants are as antagonistic as the wild-type in laboratory tests. In some trials they are more antagonistic than the wild strain, even without the application of fungicides.[93,127] Integrated control with the *Trichoderma* spp. and *Gliocladium* spp., together with selective pesticides, may offer a new and promising possibility of control in cases where no single component is effective.

V. HYPERPARASITES ON POWDERY MILDEWS

Considering the large number of plant-parasitic fungi within Erysiphaceae, the number of hyperparasites described as members of this family is surprisingly small.[48] The most common hyperparasite on powdery-mildews is the coelomycete *Ampelomyces quisqualis*. It was first described in the middle of the nineteenth century, but for many years some mycologists considered the hyperparasites pycnidia for accessory spore stages of the powdery-mildew hosts.

A. *Ampelomyces quisqualis*

A. quisqualis has many hosts within Erysiphaceae in both tropical and temperate climates. Inoculation experiments with *A. quisqualis* isolates obtained from powdery-mildews on several plant families have not given any indication of host specificity.[96] Hashioka and Nakai[46] studied the ultrastructure of growth and pycnidial development of the hyperparasite. *A. quisqualis* penetrates from cell to cell through the septal pores of the powdery-mildew, and continues growth during the gradual degeneration of the infected cells. Beuther et al.[16] did not find any evidence of toxin production by *A. quisqualis*.

Using a scanning electron microscope, Sundheim and Krekling[121] investigated the infection process of *A. quisqualis* on the cucumber powdery-mildew *Sphaerotheca fuliginea* (Fr.) Poll. Within 24 hr after inoculation the hyperparasite had germinated, and the germ tubes had developed appressorium-like structures at the point of contact with the powdery-mildew host. Within 5 days the hyperparasite had developed pycnidia with mature conidia in the powdery-mildew hyphae and conidiophores.

Yarwood[147] used conidial suspensions of *A. quisqualis* in experiments with the biological control of red-clover powdery-mildew. He reported that conidial production in the powdery-mildew host ceased within 1 week after application of the hyperparasite. Odintsova[91] controlled apple powdery-mildew (*Podosphaera leucotricha* (Ell. and Ev.) Salmon) by application of conidial suspensions of the hyperparasite.

Jarvis and Slingsby[61] obtained control of the cucumber powdery-mildew *S. fuliginea*. When conidial suspensions of the hyperparasite were applied, *A. quisqualis* reduced the powdery-mildew attack and increased cucumber yield. They also noted small, angular leaf-spots and sunken lesions on mature cucumber fruits sprayed with the hyperparasite. Damage to green plants has not been reported from other experiments with *A. quisqualis*.

Sztejnberg[123] reported on the control of *S. fuliginea* on cucumber and watermelon, and also on the control of powdery-mildews on other greenhouse crops. Under conditions of high humidity, Philipp and Crüger[96] obtained infection by the hyperparasite on powdery-mildews of greenhouse cucumber and also on powdery-mildews of crops in the open. Weekly sprays with spore suspensions controlled the powdery-mildew development during periods of humid weather. *A. quisqualis* had less of an effect on powdery-mildew in field grown crops during dry periods.

Sundheim[116] conducted control experiments with *S. fuliginea*-infected greenhouse cucumber. The yield increases over an untreated control were similar to the increases due to weekly applications of conidial suspensions of the hyperparasite and treatment with the fungicide triforine. Also, in experiments on cucumbers infected with the powdery-mildew *Erysiphe cichoracearum* DC. in commercial greenhouses, *A. quisqualis* extensively parasitized the mildew. The cucumber yields on *A. quisqualis* sprayed plots were comparable to the yields on quinomethionate-sprayed plants. Ampelomycin, a biocontrol preparation produced from a strain of *A. quisqualis*, was tested by Puzanova.[101] Ampelomycin controlled powdery-mildews on several hosts plants.

A. quisqualis tolerates several fungicides used for powdery-mildew control.[95,96,98,120] Thus, the hyperparasite should be well adapted to integrated control programs. Sundheim[116] combined the application of the hyperparasite with reduced rates of fungicides without any reduction in disease control. Philipp et al.[97] found that, when applied at normal rates, most of the acaricides, insecticides, and fungicides tested had only a slight or moderate inhibitory effect on *A. quisqualis*. In control experiments, the hyperparasite was applied at regular intervals to cover new growth of the host plant and to protect against the rapid dissemination of the powdery-mildews. Philipp et al.[97] maintained that passive transport of the hyperparasite in infected powdery-mildew conidia plays an important role in dissemination of the hyperparasite. Parasitism of powdery-mildew on unsprayed control plots also indicates the spread of airborne *A. quisqualis* inoculum.[116]

Freeze-dried spore suspensions or agar cultures of *A. quisqualis* did not have any toxic effects when subjected to a standard OECD-test for acute oral toxicity in rats. A standard eye irritation test in rabbits was also negative.[117,118] Thus, there is no evidence for the production of toxic metabolites by the hyperparasite.

B. *Tilletiopsis* Yeasts

The *Tilletiopsis* spp. are common phylloplane yeasts belonging to the family Sporobolomycetaceae. Hoch and Provvidenti[51] found strong antagonism between a *Tilletiopsis* sp. and the cucumber powdery-mildew *Sphaerotheca fuliginea*. Inoculation with a *Tilletiopsis*-spore suspension on detached, mildew infected cucumber leaves destroyed the superficial hyphae and conidial inoculum of the powdery-mildew. Development of the *Tilletiopsis* sp. on the *S. fuliginea* colonies could be observed 48 hr after inoculation. Within 5 days the powdery-mildew was eliminated. When the hyperparasite was applied at lower spore concentrations, the time needed to eradicate the powdery-mildew increased by several days.

Treatment of cucumber leaves with *Tilletiopsis* spores up to 8 days before inoculation with *S. fuliginea* prevented powdery-mildew development. The isolate used to control *S. fuliginea* also controlled apple powdery-mildew, *Podosphaera leucotricha*, and the powdery-mildew of grape, *Uncinula necator* (Schw.) Burr.[51] Hartmann et al.[45] tested the parasitism of several *Tilletiopsis* spp. on *S. fuliginea*. All suppressed the powdery-mildew colonies compared to the untreated control.

C. *Cladosporium* spp.

The *Cladosporium* spp. have been reported to parasitize several powdery-mildews. Mathur and Mukerji[85] sprayed a spore suspension of *Cladosporium spongiosum* onto leaves of *Morus alba* infected with *Phyllactinia guttata* (Wallr.) Lev. and obtained some disease control. *C. spongiosum* also parasitizes and inhibits the conidial germination of *Phyllactinia dalbergiae* Pirozynski on *Dalbergia sissoo*, while *C. cladosporioides* (Fres.) de Vries parasitizes *Erysiphe cichoracearum* on *Xanthium strumarium*.[85]

VI. HYPERPARASITES ON RUSTS

A. *Eudarluca caricis*

Eudarluca caricis (Fr.) O. Erikss., with the conidial stage *Sphaerellopsis filum* (Biv.-Bern. :F) Sutton, parasitizes many species of rust fungi. It is frequently found on the important cereal-rust species, *Puccinia coronata* Corda, *Puccinia graminis* Pers., *Puccinia recondita* Rob. ex Desm., *Puccinia sorghi* Schw., and *Puccinia striiformis* Westend.[23,68,70,104]

Kuhlman and Matthews[75] found the pine-rusts *Cronartium fusiforme* Hedgc. and Hunt. and *C. strobilinum* (Arthur) Hedgc. and Hahn. to be hosts of *E. caricis*. In a new host list, Kranz and Bradenburger[70] listed 369 rust species as hosts of *E. caricis*. The hyperparasite is common in tropical and subtropical countries, but it is also found in temperate climates.[39,63,67,89]

The conidial state of *S. filum* can be observed under low magnification as shiny, black, spherical pycnidia in clumps between the urediniospores. It also parasitizes the spermogonia, aecia, and telia of its hosts.

Carling et al.[22] characterized the *E. caricis-Puccinia graminis* relation as a destructive biotrophic relationship, and showed that penetration of the spore wall is due to a combination of mechanical and enzymatic processes. *D. filum* germinates on wheat leaves without the presence of its rust host, but urediniospores of *P. recondita* stimulate the germination of the hyperparasite.[114]

Swendsrud and Calpouzos[122] applied the hyperparasite to wheat leaves 3 days prior to inoculation with *P. recondita*. This resulted in more severe rust attacks and fewer uredinia infected than when both organisms were applied simultaneously. Application of *E. caricis* 3 days after inoculation with the rust had the same effect as simultaneous inoculation. Apparently, *E. caricis* conidia have a limited ability to survive on the wheat leaves in the absence of the rust host.

Hau and Kranz[47] assessed the effectiveness of *E. caricis* in limiting the development of

P. recondita by using a mathematical model. They concluded that the hyperparasite will reduce the final severity of leaf-rust by 60 to 80% when the pycnidia occupy 40 to 60% of the total rust-pustule surface.

Kuhlman et al.[76] conducted inoculation experiments with *C. fusiforme* and *E. caricis* on oak. The rust sori became infected by the hyperparasite, but this did not reduce the number of telia formed in the sori. They concluded that the biological control of *C. fusiforme* by *E. caricis* is not practical because of the short cycle of *C. fusiforme* on oak, the alternate host.

B. *Verticillium* spp.

Verticillium lecanii (Zimm.) Viegas is a catholic hyperparasite and parasitizes arthropods, rust fungi, powdery-mildews, and many other fungi. Spencer and Ebben[111] conducted inoculation experiments on *Sphaerotheca fuliginea*-infected greenhouse cucumber. They found that the hyperparasite did not survive for very long on the cucumber leaves. In a yield-trial cucumber plants sprayed with *V. lecanii* had less mildew than the unsprayed control, but there was no significant increase in the cucumber yield on plants sprayed with the hyperparasite.

Allen[10] inoculated rust-infected bean plants with a *V. lecanii* isolate from the aphid *Brachyocaudus helichrysi* Kltb. The aphid-isolate colonized the uredinia of the bean rust *Uromyces appendiculatus* (Pers.) Unger. Other rust hosts of *V. lecanii* are *Puccinia chrysanthemi* Roze,[66] *Hemileia vastatrix*,[81] and *Uromyces dianthi* (Pers.) Niessl.[109]

Mendgen[86] inoculated stripe-rust (*Puccinia striiformis*) infected wheat leaves with *V. lecanii*, labeled with fluorescent antibodies. He found that the urediniospore wall dissolved following infection. The hyperparasite requires air humidity above 80% for growth in the stripe-rust pustule, but the optimal condition is a relative air humidity in the range of 95 to 100%. Temperatures between 15 and 18°C allow good development of the hyperparasite on stripe-rust, and high light intensities have a positive effect on parasitism.

Spencer[109] obtained significant control of carnation-rust in greenhouse experiments using *V. lecanii* conidia. Spencer and Atkey[110] found that the number of uredinia per plant was reduced by 84 to 90% when the hyperparasite was applied together with the rust urediniospores.

Lim and Nik[80] reported *Verticillium psalliotae* Treschow growing on the uredinia of the coffee-rust *Hemileia vastatrix* Berk. and Br. in Malaysia. Some uredinia were completely covered by the white mycelium of the hyperparasite, which prevented further development of the rust. The hyperparasite penetrated living urediniospores and filled the spores with hyphae. No penetration of the vegetative rust hyphae was observed.

Grabski and Mendgen[40] used *V. lecanii* to control bean-rust (*U. appendiculatus*). The hyperparasite reduced the spread of bean-rust in greenhouse experiments, but in the field *V. lecanii* did not prevent the spread of the bean-rust fungus.

C. *Tuberculina* spp.

The hyphomycete genus *Tuberculina* includes several hyperparasites on the aecia and uredinia of rust fungi. Sundaram[115] found parasitism of *Tuberculina costaricana* Syd. on uredinia of *Puccinia penniseti* Zimm. and other *Puccinia* and *Uromyces* species in India. Sharma et al.[106] inoculated *T. costaricana* on groundnut rust (*Puccinia arachidis* Speg.) and obtained heavy parasitism of the rust urediniospores.

Tuberculina maxima Rostrup was described from cankers of the white-pine blister rust (*Cronartium ribicola* J. C. Fischer) on *Pinus strobus*. In western North America *T. maxima* is common on a number of pine rusts.[100] Kuhlman and Miller[77] reported between 15 and 20% *T. maxima* colonization of *Cronartium quercuum* (Berk.) Miyabe ex Shirai on *Pinus taeda* in North and South Carolina. There is no experimental evidence of a direct, nutritional relationship between *T. maxima* and the rust.[145]

Histological studies of the *T. maxima*-rust association indicated that *C. ribicola* does not degrade pine cells.[146] However, when *T. maxima* is present in rust-infected tissue, a rapid degradation occurs. Walls, cytoplasm, and nuclei in infected pine cells are destroyed when the *T. maxima* hyphae penetrates, but the hyperparasite is not able to attack rust-free pine tissue. Wicker and Woo[146] concluded that *T. maxima* destroys the nutritional base of the rust and suppresses production of both spermogonia and aecia. Thus, it reduces the inoculum production and delays rust damage, but *T. maxima* does not control the disease.[145]

D. Other Hyperparasites on Rusts

Kuhlman et al.[74] described *Scytalidium uredinicola* Kuhlman et al. as a new hyperparasite of *Cronartium quercuum* f. sp. *fusiforme* on loblolly pine (*Pinus taeda*) and slash pine (*P. elliottii* var. *elliottii*). Kuhlman[72,73] observed the hyperparasite in three pine-plantations during four seasons, and found this to be the most common hyperparasite on these two pines in North and South Carolina. Hiratsuka et al.[49] reported *S. uredinicola* to be common on the western gall rust *Endocronartium harknessii* (J. P. Moore) Y. Hiratsuka in Canada. Subsequent investigations showed that more than 80% of the galls were infected in some localities.[135]

S. uredinicola parasitizes the rust spores without penetration, but it causes a degradation of the spore wall and the cell contents. Tsuneda et al.[135] found hyphae of the parasite in the rust sori and in the rust hyphae within the pine tissue. The hyperparasite develops slowly and inactivation of one rust gall is usually not completed within the year of infection. Finally, the whole rust gall can be replaced by numerous arthrospores of the parasite. Cunningham and Pickard[29] identified a *S. uredinicola* metabolite, which in low concentrations inhibits germination of *E. harknessi* spores. Tsuneda et al.[135] considered *S. uredinicola* a promising biological control agent against *E. harknessii*.

Aphanocladium album (Preuss) W. Gams induces telia formation in cereal-rusts.[17] Forrer[38] found metabolic products from *A. album* which induce formation of telia in several rust fungi. Application of a purified, cell free extract induced telia production in *Puccinia graminis* f. sp. *tritici*, *Puccinia sorghi*, and *Puccinia recondita*.

A. album may parasitize urediniospores of *P. graminis* f. sp. *tritici* by penetration.[65] In infected spores, the cytoplasm disintegrates and disappears. Within a few days after application of a conidial suspension, the rust sori become completely covered by the cotton-like, white mycelium of the hyperparasite. Kog and Défago[64] inoculated 14 rust species with a conidial suspension of *A. album*, and all were heavily parasitized within 1 week after inoculation. The teliospores of three smut species were also parasitized.

Several *Cladosporium* spp. are hyperparasites of rust fungi.[105] Steyaert[113] described *Cladosporium hemileiae* Steyaert as a parasite on the coffee-rust fungus *Hemileia vastatrix* in Zaire. Tsuneda and Hiratsuka[133] described the penetration of *Cladosporium gallicola* Sutton on the western gall rust *Endocronartium harknessii*. The hyperparasite grows very rapidly and its mycelium covers *E. harknessii* within a few days after the bark ruptures and exposes the aeciospores. The mycoparasite penetrates viable rust spores and it is not known to produce antibiotics. *C. gallicola* is usually restricted to the outer rust-spore layers, and it does not affect the basal region of galls or the rust hyphae in pine tissue. Tsuneda and Hiratsuka[133] found evidence for an enzymatic action in this contact parasitism.

Traquair et al.[125] studied the parasitism of *Cladosporium uredinicola* Speg. on *Puccinia violae* (Schum.) DC, on the garden violet. The hyperparasite invades the spores from small swellings on the hyphal tips. Intercellular hyphae of *C. uredinicola* develop in the parasitized spores.

Srivastava et al.[112] compared the hyperparasites *Cladosporium sphaerospermum* Penzig, *C. uredinicola*, *Aphanocladium album*, and *Verticillium lecanii* in inoculation experiments with *Puccinia horiana* P. Henn. and three other microcyclic rusts. The two *Cladosporium* spp. were less effective than *V. lecanii* and *A. album* against the rust species tested.

VII. CONCLUSIONS

Hyperparasites are interesting disease control agents. There is strong evidence that hyperparasites may cause protection under natural conditions. This control is caused by a delicate balance between the hyperparasite and its host. This balance can be destroyed by the use of agrochemicals, both to the disfavor and favor of the hyperparasite. However, little is known about the effect of naturally occurring biological control in modern horticulture and agriculture.

There are few environmental hazards associated with the use of hyperparasites in plant protection. The hyperparasitic principle may be considered simple, but it is difficult to apply because three different organisms (the host plant, the plant pathogen, and the hyperparasite) are involved. Each are affected by agronomic factors, environment parameters, and pest control programs. These complex ecosystems require further studies. In protected cultivation the grower may be able to change the environment to favor hyperparasites. However, in field crops, the microclimate may be manipulated by cultural practices.

Wide differences in virulence between naturally occurring strains have been noted for several hyperparasites. Virulence probably has a complex genetic background. Responses to temperature and humidity, rapid spore germination, and the regulation of chitinases, proteases, and other enzymes are probably important characteristics of a successful hyperparasite. Selection of efficient strains should be combined with efforts to improve efficiency by induced mutations and genetic engineering.

REFERENCES

1. **Adams, P. B. and Ayers, W. A.,** Ecology of *Sclerotinia* species, *Phytopathology*, 69, 896, 1979.
2. **Adams, P. B. and Ayers, W. A.,** Factors affecting parasitic activity of *Sporidesmium sclerotivorum* on sclerotia of *Sclerotinia minor* in soil, *Phytopathology*, 70, 366, 1980.
3. **Adams, P. B. and Ayers, W. A.,** *Sporidesmium sclerotivorum*: distribution and function in natural biological control of sclerotial fungi, *Phytopathology*, 71, 90, 1981.
4. **Adams, P. B. and Ayers, W. A.,** Biological control of *Sclerotinia* lettuce drop in the field by *Sporidesmium sclerotivorum*, *Phytopathology*, 72, 485, 1982.
5. **Adams, P. B. and Ayers, W. A.,** Histological and physiological aspects of infection of sclerotia of *Sclerotinia* species by two mycoparasites, *Phytopathology*, 73, 1072, 1983.
6. **Adams, P. B., Marios, J. J., and Ayers, W. A.,** Population dynamics of the mycoparasite, *Sporidesmium sclerotivorum*, and its host, *Sclerotinia minor*, in soil, *Soil Biol. Biochem.*, 16, 627, 1984.
7. **Ahmed, A. H. M., and Tribe, H. T.,** Biological control of white rot of onion (*Sclerotium cepivorum*) by *Coniothyrium minitans*, *Plant Pathol.*, 26, 75, 1977.
8. **Ahrens, W.,** Einfluss des Mykoparasiten *Pythium oligandrum* auf den Aufgang von Zuckerrueben, *Med. Fac. Landbouwv.*, Rijksuniv. Gent, 47, 811, 1982.
9. **Al-Hamdani, A. M. and Cooke, R. C.,** Effects of the mycoparasite *Pythium oligandrum* on cellulolysis and sclerotium production by *Rhizoctonia solani*, *Trans. Br. Mycol. Soc.*, 81, 619, 1983.
10. **Allen, D. J.,** *Verticillium lecanii* on the bean rust fungus, *Uromyces appendiculatus*, *Trans. Br. Mycol. Soc.*, 79, 362, 1982.
11. **Ayers, W. A. and Adams, P. B.,** Mycoparasitism of sclerotia of *Sclerotinia* and *Sclerotium* species by *Sporidesmium sclerotivorum*, *Can. J. Microbiol.*, 25, 17, 1979.
12. **Ayers, W. A. and Lumsden, R. D.,** Mycoparasitism of oospores of *Pythium* and *Aphanomyces* species by *Hyphochytrium catenoides*, *Can. J. Microbiol.*, 23, 38, 1976.
13. **Aytoun, R. S. C.,** The genus *Trichoderma*: its relationship with *Armillaria mellea* (Vahl ex Fries) Quel. and *Polyporus schweinitzii* Fr., together with preliminary observations on its ecology in woodland soils, *Trans. Bot. Soc. Edinburgh*, 36, No. 2, 99, 1953.
14. **Barak, R., Elad, Y., Mirelman, D., and Chet, I.,** Lectins: a possible basis for specific recognition in the interaction of *Trichoderma* and *Sclerotium rolfsii*, *Phytopathology*, 75, 458, 1985.
15. **Barnett, H. L. and Binder, F. L.,** The fungal host-parasite relationship, *Annu. Rev. Phytopathol.*, 11, 273, 1973.

16. **Beuther, E., Phillipp, W.-D., and Grossmann, F.**, Untersuchungen zum Hyperparasitismus von *Ampelomyces quisqualis* auf Gurkenmehltau (*Sphaerotheca fuliginea*), *Phytopathol. Z.*, 101, 265, 1981.

17. **Biali, M., Dinoor, A., Eshed, H., and Kenneth, R.**, *Aphanocladium album*, a fungus inducing teliospore production in rusts, *Ann. Appl. Biol.*, 72, 37, 1972.

18. **Blakeman, J. P. and Fokkema, N. J.**, Potential for biological control of plant diseases on the phylloplane, *Annu. Rev. Phytopathol.*, 20, 167, 1982.

19. **Boosalis, M. G.**, Effect of soil temperature and green-mature amendment of unsterilized soil on parasitism of *Rhizoctonia solani* by *Penicilliium vermiculatum* and *Trichoderma* sp., *Phytopathology*, 46, 473, 1956.

20. **Burdsall, H. H., Hoch, H. C., Boosalis, M. G., and Setliff, E. C.**, *Laetisaria arvais* (Apyllophorales, Corticiaceae): A possible biological control agent for *Rhizoctonia solani* and *Pythium* species, *Mycologia*, 72, 728, 1980.

21. **Campbell, W. A.**, A new species of *Coniothyrium* parasitic on sclerotia, *Mycologia*, 39, 190, 1947.

22. **Carling, D. E., Brown, M. F., and Millikan, D. E.**, Ultrastructural examination of the *Puccinia graminis* — *Darluca filum* host-parasite relationship, *Phytopathology*, 66, 419, 1976.

23. **Chaika, M. N.**, Development of *Darluca filum* hyperparasite on different rust (*Puccinia*) species under conditions of artificial infection, *Vestnik Moskovskogo Universiteta. Seriia XVI: Biologiia Moskovskii Universitet*, 19, 1978.

24. **Coley-Smith, J. R.**, Studies on the biology of *Sclerotium cepivorum* Berk. III. Host range; persistence and viability of sclerotia, *Ann. Appl. Biol.*, 47, 511, 1959.

25. **Coley-Smith, J. R. and Cooke, R. C.**, Survival and germination of fungal sclerotia, *Annu. Rev. Phytopathol.*, 9, 65, 1971.

26. **Cook, R. J.**, Biological control of plant pathogens: Theory to application, *Phytopathology*, 75, 25, 1985.

27. **Cook, R. J. and Baker, K. F.**, *The Nature and Practice of Biological Control of Plant Pathogens*, American Phytopathological Society, St. Paul, Minn., 1983, 539.

28. **Corke, A. T. K. and Rishbeth, J.**, Use of microorganisms to control plant diseases, in *Microbial Control of Insects, Mites and Plant Diseases 1970—1980*, Burges, H. D., Ed., Academic Press, London, 1981, 717.

29. **Cunningham, J. E. and Pickard, M. A.**, Maltol, a metabolite of *Scytalidium uredinicola* which inhibits spore germination of *Endocronartium harknessii*, the western gall rust, *Rev. Plant Pathol.*, 1986.

30. **Deacon, J. W.**, Studies on *Pythium oligandrum*, an aggressive parasite of other fungi, *Trans. Br. Mycol. Soc.*, 66, 383, 1976.

31. **Dennis, C. and Webster, J.**, Antagonistic properties of species-groups of *Trichoderma*. III. Hyphal interaction, *Trans. Br. Mycol. Soc.*, 57, 363, 1971.

32. **Dubos, B., Roudet, J., Bulit, J., and Bugaret, Y.**, L'utilisation du *Trichoderma harzianum* Rifai dans la pratique vitiole pour luttre contre la pourriture grise (*Botrytis cinerea* Pers.), *Les Colloques de l'INRA*, 18, 289, 1983.

33. **Durell, L. M.**, Hyphal invasion by *Trichoderma viride*, *Mycopath. Mycol. Appl.*, 35, 138, 1966.

34. **Elad, Y., Barak, R., Chet, I., and Henis, Y.**, Ultrastructural studies of the interaction between *Trichoderma* spp. and plant pathogenic fungi, *Phytopathol. Z.*, 107, 168, 1983.

35. **Elad, Y., Barak, R., and Chet, I.**, Possible role of lectins in mycoparasitism, *J. Bacteriol.*, 154, 1431, 1983.

36. **Elad, Y., Chet, I., Boyle, P., and Henis, Y.**, Parasitism of *Trichoderma* spp. on *Rhizoctonia solani* and *Sclerotium rolfsii* — scanning electron microscopy and fluorescence microscopy, *Phytopathology*, 73, 85, 1983.

37. **Filonow, A. B. and Lockwood, J. L.**, Evaluation of several actinomycetes and the fungus, *Hyphochytrium catenoides*, as biocontrol agents for *Phytophthora* root rot of soybean, *Plant Dis.*, 69, 1033, 1985.

38. **Forrer, H. R.**, Der Einfluss von Stoffwechselprodukten des Mycoparasiten *Aphanocladium album* auf die Teleutosporenbildung von Rostpilzen, *Phytopathol. Z.*, 88, 306, 1977.

39. **Gonzalez, Avila M. and Castellanos, J. J.**, Presencia del microparasito *Darluca filum* sobre uredosporus *Uromyces paseoli* var. *typica*, *Cien. Agric.*, 3, 119, 1978.

40. **Grabski, G. C. and Mendgen, K.**, Einsatz von *V. lecanii* als biologisches Schädlingsbekampfungsmittel gegen den Bohnenrostpilz *U. appendiculatus* var. *appendiculatus* im Feld und im Gewächschaus, *Phytopathol. Z.*, 113, 243, 1985.

41. **Griffin, D. M.**, *Ecology of Soil Fungi*, Chapman and Hall, London, 1972, 193.

42. **Grosclaude, C., Richard, J., and Dubos, B.**, Inoculation of *Trichoderma viride* spores via pruning shears for biological control of *Stereum purpureum* on plum tree wounds, *Plant Dis. Rep.*, 57, 25, 1973.

43. **Gulino, M. L., Mezzalama, M., and Garibaldi, A.**, Biological and integrated control of *Botrytis cinerea* in Italy: experimental results and problems, *Quaderni della Scuola de Specializzazione in Virticultura ed Enologia*, Universita di Torino, 9, 299, 1985.

44. **Hadar, Y., Chet, I., and Henis, Y.**, Biological control of *Rhizoctonia solani* damping off with wheat bran culture of *Trichoderma harzianum*, *Phytopathology*, 69, 64, 1979.

45. **Hartmann, H., Riggs, W. A., and Hall, J. W.,** Screening for biological control agents of powdery mildew (*Sphaerotheca fuliginea*) on cucumbers, *Phytopathology*, 74, 864, 1984.

46. **Hashioka, Y. and Nakai, Y.,** Ultrastructure of pycnidial development and mycoparasitism of *Ampelomyces quisqualis* parasitic on Erysiphales, *Trans. Mycol. Soc. Japan*, 21, 329, 1980.

47. **Hau, B. and Kranz, J.,** Modellrecknungen zur Wirkung des Hyperparasiten *Eudarluca caricis* auf Rostepidemien, *Z. Pflanzenkr. Pflanzenschutz*, 85, 131, 1978.

48. **Hijwegen, T. and Buchenauer, H.,** Isolation and identification of hyperparasitic fungi associated with Erysiphaceae, *Neth. J. Plant Pathol.*, 90, 79, 1984.

49. **Hiratsuka, Y., Tsuneda, A., and Sigler, L.,** Occurrence of *Scytalidium uredinicola* on *Endocronartium harknessii* in Alberta, Canada, *Plant Dis. Rep.*, 63, 512, 1979.

50. **Hoch, H. C. and Abawi, G. S.,** Biological control of *Pythium* root rot of table beet with *Corticium* sp., *Phytopathology*, 69, 417, 1979.

51. **Hoch, H. C. and Provvidenti, R.,** Mycoparasitic relationships: cytology of the *Sphaerotheca fuliginea-Tilletiopsis* sp. interaction, *Phytopathology*, 69, 359, 1979.

52. **Howell, C. R.,** Effect of *Gliocladium virens* on *Pythium ultimum*, *Rhizoctonia solani*, and Damping-off of cotton seedlings, *Phytopathology*, 72, 496, 1982.

53. **Hsu, S. C. and Lockwood, J. L.,** Biological control of *Phytophthora* root rot of soybean by *Hyphochytrium catenoides* in greenhouse tests, *Phytopathol. Z.*, 109, 139, 1984.

54. **Huang, H. C.,** *Gliocladium catenulatum*: hyperparasite of *Sclerotinia sclerotiorum* and *Fusarium* species, *Can. J. Bot.*, 56, 2243, 1978.

55. **Huang, H. C.,** Control of *Sclerotinia* wilt of sunflower by hyperparasites, *Can. J. Plant Pathol.*, 2, 26, 1980.

56. **Huang, H. C.,** Distribution of *Coniothyrium minitans* in Manitoba sunflower fields, *Can. J. Plant Pathol.*, 3, 219, 1981.

57. **Huang, H. C. and Hoes, J. A.,** Penetration and infection of *Sclerotinia sclerotiorum* by *Coniothyrium minitans*, *Can. J. Bot.*, 54, 406, 1976.

58. **Hunter, W. E., Duniway, J. M., and Butler, E. E.,** Influence of nutrition, temperature, moisture, and gas composition on parasitism of *Rhizopus oryzae* by *Syncephalis californica*, *Phytopathology*, 67, 664, 1977.

59. **Iwasaki, T., Hayashi, K., and Funatsu, M.,** Purification and characterization of two types of cellulase from *Trichoderma koningii*, *J. Biochem. (Tokyo)*, 55, 209, 1964.

60. **Jarvis, W. R.,** Progress in the biological control of plant diseases, Proc. 10th Int. Congr. Plant Protection, Brighton, U.K., 1983, 1095.

61. **Jarvis, W. R. and Slingsby, K.,** The control of powdery mildew of greenhouse cucumber by water sprays and *Ampelomyces quisqualis*, *Plant Dis. Rep.*, 61, 728, 1977.

62. **Jones, D. and Watson, D.,** Parasitism and lysis by soil fungi of *Sclerotinia sclerotiorum* (Lib.) de Bary, a phytopathogenic fungus, *Nature (London)*, 224, 287, 1969.

63. **Kala, S. P. and Gaur, R. D.,** New host records for *Eudarluca caricis* from India, *Indian Phytopathol.*, 36, 408, 1983.

64. **Koc, N. K. and Défago, G.,** Studies on the host range of the hyperparasite *Aphanocladium album*, *Phytopathol. Z.*, 107, 214, 1983.

65. **Koc, N. K., Forrer, H. R., and Défago, G.,** Hyperparasitism of *Aphanocladium album* on aecidiospores and teliospores of *Puccinia graminis* f. sp. *tritici*, *Phytopathol. Z.*, 107, 219, 1983.

66. **Kotthoff, P.,** *Verticillium coccorum* (Petch) Westerdijk als Parasit auf *Puccinia chrysanthemi* Roze, *Angew. Bot.*, 19, 127, 1937.

67. **Kranz, J.,** Zur naturliche Verbreitung des Rostparasiten *Eudarluca caricis* (Fr.) O. Eriks., *Phytopathol. Z.*, 65, 43, 1969.

68. **Kranz, J.,** A host list of the rust parasite *Eudarluca caricis* (Fr.) O. Eriks., *Nova Hedwigia*, 24, 169, 1973.

69. **Kranz, J.,** Hyperparasitism of biotrophic fungi, in *Microbial Ecology of the Phylloplane*, Blakeman, J. P., Ed., Academic Press, London, 1981, 327.

70. **Kranz, J. and Brandenburger, W.,** An amended host list of the rust parasite *Eudarluca caricis*, *Z. Pflanzenkr. Pflanzenschutz*, 88, 682, 1981.

71. **Kuhlman, E. G.,** Hypovirulence and hyperparasitism, in *Plant Disease*, Vol. 5, Horsfall, J. G. and Cowling, E. B., Eds., Academic Press, New York, 1980, 363.

72. **Kuhlman, E. G.,** Parasite interaction with sporulation by *Cronartium querquum* f. sp. *fusiforme* on loblolly and slash pine, *Phytopathology*, 71, 348, 1981.

73. **Kuhlman, E. G.,** Mycoparasitic effects of *Scytalidium uredinicola* on aeciospore production and germination of *Cronartium quercuum* f. sp. *fusiforme*, *Phytopathology*, 71, 186, 1981.

74. **Kuhlman, E. G., Carmichael, J. W., and Miller, T.,** *Scytalidium uredinicola*, a new mycoparasite of *Cronartium fusiforme* on *Pinus*, *Mycologia*, 68, 1188, 1976.

75. **Kuhlman, E. G. and Matthews, F. R.,** Occurrence of *Darluca filum* on *Cronartium strobilinum* and *Cronartium fusiforme* infecting oak, *Phytopathology*, 66, 1195, 1976.

76. **Kuhlman, E. G., Matthews, F. R., and Tillerson, H. P.,** Efficacy of *Darluca filum* for biological control of *Cronartium fusiforme* and *C. strobolinum*, *Phytopathology*, 68, 507, 1978.

77. **Kuhlman, E. G. and Miller, T.,** Occurrence of *Tuberculina maxima* on fusiform rust galls in the southeastern United States, *Plant Dis. Rep.*, 60, 627, 1976.

78. **Larsen, H. J., Boosalis, M. G., and Kerr, E. D.,** Temporary depression of *Rhizoctonia solani* field populations by soil amendments with *Laetisaria arvalis*, *Plant Dis.*, 69, 347, 1985.

79. **Lewis, J. A. and Papavizas, G. C.,** Integrated control of *Rhizoctonia* fruit rot of cucumber, *Phytopathology*, 70, 85, 1980.

80. **Lim, T. K. and Nik, W. Z.,** Mycoparasitism of the coffee rust pathogen, *Hemileia vastatrix*, by *Verticillium psalliotae* in Malaysia, *Pertanika*, 6, 23, 1983.

81. **Locci, R., Ferrante, G. M., and Rodrigues, C. J.,** Studies by transmission and scanning electron microscopy on the *Hemileia vastatrix* — *Verticillium hemileiae* association, *Riv. Patol. Veg.*, 7, 127, 1971.

82. **Lumsden, R. D.,** Ecology of mycoparasitism, in *The Fungal Community. Its Organization and Role in the Ecosystem*, Wicklow, D. T. and Carroll, G. C., Eds., Marcel Dekker, New York, 1981, 295.

83. **Martin, S. B., Abawi, G. S., and Hoch, H. C.,** Influence of the antagonist *Laetisaria arvalis* on infection of table beets by *Phoma betae*, *Phytopathology*, 74, 1092, 1984.

84. **Martin, S. B., Hoch, H. C., and Abawi, G. S.,** Population dynamics of *Laetisaria arvalis* and low-temperature *Pythium* spp. in untreated and pasturized beet field soils, *Phytopathology*, 73, 1445, 1983.

85. **Mathur, M. and Mukerji, K. G.,** Antagonistic behaviour of *Cladosporium spongiosum* against *Phyllactinia dalbergiae* on *Dalbergia sissoo*, *Angew. Bot.*, 55, 75, 1981.

86. **Mendgen, K.,** Growth of *Verticillium lecanii* in pustules of stripe rust (*Puccinia striiformis*), *Phytopathol. Z.*, 102, 301, 1981.

87. **Merriman, P. R.,** Survival of sclerotia of *Sclerotinia sclerotiorum* in soil, *Soil Biol. Biochem.*, 7, 385, 1976.

88. **Moody, A. R. and Gindrat, D.,** Biological control of cucumber black root rot by *Gliocladium roseum*, *Phytopathology*, 67, 1159, 1977.

89. **Morelet, M. and Pinon, J.,** *Darluca filum* hyperparasite du genre *Melampsora* sur peuplier et saule, *Rev. For. Fr.*, 25, 378, 1973.

90. **Nomura, K., Yasui, T., Kiyooka, S., and Kobayashi, T.,** Xylanases of *Trichoderma viride*. Some properties of enzyme reactions and a preliminary experiment of xylan hydrolysis, *J. Ferment. Tech.*, 46, 634, 1968.

91. **Odintsova, O.,** Role of a hyperparasite *Cicinnobolus cesatii* D. By. in suppressing powdery mildew on apple trees, *Mikol. Fitopatol.*, 9, 337, 1975.

92. **Papavizas, G. C.,** *Trichoderma* and *Gliocladium*: biology, ecology, and potential for biocontrol, *Annu. Rev. Phytopathol.*, 23, 23, 1985.

93. **Papavizas, G. C. and Lewis, J. A.,** Physiological and biocontrol characteristics of stable mutants of *Trichoderma viride* resistant to MBC fungicides, *Phytopathology*, 73, 407, 1983.

94. **Papavizas, G. C. and Lumsden, R. D.,** Biological control of soilborne fungal propagules, *Annu. Rev. Phytopathol.*, 18, 389, 1980.

95. **Philipp, W-D., Beuther, E., and Grossman, F.,** Untersuchungen über den Einfluss von Fungiziden auf *Ampelomyces quisqualis* im Hinblick auf eine integrierte Bakampfung von Gurkenmehltau unter Glas, *Z. Pflanzenkr. Pflanzenschutz*, 89, 575, 1982.

96. **Philipp, W-D. and Crüger, G.,** Parasitismus von *Ampelomyces quisqualis* auf Echten Mehltaupilzen an Gurken und anderen Gemüsearten, *Z. Pflanzenkr. Pflanzenschutz*, 86, 129, 1979.

97. **Philipp, W-D., Grauer, U., and Grossmann, F.,** Ergänzende Untersuchungen zur biologischen und integrierten Bekämpfung von Gurkenmehltau unter Glas durch *Ampelomyces quisqualis*, *Z. Pflanzenkr. Pflanzenschutz*, 91, 438, 1984.

98. **Philipp, W-D. and Kirchhoff, J.,** Wechselwirkungen zwischen Triadimefon und dem Mehltauparasiten *Ampelomyces quisqualis* in vitro, *Z. Pflanzenkr. Pflanzenschutz*, 90, 68, 1983.

99. **Pistole, T. G.,** Interaction of bacteria and fungi with lectins and lectin-like substances, *Annu. Rev. Microbiol.*, 35, 85, 1981.

100. **Powell, J. M.,** Incidence and effect of *Tuberculina maxima* on cankers of the pine stem rust, *Cronartium comandrae*, *Phytoprotection*, 52, 104, 1971.

101. **Puzanova, L. A.,** Hyperparasites of *Ampelomyces* Ces. ex Schlecht. and their possible application to biological control of powdery mildew, *Mikol. Fitopatol.*, 18, 333, 1984.

102. **Ricard, J. L., Grosclaude, C., and Ale-Agha, N.,** Antagonism between *Eutypa armeniacea* and *Gliocladium roseum*, *Plant Dis. Rep.*, 58, 983, 1974.

103. **Roth, G.,** *Trichoderma viride* as a hyperparasite of stem rot disease on sugarcane, *Phytopathol. Z.*, 65, 176, 1969.

104. **Schroeder, H. von and Hassebrauk, K.,** Beiträge zur Biologie von *Darluca filum* (Biv.) Cast. und einigen anderen auf Uredineen beobachteten Pilzen, *Zentralbl. Bakteriol. Parasitkd.,* Abteilung 2, 110, 676, 1957.
105. **Sharma, I. K. and Heather, W. A.,** Hyperparasitism of *Melampsora larici-populina* by *Cladosporium herbarum* and *Cladosporium* tenuissimum, *Indian Phytopathol.,* 34, 395, 1981.
106. **Sharma, N. D., Vyas, S. C., and Jain, A. C.,** *Tuberculina costaricana* Syd.: a new hyperparasite on groundnut rust (*Puccinia arachidis* Speg.), *Curr. Sci.,* 46, 311, 1977.
107. **Sneh, B., Humble, S. J., and Lockwood, L. J.,** Parasitism of oospores of *Phytophthora megasperma* var. *sojae, P. cactorum, Pythium* sp., and *Aphanomyces euteiches* in soil by oomycetes, chytridiomycetes, hyphomycetes, actinomycetes, and bacteria, *Phytopathology,* 67, 622, 1977.
108. **Snyder, W. C., Wallis, G. W., and Smith, S. N.,** Biological control of plant pathogens, in *Theory and Practice of Biological Control,* Huffaker, C. B. and Messenger, P. S., Eds., Academic Press, New York, 1976, 521.
109. **Spencer, D. M.,** Parasitism of carnation rust (*Uromyces dianthi*) by *Verticillium lecanii, Trans. Br. Mycol. Soc.,* 74, 191, 1980.
110. **Spencer, D. M. and Atkey, P. T.,** Parasitic effects of *Verticillium lecanii* on two rust fungi, *Trans. Br. Mycol. Soc.,* 77, 535, 1981.
111. **Spencer, D. M. and Ebben, M. H.,** Biological control of cucumber powdery mildew, *Annual Report Glasshouse Crops Research Institute 1981,* Littlehampton, West Sussex, U.K., 1983, 128.
112. **Srivastava, A. K., Defago, G., and Kern, H.,** Hyperparasitism of *Puccinia horiana* and other microcyclic rusts, *Phytopathol. Z.,* 114, 73, 1985.
113. **Steyaert, R. L.,** *Cladosporium hemileiae* n. sp. Un parasite de l'*Hemileia vastatrix* Berk. et Br., *Bull. Soc. R. Bot. Belg.,* 63, 46, 1930.
114. **Stahle, U. and Kranz, J.,** Interactions between *Puccinia recondita* and *Eudarluca caricis* during germination, *Trans. Br. Mycol. Soc.,* 82, 562, 1984.
115. **Sundaram, N. V.,** Studies on parasites of the rusts, *Indian J. Agric. Sci.,* 32, 266, 1962.
116. **Sundheim, L.,** Control of cucumber powdery mildew by the hyperparasite *Ampelomyces quisqualis* and fungicides, *Plant Pathol.,* 31, 209, 1982.
117. **Sundheim, L.,** The hyperparasite *Ampelomyces quisqualis* in biological control of cucumber powdery mildew, *Proc. 10th Int. Congr. Plant Protection (Abstr.),* Brighton, U.K., 1983, 1110.
118. **Sundheim, L.,** L'hyperparasite *Ampelomyces quisqualis* dans la lutte contre l'oidium du concombre, *Les Colloques de l'INRA,* 18, 145, 1984.
119. **Sundheim, L.,** Use of hyperparasites in biological control of biotrophic plant pathogens, in *Microbiology of the Phylloplane,* Fokkema, N. and van den Heuvel, J., Eds., Cambridge University Press, London, 1986, 333.
120. **Sundheim, L. and Amundsen, T.,** Fungicide tolerance in the hyperparasite *Ampelomyces quisqualis* and integrated control of cucumber powdery mildew, *Acta Agric. Scand.,* 32, 349, 1982.
121. **Sundheim, L. and Krekling, T.,** Host-parasite relationships of the hyperparasite *Ampelomyces quisqualis* and its powdery mildew host *Sphaerotheca fuliginea, Phytopathol. Z.,* 104, 202, 1982.
122. **Swendsrud, D. P. and Calpouzos, L.,** Effect of inoculation sequence and humidity on infection of *Puccinia recondita* by the mycoparasite *Darluca filum, Phytopathology,* 62, 931, 1972.
123. **Sztejnberg, A.,** Biological control of powdery mildews by *Ampelomyces quisqualis, Phytopathology,* 69(Abstr.), 1047, 1979.
124. **Toyama, N. and Ogawa, K.,** Purification and properties of *Trichoderma viride* mycolytic enzymes, *J. Ferment. Tech.,* 46, 626, 1968.
125. **Traquair, J. A., Meloche, R. B., Jarvis, W. R., and Baker, K. W.,** Hyperparasitism of *Puccinia violae* by *Cladosporium uredinicola, Can. J. Bot.,* 62, 181, 1984.
126. **Tribe, H. T.,** On the parasitism of *Sclerotinia trifoliorum* by *Coniothyrium minitans, Trans. Br. Mycol. Soc.,* 40, 489, 1957.
127. **Tronsmo, A.,** Muligheter for integrert bekjempelse av soppsykdommer, Informasjonsmøte i plantevern 1985, *Aktuelt fra Statens Fagtjeneste for Landbruket,* Ås, Norway Nr. 2, 1985, 107.
128. **Tronsmo, A.,** Use of *Trichoderma* spp. in biological control of necrotrophic pathogens, in *Microbiology of the Phyllophlane,* Fokkema, N. and van den Heuvel, J., Eds., Cambridge University Press, London, 1986, 348.
129. **Tronsmo, A. and Dennis, C.,** The use of *Trichoderma* species to control strawberry fruit rots, *Neth. J. Plant Pathol.,* 83, 449, 1977.
130. **Tronsmo, A. and Dennis, C.,** Effect of temperature on antagonistic properties of *Trichoderma* species, *Trans. Br. Mycol. Soc.,* 71, 469, 1978.
131. **Tronsmo, A. and Raa, J.,** Antagonistic action of *Trichoderma pseudokoningii* against the apple pathogen *Botrytis cinerea, Phytopathol. Z.,* 89, 216, 1977.
132. **Tronsmo, A. and Ystaas, J.,** Biological control of *Botrytis cinerea* on apple, *Plant Dis.,* 64, 1009, 1980.
133. **Tsuneda, A. and Hiratsuka, Y.,** Mode of parasitism of a mycoparasite, *Cladosporium gallicola* on western gall rust, *Endocronartium harknessii, Can. J. Plant Pathol.,* 1, 31, 1979.

134. **Tsuneda, A. and Hiratsuka, Y.,** Biological control of pine stem rusts by mycoparasites, Proc. Japan Academy, Series B, *Phys. Biol. Sci.,* 57, 337, 1981.

135. **Tsuneda, A., Hiratsuka, Y., and Maruyama, P. J.,** Hyperparasitism of *Scytalidium uredinicola* on western gall rust, *Endocronartium harknessii, Can. J. Bot.,* 58, 1154, 1980.

136. **Tu, J. C.,** *Gliocladium virens,* a destructive mycoparasite of *Sclerotinia sclerotiorum, Phytopathology,* 70, 670, 1980.

137. **Tu, J. C. and Vaartaja, O.,** The effect of the hyperparasite (*Gliocladium virens*) on *Rhizoctonia* root rot of white beans, *Can. J. Bot.,* 59, 22, 1981.

138. **Turner, G. J. and Tribe, H. T.,** On *Coniothyrium minitans* and its parasitism of *Sclerotinia* species, *Trans. Br. Mycol. Soc.,* 66, 97, 1976.

139. **Uecker, F. A., Ayers, W. A., and Adams, P. B.,** A new hyphomycete on sclerotia of *Sclerotinia sclerotiorum, Mycotaxon,* 7, 275, 1978.

140. **Upadhyay, R. S. and Rai, B.,** Mycoparasitism with reference to biological control of plant diseases, in *Recent Advances in Plant Pathology,* Husain, A., Ed., Lucknow, Print House, Lucknow, India, 1983, 48.

141. **Veselý, D.,** Potential biological control of damping-off pathogens in emerging sugar beet by *Pythium oligandrum, Phytopathol. Z.,* 90, 113, 1977.

142. **Veselý, D.,** Parasitic relationships between *Pythium oligandrum* Drechsler and some other species of the Oomycetes class, *Zentralbl. Bakteriol. Parasitkd., Abteiling* 2, 133, 341, 1978.

143. **Veselý, D. and Hejdánek, S.,** Microbial relations of *Pythium oligandrum* and problems in the use of this organism for the biological control of damping-off in sugar beet, *Zentralbl. Mikrobiol.,* 139, 257, 1984.

144. **Walker, J. A. and Maude, R. B.,** Natural occurrence and growth of *Gliocladium roseum* on the mycelium and sclerotia of *Botrytis allii, Trans. Br. Mycol. Soc.,* 65, 335, 1975.

145. **Wicker, E. F.,** Natural control of white pine blister rust by *Tuberculina maxima, Phytopathology,* 71, 997, 1981.

146. **Wicker, E. F. and Woo, J. Y.,** Histology of blister rust cankers parasitized by *Tuberculina maxima, Phytopathol. Z.,* 76, 356, 1973.

147. **Yarwood, C. E.,** *Ampelomyces quisqualis* on clover mildew, *Phytopathology,* 22, 31, 1932.

Chapter 5

TRICHODERMA AS A BIOCONTROL AGENT*

Homer D. Wells

TABLE OF CONTENTS

---

*   This chapter was prepared by a United States Government employee as a part of his official duties and legally cannot be copyrighted. Mention of a trademark or proprietary product does not constitute a guarantee or warranty of the product by the U.S. Department of Agriculture and does not imply approval to the exclusion of other products that may also be suitable.

I. INTRODUCTION

Bissett[8] characterized the genus *Trichoderma* as "rapidly growing colonies bearing tufted or pustulate, repeatedly branched conidiophores with lageniform phialids and hyaline or green conidia born in slimy heads." Teleomorphs, where known, belong to *Hypocrea* or a related genera. Rifai[53] monographed the genus in 1969 and settled on nine species aggregates: (1) *Trichoderma piluliferum*, Webster and Rifai, (2) *T. polysporum* (Link ex Pers.) Rifai, (3) *T. hamatum* (Bon.) Bain., (4) *T. koningii* Oud., (5) *T. aureoviride* Rifai, (6) *T. harzianum* Rifai, (7) *T. longibrachiatum* Rifai, (8) *T. pseudokoningii* Rifai, and (9) *T. viride* Pers. ex S. F. Gray. He said this was a highly artificial classification, and a more satisfactory classification would depend on defining criteria by which additional species could be separated and identified. In 1984 Bissett[8] made a partial revision of the genus *Trichoderma*, and he established *Longibrachiatum* as a section of the genus in which he included: *T. viride, T. koningii, T. pseudokoningii,* and *T. longibrachiatum* and added two new species *T. citrinoviride* Bissett and *T. atroviride* Bissett.

In this paper I will follow the species-aggregate concept of Rifai.[53] This is because the work being discussed was done, for the most part, without the benefit of Bissett's input. With the wide utilization of *Trichoderma* in research and industry, it is essential that the best possible classification of the genus be made as soon as feasible. Therefore, it is essential that Bissett complete his work on the genus at an early date.

Work on biocontrol prior to 1969 was carried out using Gilman and Abbot's classification (1927)[26] or Bisby's conclusion (1939)[7] that *Trichoderma* was a monotypic genus and all green spored *Trichoderma* isolates were indiscriminately identified as *T. viride*, or *T. lignorum* (Tode) Harz, a synonym of *T. viride*. Mycologists in the U.S.S.R. accepted the classification of *T. lignorum [viride]* and *T. koningii*[6] as representing the genus. In recent years there has been a general tendency for all workers to accept Rifai's species-aggregate concept.

In addition to the confusion in species concepts in *Trichoderma*, there are also considerable indications that biocontrol workers have been confused in separating *Gliocladium* and *Trichoderma*.[14,68] While this major factor cannot be ignored, the authors' assigned nomenclature will be accepted, except in cases where the preponderance of evidence indicates the assigned name is incorrect. For the most part this will consist of changing *T. lignorum* to *T. viride*.

Mycoparasitism has long been known to occur in a wide range of fungi, including some *Hypocrea*; however, it was not until Weindling's paper in 1932[66] that the potential value of *Trichoderma* as a biocontrol agent was recognized. Weindling observed *T. lignorum [viride]* as a parasite with hyphae coiling around and killing *Rhizoctonia solani* Kuhn. He also found that species of *Phytophthora, Pythium,* and *Rhizopus,* and *Sclerotium rolfsii* Sacc. were susceptible to mycoparasitism by *T. viride*. Based on these results Weindling suggested that abundant inoculations with *T. viride* might be used to control certain pathogens. Since 1932, Weindling and numerous other workers have been involved, with varying degrees of success, in elucidating principals and mechanisms of biocontrol of plant diseases and the potential application to production situations with *Trichoderma*.

In discussing the role of *Trichoderma* as a biocontrol agent I will use the broad concept that biocontrol of plant diseases involves all the interactions of plant pathogens and their antagonists, including indirect interventions into the environment (such as crop rotations, etc.), organic amendments, chemical pesticide intervention, application of an antagonist, and augmentation with a food base. Currently the activity on *Trichoderma* has reached an intensity sufficient for the Henry Doubleday Research Association to issue a *Trichoderma Newsletter*.

Trichoderma has been an exceptionally good model with which to study biocontrol because it is ubiquitous, easy to isolate and culture, grows rapidly on many substrates, affects a wide range of plant pathogens, is rarely pathogenic on higher plants, acts as a mycoparasite, competes well for food and site, produces antibiotics, and has an enzyme system capable of attacking a wide range of plant pathogens. While *Trichoderma* is recognized primarily as an antagonist against pathogenic fungi, it also has been reported to be antagonistic to certain bacterial pathogens. *Trichoderma* has also been widely studied for its antibiotic properties for medicine and its enzyme systems in food processing and biodegradation. As a general indicator of the attention attracted by *Trichoderma*, it has been discussed in more than 1700 scientific papers in the last 10 years, and in more than 2300 papers in the last 50 years. For convenience this chapter will be divided into sections dealing with *Trichoderma* as a biocontrol agent for plant pathogens by mycoparasitism, antibiosis, and competition, and a section on biocontrol strategies. These factors, however, are probably never mutually exclusive. Thus, in instances, where needed for clarity or emphasis, I may duplicate a limited amount of discussion. I will certainly not attempt to cover all the work on *Trichoderma* as a biocontrol agent, but will try to cite a few of the major milestones that have led to new advances in biocontrol. Citation of review papers and textbooks will be kept to a minimum and used only to reinforce a point under consideration. In a number of instances numerous other publications could have been cited to make the same point as appears in the manuscript, but it is my feeling that too many citations may only obscure the principals being presented.

II. MYCOPARASITISM

Ignoring the fact that there is likely to be antibiosis and other factors as a precondition for mycoparasitism, I will, nevertheless, discuss the mycoparasitic activity of *Trichoderma* as a separate entity. As mentioned previously, Weindling[66] was the first to report *Trichoderma* as a mycoparasite of *R. solani, S. rolfsii,* the *Phytophthora* sp., the *Pythium* spp., and the *Rhizopus* spp. Other workers, since that time, have observed its parasitism on numerous other phycomycetes, fungi imperfecti, ascomycetes, and basidiomycetes. In some instances *Trichoderma* has been a serious pest on edible mushrooms warranting intervention with control procedures. Weindling[66] pictured *Trichoderma* mycelia growing adjacent to *Rhizoctonia* hyphae, coiling around, attaching to, and growing within the *Rhizoctonia* hyphae. The hyphae of the two organisms did not have to make contact for the host hyphae to lose the integrity of its protoplasm. Later the host hyphae would consist of only empty cells. Thus, *Trichoderma* is able to act as a parasite of certain other fungi without necessarily making physical contact. A partial listing of the genera of plant pathogenic fungi that *Trichoderma* has been shown to parasitize includes: *Rhizoctonia, Sclerotium, Sclerotinia, Helminthosporium, Fusarium, Armillaria, Colletotrichum, Verticillium, Venturia, Endothia, Pythium, Phytophthora, Rhizopus, Diaporthe,* and *Fusicladium.*

Dennis and Webster[16] observed major differences in the ability of the *Trichoderma* spp. to parasitize a number of different soil-borne pathogens. Some isolates did not react with any hyphae of *Fusarium oxysporum* Schlecht, ex. Fr., and 10 of 80 *Trichoderma* isolates did not show coiling around the hyphae of any test fungi. Ability to produce antibiotics was not related to parasitism. In some instances there appeared to be a negative correlation between the ability of *Trichoderma* isolates to produce antibiotics, and mycoparasitism. The production of antibiotics often caused a visual change in the growth of the leading edges of the test fungus and *Trichoderma*, indicating that the plant pathogen may also be utilizing some defense mechanism against the *Trichoderma*. Their studies indicated no differences in parasitism between pH 4.0 and 6.5. In experiments using plastic threads they were not able to show coiling by *Trichoderma*, indicating that the coiling was not a contact stimulus. Their studies also showed that with a number of isolates the mycelium of the plant pathogen began to lose its contents before coming in contact with the hyphae of *Trichoderma*, thus indicating that the *Trichoderma* could affect the plant pathogen before the two organisms came together.

Most studies on parasitism are essentially in agreement with Weindling's[66] initial report. Weindling did not discuss the hyphal clamps of *Trichoderma* attached to the host. These, however, were evident in some of his illustrations demonstrating mycoparasitism. In 1981 Chet et al.[12] found that the hyphae of *T. hamatum* grew directly toward hyphae of *R. solani,* indicating that this was not a random phenomenon.

The scanning electron microscopy and fluorescence microscopy work of Elad et al.[19] in 1983 significantly enhanced our understanding of the nature of parasitism. *Trichoderma* entered *R. solani* or *S. rolfsii* by dissolving holes in the host hyphae. Removal of the coiled hyphae also showed partial lysis outlining the area of former contact. Their studies demonstrated that adding cycloheximide to a solid growth medium prevented *T. harzianum* from invading colonies of *R. solani* and *S. rolfsii*. Enzymatic activity was also reduced in the presence of cycloheximide. Intense fluorescence was observed in the coiled hyphae zones. These fluorescence sites appeared to be present in areas where enzymatic activity of *Trichoderma* was greatest. Their work indicated β-1,3-glucanase was excreted by *Trichoderma* at the contact sites.

In addition to parasitism of fungal hyphae, *Trichoderma* will also attack rhizomorphs, sclerotia, and fruiting structures of numerous fungi. A number of workers have discussed mycoparasitism of rhizomorphs and sclerotia of several soil fungi by *Trichoderma*. Rhizomorphs of *Armillaria mellea* (Vahl) Quel. have a pseudosclerotial covering through which *Trichoderma* must penetrate before the mycelium can be attacked. Bliss[9] indicated this wall had to be weakened before *Trichoderma* could penetrate in natural, untreated soils. However, in sterilized soils in which both *Trichoderma* and *A. mellea* rhizomorphs were added, *Trichoderma* could penetrate the pseudosclerotial covering in the absence of natural soil fungistasis. He indicated that the lack of fungistasis may have allowed the *Trichoderma* population to increase and become more aggressive.

A considerable number of references have indicated that *Trichoderma* attacks and destroys sclerotia, but there has been a lack of detail on the histology of parasitism. Hennis et al.[32] studied the parasitism of 28 isolates of *Trichoderma* on *S. rolfsii*. They found that isolates varied in their ability to penetrate the sclerotia of *S. rolfsii*. In stained sections *Trichoderma* was observed in the cortex and medulla. Degradation was not observed in the rind or cortex in initial stages of penetration. However, at later stages sclerotia often disintegrated under slight pressure. The medulla cells appeared lysed and *Trichoderma* hyphae formed chlamydospores within it. Conidia were not observed within the disintegrated sclerotia. In degraded sclerotia the cortex tissue was replaced by *Trichoderma* which grew out through the rind.

III. ANTIBIOSIS

In this paper antibiosis includes all antagonistic chemical products produced and released into the environment by living *Trichoderma*. This includes antibiotic compounds and extracellular enzyme systems that damage plant pathogens, as well as compounds that may enhance *Trichoderma* in maintaining a favorable balance as a portion of the biota.

Antibiotic antibiosis was first reported by Weindling[67] and Weindling and Emerson.[69] The isolates that produced gliotoxin, however, were later identified as a *Gliocladium*.[14,65,68] Also, the isolates that produced gliotoxin and viridin for Brian and co-workers,[10,11] were later shown to be *Gliocladium*.[65] No existing cultures of *Trichoderma* have been shown to produce these two compounds. However, Webster and Lomas[65] pointed out that many isolates recovered from the soil had been shown to produce gliotoxin and viridin, and it could not be ruled out that many of these were not *Trichoderma*. Since the more recent work does not report gliotoxin and viridin as a product of *Trichoderma*, I will minimize these two antibiotics as a major avenue of antibiosis by *Trichoderma*. Trichodermin, a sequisterpene, was de-

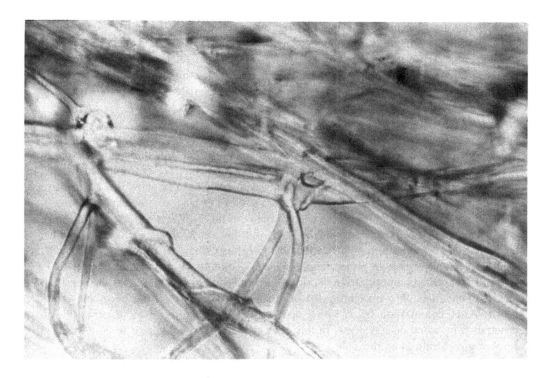

FIGURE 1. *Trichoderma harzianum* hyphae parasitizing hyphae of *Sclerotium rolfsii* showing coiling around and clamps on *S. rolfsii* hyphae and a strand of *T. harzianum* hyphae apparently within the *S. rolfsii* hypha.

scribed as an antibiotic produced by *Trichoderma* by Godtfredsen and Vangedal.[28] Dermadine, a product of *Trichoderma*,[50] is an unsaturated monobasic acid active against a wide range of fungi and both Gram positive and Gram negative bacteria. Suzukacillin® and Alamethicine® are peptides produced by *Trichoderma* with both antifungal and antibacterial properties.[39,44,51] Dennis and Webster[15] reported that acetaldehyde was the major volatile antibiotic produced by *Trichoderma*.

The production of antibiotics by *Trichoderma* is not highly correlated with mycoparasitism, therefore other systems of antibiosis are involved. Enzymes are important factors facilitating the ability of *Trichoderma* to compete for substrate as well as to directly attack a wide array of plant pathogens. Lytic enzymes are most often implicated as major features in nonantibiotic antibiosis of plant pathogens. Rodriguez-Kabana et al.[54,55,57] have demonstrated that proteolytic enzymes play a significant role in the destruction of the enzymatic activity of *S. rolfsii*. They demonstrated that the proteolytic enzyme activity was greatest at or near a neutral pH. Much of the recent work has centered on enzymes produced by *Trichoderma* that attack the cell walls of various fungi. These are primarily β-1,3-glucanase, chitinase, and cellulase. By the use of fluorescent microscopy Elad et al.[19] were able to demonstrate increased enzymatic activity at the point where *T. harzianum* attached to *S. rolfsii* or *R. solani*, with clamps or by coiling (Figure 1). Harman et al.,[31] in work with *T. hamatum* as a biocontrol for the *Pythium* spp. and *R. solani*, were able to increase the efficacy of *T. hamatum* by the addition of chitin, presumably by enhancing the chitinase activity of *Trichoderma*. Work by Sivan et al.[61] indicates that not all of the biocontrol activity of some of the best *Trichoderma* isolates can be accounted for by presently known enzymatic activity. Thus, it is very likely that sometimes both antibiotic and enzymatic properties can be combined for broad spectrum biocontrol. The extreme variability in ability of different isolates of *Trichoderma* to attack different pathogens and the high degree of interaction of

Trichoderma and plant pathogens[5] makes it highly unlikely that the relatively few compounds already defined are the only compounds active in antibiosis. There is a considerable need for more research on the role of antibiosis in *Trichoderma* as related to its ability to serve as a biocontrol agent. This may be especially important in determining the type of food base that should be supplied with *Trichoderma* for the control of specific plant pathogens.

IV. COMPETITION

The omnipresence of *Trichoderma* in agricultural and natural soils throughout the world is prima facie evidence that it must be an excellent competitor for space and nutritional resources. This section, however, will deal only with examples that show this competitive ability directly affecting suppression of pathogen activity or plant disease expression.

Wood rotting fungi have been controlled in many instances by the presence of or use of the *Trichoderma* spp. Hulme and Shields[35] showed that *Trichoderma viride* could protect birch (*Betula papyrifera*) from rot caused by *Polyporus hirsutus, P. versicolor,* and *P. adustus.* They were able to protect logs for up to 7 months by inoculating them with *T. viride.* This resulted primarily from the ability of *Trichoderma* to utilize all of the nonstructured carbohydrates thus preventing the destructive basidiomycete from becoming established. Also, colonization by *T. viride* appeared to leave the wood more permeable to penetration by wood preservatives. In this case, antibiosis did not appear to be a factor in *Trichoderma* acting as a protectant.

Toole[63] showed that *T. viride* protected southern pine wood from rot by *Poria monticola* Muss. without adversely affecting the wood quality. There was no indication that factors other than competition were involved. Pottel and co-workers[48,49] used *T. harzianum* to effectively control rots caused by hymenomycetes (primarily *Coriolus versicolor* (L. ex Fries) Quel on *Acer rubrum* L.) for up to 2 years. Since the *Trichoderma* remained superficial, they were not able to determine the mechanism of protection. Smith et al.[62] suggested that by using *T. harzianum* to treat red maple (*A. rubrum*) wounds the *Trichoderma* preempted *Phialophora melinii* (Nannf.) Covart, a pioneer colonizer of red maple wounds, and prevented colonizing by the rotting fungus *Fomes connatus* (Weinn.) Gill. A part of this control appeared to be related to the removal of phenols by *P. melinii,* whereas the *Trichoderma* isolates allowed the normal accumulation of plant phenols which are toxic to *F. connatus.* Thus, the competition was not directly against the pathogen, but against the primary colonizer which ordinarily detoxified the site for invasion by the rotting organism.

The use of carbon disulfide as a means of partial sterilization of soil has been practiced in Europe for the control of *Armillaria mellea* for nearly a century.[27] This was assumed to be primarily a chemical effect until Bliss[9] showed that sublethal doses of the chemical gave *Trichoderma* a competitive advantage in the environment as an antagonist to the *A. mellea.* It was hypothesized that the additional competitive ability of the *Trichoderma* in fumigated soil allowed it to overcome the protection mechanisms of the pseudosclerotial phase of *A. mellea,* thus effecting disease control. Garrett[24,25] substantiated most of the work reported by Bliss and believed competition by *Trichoderma* was the primary factor resulting in increased control of *A. mellea.* Saksena[58] attributed most of the effect of partial sterilization with carbon disulfide to a better tolerance to the fumigant by *Trichoderma,* allied with its intrinsically high growth rate. Moubasher[40] showed that *T. viride* had only moderate tolerance to carbon disulfide but this was combined with its high recolonizing ability after the optimum dose of carbon disulfide. Ohr et al.,[43] in studying soils modified by treatments with sublethal doses of methyl bromide, found a negative linear regression of *Trichoderma* populations and *A. mellea* in nonsterile soil. Thus, the more *Trichoderma* available the lower the survival of *A. mellea* root rot fungus. Munnecke et al.,[41] using both carbon disulfide and methyl bromide as soil fumigants, postulated that the *A. mellea* was weakened by the fumigant, and it is during this weakened period that *T. viride* exerts its competitive and antagonistic ability. Irrespective of all the proposed modes of action it appears that the normal quotient of *Trichoderma* in the soil microflora will not control *A. mellea*, but that an intervention must be made through partial sterilization to allow a temporary competitive advantage for the *Trichoderma*. The resulting population increase allows sufficient antibiosis and parasitism to control the pathogen. Most likely competition is relative only in that adequate *Trichoderma* must be present and actively releasing metabolites to overcome defense mechanisms of *A. mellea*.

The competitive ability of *Trichoderma*, and thus its biocontrol potential, has been shown to be affected by soil properties. Weindling[67] noted that pH affected the relative value of *Trichoderma* as an antagonist to *R. solani* by affecting the growth and development of each fungus. Since that time numerous other workers have studied similar phenomena. The availability of iron is important for the growth and competitive ability of *Trichoderma*. Hubbard et al.[34] found that in soils with low iron, bacteria belonging to the fluorescence-type pseudomonads tied-up available iron by production of siderophores and thus reduced the ability of *T. hamatum* to act as a biocontrol. This may be tied indirectly to some of the observations on pH in that higher pH may make iron less available. Warcup[64] showed that *Trichoderma* preferred moist soils over arid soils. This phenomena has also been reported many times.

Backman et al.[4] showed that when fungicides that were toxic to *Trichoderma* were applied for peanut leafspot control, there was an increase in disease caused by *S. rolfsii*, a pathogen susceptible to biocontrol by *Trichoderma*. Thus, there are a wide array of factors that affect *Trichoderma*'s competitive ability to be present and active as a biocontrol agent. Rodriguez-Kabana and Curl[56] presented an extensive review on the effects of nontarget pesticides on soil-borne fungi.

Papavizas, in a review,[45] indicates soil antagonism to *Trichoderma* is a major factor regulating the potential inoculum level of *Trichoderma* in the soil. Thus soil antagonism plays a direct role in reducing the efficacy of *Trichoderma* as a biocontrol agent. The entire soil biota and physical/chemical nature most likely play a major role in reducing the number of propagules of *Trichoderma* remaining in the soil following the introduction of *Trichoderma* as a biocontrol agent.

V. BIOCONTROL STRATEGIES

When Weindling[66] first discovered the biocontrol potential of *Trichoderma*, his primary inclination was to effect biocontrol by introducing *Trichoderma* inoculum into the soil. Some success was obtained in sterilized soil, however, practical use of this method was not realized.[2,70]

Antibiotic properties of *Trichoderma*[10,14-16,28,67] have been reported many times but this concept is not directly utilized as a biocontrol procedure (Figure 2). Numerous Russian workers have reported significant benefit from the use of Trichodermin-1, Trichodermin-2, Trichodermin-3, and Trichodermin-4 for the control of several diseases.[21,22,37,47] These Trichodermin materials are crude preparations containing most of the substrate within which they are grown. Thus, their effectiveness may be due to a number of antibiotic and enzymatic compounds. These Trichodermin materials should not be confused with the sequisterpene antibiotic, Trichodermin, described by Godtfredsen and Vangedal.[28] The dosage rates of these preparations of Trichodermin-1, -2, -3, and -4 needed for adequate disease control may be too high for practical application in agriculture.

The first practical control of plant diseases with *Trichoderma* were the results obtained by partial soil sterilization in the control of *Armillaria mellea* root rot in citrus. It was shown that the partial sterilization with selective fumigants weakened the growth of the plant

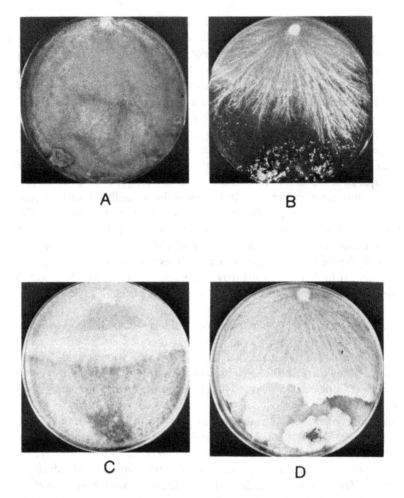

FIGURE 2.   Antagonism after 7 days of 4 different isolates of *Trichoderma* against a single isolate of *Sclerotium rolfsii* on 20% V-8 juice agar in 9 cm culture dishes. (A) *Trichoderma* has obliterated essentially all *S. rolfsii*. (B) *Trichoderma* has limited the surface and subsurface growth of *S. rolfsii* but *S. rolfsii* has produced aerial rhizomorphs that are growing across the *Trichoderma*. (C) *Trichoderma* and *S. rolfsii* mycelium have met and appear to be growing through each other without significant adverse effects to either fungus. (D) The *S. rolfsii* mycelium has become dense and is growing over the *Trichoderma*.

pathogen and allowed a bloom of *Trichoderma* which was capable of invading and killing *A. mellea*.[9,20,24] This is a practice still under study and utilization.

The use of *Trichoderma* as a preventative of infections following wounding or pruning of trees has had considerable success.[17,18,29,48,49,52,59,60,62] Commercial preparations of *Trichoderma* are available for treating above-ground portions of trees against a number of wood rotting diseases.

Smith et al.[62] found that the pioneering fungus, *Phialophora melinii*, normally invaded red maple wounds first. It utilized the gallic acid that is the major phenolic constituent of red maple involved in giving the wound resistance to *F. connatus*. The *T. harzianum* was tolerant to, but did not remove, the gallic acid. Therefore, by introducing *T. harzianum* as the primary colonizer the food base was utilized without removing the tree's gallic acid product, resulting in effective biocontrol.

In 1982 Wells et al.[71] and Baker and Cook[13] first demonstrated under field conditions that an effective *Trichoderma* isolate (*T. harzianum*), combined with a food base, had highly significant potential as a biocontrol agent of the soil-borne plant disease caused by *S. rolfsii*. Since that time much work has been done finding highly effective *Trichoderma* biotypes and developing food bases and delivery systems for enhanced biocontrol of numerous pathogens. In 1975 Backman and Rodriguez-Kabana[3] developed a growth and delivery system using a diatomaceous earth granule impregnated with 10% molasses. This substrate and media served as a food base for growing *Trichoderma* inoculum and as an additional food base to be applied with the *Trichoderma* to control *S. rolfsii* in peanuts. In 1979 Hadar et al.[30] demonstrated that wheat bran was an effective media for growth and carrier for *Trichoderma* as a biocontrol for soil-borne diseases. Harman et al., in 1981,[31] showed that the addition of chitin or fungus cell walls with inoculum as a seed coating increased the efficacy of *T. hamatum* as a seed treatment for peas and radish. Nelson et al., in 1983,[42] and Hoitnik and Kuter, in 1985,[33] demonstrated composted hardwood bark as an excellent vehicle for growth and delivery of suppressive *Trichoderma* spp. Fisher et al.[23] proposed fluid drilling as a delivery system for biocontrol with small seeded vegetables. In 1984 Jones et al.[36] obtained excellent results with lignite and stillage as substrates and carriers for *Trichoderma* as a biocontrol agent. In 1985 Lewis and Papavizas[38] showed that young *Trichoderma* mycelia delivered in a partially consumed food base was the most effective method of delivery of *Trichoderma* as a biocontrol agent. They theorized this was because mycelia is much more resistant to soil fungistasis than conidia. Therefore, they hypothesized, it may be more efficient to apply *Trichoderma* with a partially consumed food base than to augment *Trichoderma* with an additional food base at the time of delivery. Sivan et al., in 1984,[61] found wheat-bran plus peat was the superior mixture for survival of *Trichoderma* as a biocontrol inoculum source for soil and potting mixtures.

General-purpose pesticides have effects on *Trichoderma* and thus affect its biocontrol potential.[4,56] Abed-El Moity et al.[1] and Papavizas et al.[46] have developed some fungicidal tolerance by the use of mutagens and selection technique on *Trichoderma*. Until more is known about the need and use of pesticides on nontarget organisms and their effects on the biocontrol potential of *Trichoderma*, this will remain a major obstacle to the effective use of *Trichoderma* as a biocontrol organism. This will be especially limiting on crops requiring numerous applications of pesticides to control foliar diseases.

Judging from the direction of current publications, effective biocontrol strategies using *Trichoderma* for the control of soil-borne diseases should be available for use by the general public in the foreseeable future. These strategies will no doubt first be tailored for high-value intensive crops such as ornamentals, turf, and vegetables. Discussions with various individuals indicate that strains of *Trichoderma* and inoculum growth and delivery systems are being patented in a number of countries. This will not only heighten the intensity of research, but should also stimulate the use of biocontrol as a main strategy of disease control. If biocontrol receives favorable treatment by environmental regulators, media, and the public as compared to chemical pesticides, this should further enhance the research and use of *Trichoderma* as a biocontrol agent.

REFERENCES

1. **Abed-El Moity, T. H., Papavizas, G. C., and Shatla, M. N.,** Induction of new isolates of *Trichoderma harzianum* tolerant to fungicides and their experimental use for control of white rot of onion, *Phytopathology*, 72, 396, 1982.
2. **Allen, M. C. and Haenseler, C. M.,** Antagonistic action of *Trichoderma* on *Rhizoctonia* and other soil fungi, *Phytopathology*, 25, 244, 1935.

3. **Backman, P. A. and Rodriguez-Kabana, R.,** A system for growth and delivery of biocontrol agents to the soil, *Phytopathology*, 65, 819, 1975.

4. **Backman, P. A., Rodriguez-Kabana, R., and Williams, J. C.,** The effect of peanut leafspot fungicides on the non-target pathogen, *Sclerotium rolfsii, Phytopathology*, 65, 773, 1975.

5. **Bell, D. K., Wells, H. D., and Markham, C. R.,** In vitro antagonism of *Trichoderma* species against six fungal plant pathogens, *Phytopathology*, 72, 379, 1982.

6. **Bilai, V. I.,** *Antibiotic-Producing Microscopic Fungi,* Elsevier, Amsterdam, 1963, 115.

7. **Bisby, G. R.,** *Trichoderma viride* Pers. ex Fr. and notes on *Hypocrea, Trans. Br. Mycol. Soc.,* 23, 149, 1939.

8. **Bissett, J.,** A revision of the genus *Trichoderma.* L. Section *Lonibrachiatum* Sect. Nov., *Can. J. Bot.,* 62, 924, 1984.

9. **Bliss, D. E.,** The destruction of *Armillaria* in citrus soils, *Phytopathology,* 41, 665, 1951.

10. **Brian, P. W., Curtis, P. J., Hemming, H. G., and McGowan, J. C.,** The production of viridin by pigment-forming strains of *Trichoderma viride, Ann. Appl. Biol.,* 33, 190, 1946.

11. **Brian, P. W. and Hemming, H. G.,** Gliotoxin, a fungastatic metabolic product of *Trichoderma viride, Ann. Appl. Biol.,* 32, 214, 1945.

12. **Chet, I., Harman, G. E., and Baker, R.,** *Trichoderma hamatum:* its hyphal interaction with *Rhizoctonia solani* and *Phythium* sp., *Microbiology,* 7, 29, 1981.

13. **Cook, R. J. and Baker, K. F.,** *The Nature and Practice of Biocontrol of Plant Pathogens,* American Phytopathological Society, St. Paul, Minn. 1983, 539.

14. **Dennis, C. and Webster, J.,** Antagonistic properties of species-groups of *Trichoderma* I. Production of volatile antibiotics, *Trans. Br. Mycol. Soc.,* 157, 25, 1971.

15. **Dennis, C. and Webster, J.,** Antagonistic properties of species-groups of *Trichoderma* II. Production of volatile antibiotics, *Trans. Br. Mycol. Soc.,* 157, 41, 1971.

16. **Dennis, C. and Webster, J.,** Antagonistic properties of species-groups of *Trichoderma* III. Hyphal interaction, *Trans. Br. Mycol. Soc.,* 57, 363, 1971.

17. **Dubos, B. and Ricard, J. L.,** Curative treatment of peach trees against silver leaf disease (*Sternum perpureum*) with *Trichoderma viride* preparations, *Plant. Dis. Rep.,* 58, 157, 1974.

18. **Duncan, C. G. and Deverall, F. J.,** Degradation of wood preservatives by fungi, *Appl. Microbiol.,* 12, 57, 1964.

19. **Elad, Y., Chet, I., Boyle, P., and Hennis, Y.,** Parasitism of *Trichoderma* spp. on *Rhizoctonia solani* and *Sclerotium rolfsii* scanning electron microscopy and fluorescence microscopy, *Phytopathology,* 73, 85, 1983.

20. **Evans, E.,** Survival and recolonization by fungi in soil treated with formalin or carbon disulfide, *Trans. Br. Mycol. Soc.,* 38, 335, 1955.

21. **Fedorinck, N. S. and Chardymskaya, A. V.,** I Spytanie biopreparta Trikhodermin-3 v bor'be s Korneedom Vskhodov Svekly, *Byull. Vses. Nauchno-issled. Inst. Zashch. Rast.,* 1, 3, 1968.

22. **Fedorinck, N. S., Tarunina, T. A., Tyutyunnikov, M. G., and Kudryavtseva, K. I.,** Trichodermin-4, a new biological preparation for plant disease control, *8th Int. Congr. Plant Prot.,* 3, 67, 1975.

23. **Fisher, C. G., Conway, K. E., and Motes, J. E.,** Fluid drilling: a potential delivery system for fungal biocontrol agents with small seeded vegetables, *Proc. Okla. Acad. Sci.,* 63, 100, 1983.

24. **Garrett, S. D.,** Effect of soil microflora selected by carbon disulfide fumigation on survival of *Armillaria mellea* in woody host tissue, *Can. J. Microbiol.,* 3, 135, 1957.

25. **Garrett, S. D.,** *Pathogenic Root-Infecting Fungi,* Cambridge University Press, London, 1970, 294.

26. **Gilman, J. C. and Abbott, E. V.,** A summary of the soil fungi, *Iowa St. Coll. J. Sci.,* 1, 225, 1927.

27. **Girard, A.,** Reserches Sur l'augmentation des recolter par l'injection dans de sol de sulfure de carbone a'doses massives, *Soc. Natl. d'agr. de France, Bul.,* 54, 356, 1894.

28. **Godtfredsen, W. O. and Vangedal, S.,** Trichodermin, a new sesquiterpene antibiotic, *Acta. Chem. Scand.,* 19, 1088, 1965.

29. **Grosclaude, C.,** Premiers essais de protection biologique des blessures de taille vis-a-vis du *Stereum purpureum* Press, *Ann. Phytopathol.,* 2, 507, 1970.

30. **Hadar, Y., Chet, I., and Henis, Y.,** Biological control of *Rhizoctonia solani* damping-off with wheat-bran culture of *Trichoderma harzianum, Phytopathology,* 69, 64, 1979.

31. **Harman, G. E., Chet, I., and Baker, R.,** Factors affecting *Trichoderma hamatum* applied to seeds as a biocontrol agent, *Phytopathology,* 71, 569, 1981.

32. **Hennis, Y., Adams, P. B., Lewis, J. A., and Papavizas, G. C.,** Penetration of sclerotia of *Sclerotium rolfsii* by *Trichoderma* spp., *Phytopathology,* 73, 1043, 1983.

33. **Hoitink, H. A. J. and Kuter, G. A.,** Effect of composts in container media on diseases caused by soilborne plant pathogens, *Acta. Hortic.,* 172, 191, 1985.

34. **Hubbard, J. P., Harman, G. E., and Hadar, Y.,** Effect of soilborne *Pseudomonas* spp. on the biocontrol agent, *Trichoderma hamatum* on pea seeds, *Phytopathology,* 73, 655, 1983.

35. **Hulme, M. A. and Shields, J. K.**, Biocontrol of decay fungi in wood by competition for non-structured carbohydrates, *Nature (London)*, 227, 300, 1970.

36. **Jones, R. W., Pettit, R. E., and Taber, R. A.**, Lignite and stillage: carrier and substrate for application of fungal biocontrol agents to soil, *Phytopathology*, 74, 1167, 1984.

37. **Kozlova, L. N., Zlothina, G. D., and Bashmakova, V. I.**, Rezultaty izucheniya metodor borby s uvydanium khlopchatnika V tadzhikistane, *Bvull. Nauchno Tekh. Inf.*, Tadzh. *Nauchno Issled. Inst. Selsk. Khoz.*, 4, 95, 1966.

38. **Lewis, J. A. and Papavizas, G. C.**, Effect of mycelial preparations of *Trichoderma* and *Gliocladium* on populations of *Rhizoctonia solani* and the incidence of damping-off, *Phytopathology*, 75, 812, 1985.

39. **Meyer, C. E. and Reusser, F.**, A polypeptide anti-bacterial agent isolated from *Trichoderma viride*, *Experienta*, 23, 85, 1967.

40. **Moubasher, A. H.**, Selective effects of fumigation with carbon disulfide on the soil fungus flora, *Trans. Br. Mycol. Soc.*, 46, 338, 1963.

41. **Munnecke, D. E., Kolbezen, M. J., and Wilber, W. D.**, Effect of methyl bromide or carbon disulfide on *Armillaria* and *Trichoderma* growing on agar medium and relation to survival of *Armillaria* in soil following fumigation, *Phytopathology*, 63, 1352, 1973.

42. **Nelson, E. B., Kuter, G. A., and Hoitinik, H. A. J.**, Effects of fungal antagonists and compost age on suppression of *Rhizoctonia* damping-off in container media amended with composted hardwood bark, *Phytopathology*, 73, 1457, 1983.

43. **Ohr, H. D., Munnecke, D. E., and Bricker, J. L.**, The interaction *Armillaria mellea* and *Trichoderma* spp. as modified by methyl bromide, *Phytopathology*, 63, 965, 1973.

44. **Ooka, T., Shimojima, Y., Akimoto, T., Senoh, S., and Abe, J.**, A new antibacterial peptide "Suzukacillin", *Agric. Biol. Chem.*, 30, 700, 1966.

45. **Papavizas, G. C.**, *Trichoderma* and *Gliocladium*: biology, ecology, and potential for biocontrol, *Annu. Rev. Phytopathol.*, 23, 23, 1985.

46. **Papavizas, G. C., Lewis, J. A., and Abed-El Moity, T. H.**, Evaluation of new biotypes of *Trichoderma harzianum* for tolerance to benomyl and enhanced biocontrol capabilities, *Phytopathology*, 72, 126, 1982.

47. **Ponomareva, G. Ya.**, Khimicheskie i biologicheskie metody zashchity-rostenii, *Trudy Vses. Inst. Rast.*, 24, 182, 1965.

48. **Pottel, H. W. and Shigo, A. L.**, Treatment of wounds of *Acer rubrum* with *Trichoderma viride*, *Eur. J. For. Pathol.*, 5, 274, 1975.

49. **Pottel, H. W., Shigo, A. L., and Blanchard, R. O.**, Biological control of wood hymenomycetes by *Trichoderma*, *Plant Dis. Rep.*, 61, 687, 1977.

50. **Pyke, T. R. and Dietz, A.**, U-21,963, a new antibiotic I. Discovery and biological activity, *Appl. Microbiol.*, 14, 506, 1966.

51. **Reusser, F.**, Biosynthesis of antibiotic U-22,324, a cyclic polypeptide, *J. Biol. Chem.*, 242, 243, 1967.

52. **Ricard, J. L.**, Biological control of *Fomes annosus* in Norway Spruce (*Picca abies*) with immunizing commensation, *Stud. For. Suec.*, 84, 50, 1970.

53. **Rifai, M. A.**, A revision of the genus *Trichoderma*, *Mycol. Papers, Imp. Mycol. Inst.*, 116, 1, 1969.

54. **Rodriguez-Kabana, R.**, Enzymatic interactions of *Sclerotium rolfsii* and *Trichoderma viride* in mixed soil culture, *Phytopathology*, 59, 910, 1969.

55. **Rodriguez-Kabana, R. and Curl, E. A.**, Saccharase activity of *Sclerotium rolfsii* in soil and the mechanism of antagonistic action by *Trichoderma viride*, *Phytopathology*, 58, 985, 1968.

56. **Rodriguez-Kabana, R. and Curl, E. A.**, Nontarget effects of pesticides on soilborne pathogens and disease, *Annu. Rev. Phytopathol.*, 18, 311, 1980.

57. **Rodriguez-Kabana, R., Kelley, W. D., and Curl, E. A.**, Proteolytic activity of *Trichoderma viride* in mixed culture with *Sclerotium rolfsii* in soil, *Can. J. Microbiol.*, 24, 487, 1978.

58. **Saksena, S. B.**, Effects of carbon disulfide fumigation on *Trichoderma viride* and other soil fungi, *Trans. Br. Mycol. Soc.*, 43, 111, 1960.

59. **Schortle, W. C.**, Mechanism of compartmentalization of decay in living trees, *Phytopathology*, 69, 1147, 1979.

60. **Schortle, W. C. and Cowling, E. B.**, Interaction of live sapwood and fungi commonly found in discolored wood and decayed wood, *Phytopathology*, 68, 617, 1978.

61. **Sivan, A., Elad, Y., and Chet, T.**, Biological control effects of a new isolate of *Trichoderma harzianum* on *Pythium aphanidermatum*, *Phytopathology*, 74, 498, 1984.

62. **Smith, K. T., Blanchard, R. O., and Schortle, W. C.**, Postulated mechanism of biocontrol of decay fungi in red maple wounds treated with *Trichoderma harzianum*, *Phytopathology*, 71, 496, 1981.

63. **Toole, E. R.**, Interaction of mold and decay fungi on wood in laboratory tests, *Phytopathology*, 61, 124, 1971.

64. **Warcup, J. H.**, Soil steaming: a selective method for the isolation of ascomycetes from soil, *Trans. Br. Mycol. Soc.*, 34, 515, 1951.

65. **Webster, J. and Lomas, N.,** Does *Trichoderma viride* produce gliotoxin and viridin, *Trans. Br. Mycol. Soc.,* 47, 535, 1964.
66. **Weindling, R.,** *Trichoderma lignorum* as a parasite of other soil fungi, *Phytopathology,* 22, 837, 1932.
67. **Weindling, R.,** Studies on a lethal principle effective in the parasitic action of *Trichoderma lignorum* on *Rhizoctonia solani* and other soil fungi, *Phytopathology,* 24, 1153, 1934.
68. **Weindling, R.,** Isolation of toxic substances from the cultural filtrates of *Trichoderma* and *Gliocladium, Phytopathology,* 27, 1175, 1937.
69. **Weindling, R. and Emerson, O. H.,** The isolation of a toxic substance from the cultural filtrate of *Trichoderma, Phytopathology,* 26, 1068, 1936.
70. **Weindling, R. and Fawcett, H. S.,** Experiments in the control of *Rhizoctonia* damping-off of citrus seedlings, *Hilgardia,* 10, 1, 1936.
71. **Wells, H. D., Bell, D. K., and Jaworski, C. A.,** Efficacy of *Trichoderma harzianum* as a biocontrol for *Sclerotium rolfsii, Phytopathology,* 62, 442, 1972.

Chapter 6

MANAGEMENT OF BACTERIAL POPULATIONS FOR FOLIAR DISEASE BIOCONTROL*

Guy R. Knudsen and Harvey W. Spurr, Jr.

TABLE OF CONTENTS

*   Cooperative investigations of the Agricultural Research Service, U.S. Department of Agriculture, and North Carolina State University, Department of Plant Pathology, Raleigh. Paper No. 10297 of the Journal Series of the North Carolina Agricultural Research Service, Raleigh, N.C. 27695-7601. The use of trade names in this publication does not imply endorsement by the USDA or the North Carolina Agricultural Research Service of the products named, nor criticism of similar ones not mentioned.

I. INTRODUCTION

Recent events have been both exciting and discouraging for researchers trying to control foliar plant diseases by managing antagonistic bacteria. Exciting because of the discovery of new potential control agents and because of the elucidation of antagonistic mechanisms that confer control.[5,9] Also, advances in recombinant DNA technology offer possibilities for "customizing" new microbes for use in biocontrol.[23,24] Discouragement, however, inevitably accompanies the failure of bacterial antagonists to adequately control plant disease in the field. Numerous potential biocontrol agents have fallen by the wayside following unsuccessful field trials. Spurr and Knudsen[35] used the term "silver bullets" to describe these potential antagonists: the analogy is with the legendary silver bullet required to kill a werewolf. Biocontrol researchers have spent considerable time and effort searching for an elusive silver bullet — the ideal biocontrol agent. Gnotobiotic laboratory studies have dominated the screening and preliminary evaluation process for potential control agents. Because many strains of bacteria inhibit fungi or other strains of bacteria in vitro, bioassays of this type tend to generate enthusiasm and a quick premature progression, to field trials.

The development of a methodology to manage foliar biocontrol agents appears to be slow when compared with the dramatic successes in early work with chemical control measures. However, biocontrol requires a synthesis of knowledge from many sciences. Insight comes from systems analysis, epidemiology, microbial ecology, and genetics. Our purpose is to discuss how current research in these areas contributes to a systematic approach to the management of epiphytic bacterial populations. Our definition of management is broad. We include those preliminary activities used to find a bacterial strain with good potential to be an effective agent, ways to improve agent performance on the foliar surface, and the use of certain management tools to predict and evaluate field performance. We suggest some profitable directions for future research.

II. FINDING AND IMPROVING BIOCONTROL AGENTS

Two general approaches may lead to the development of successful biocontrol agents. First, one may search in the natural environment for organisms having the survival and antagonistic characteristics necessary to be promising candidates. Second, it is possible to alter the survival and/or efficacy of strains, either through improved methods of formulation or by genetic manipulation. Since identification of those natural properties that allow certain bacteria to compete and maintain high densities in the natural environment may provide some insight into the desirable attributes of a biocontrol agent, we will first consider that approach.

Certain groups of bacteria, including fluorescent pseudomonads, xanthomonads, and the *Erwinia* spp., are in relatively high densities on leaf surfaces.[6,17] Nonpathogenic members of these groups have been recognized as potential biocontrol agents.[5,6,22] Leben[21] contrasted "resident" and "casual" epiphytic bacteria: residents are part of the permanent, reproducing microflora of the leaf surface, whereas casuals are not indigenous, but have arrived from some other milieu. Similar terminology used by Lynch et al.[25] described microbial populations as either indigenous (autochthonous), i.e., growing in and contributing to the overall metabolic activity within an environment, or exotic (allochthonous). There is little qualitative or quantitative information available to distinguish between resident and casual components of the phyllosphere microflora. Doetsch and Cook[10] offered four criteria to identify members of an indigenous microbial population: the organism should always be observable in that particular environment, it must be able to reproduce and metabolize substrates normally found there, it must produce chemical changes known to occur in that environment, and it must be present in numbers adequate to produce those changes. Do these criteria apply to the phyllosphere ecosystem? Since the physical microenvironment of the leaf surface is more variable than that of the rumen or rhizosphere, phyllosphere residents may be more transient than those in other habitats. Considerable temporal variation has been observed for different groups of organisms: pigmented rod-shaped bacteria, nonpigmented rods, and yeasts.[17] A regular, annual succession of microorganisms apparently occurs on some leaf surfaces.[17] Some species may be difficult to find at certain times of the year, although their presence at other times is fairly predictable. Ephemeral sources of nutrients, such as the pollen produced by different plant species at different times of the year, probably contribute to variability in the composition of the phyllosphere microflora.

Blakeman and Brodie[5] and Morris and Rouse[26] considered the availability of nutrients in the phyllosphere, and utilization of these nutrients by epiphytic bacteria. They noted that chemical or nutritional variables can be more easily manipulated than the physical microenvironment. Thus, populations of epiphytes might be increased by applying specific nutrients at the right time. Morris and Rouse[26] determined the ability of different strains of epiphytic bacteria from snap bean leaves to grow on a variety of single-carbon source media. By applying combinations of nutrients to foliage in field plots, they modified the composition of the bacterial community, altered the population size of fluorescent pseudomonads, and in some cases reduced disease caused by *Pseudomonas syringae*.

Future efforts to identify strains that selectively utilize specific combinations of nutrients may provide a powerful tool for management of epiphytic populations. Potentially useful nutrient sources need not be limited to those normally found on the leaf surface: the chemical industry has produced a variety of compounds not previously found in nature, and organisms capable of metabolizing some of these novel compounds might be given a selective advantage in their presence. The use of selective media, enrichment culture, and mutagenesis may help to obtain strains carrying the appropriate catabolic genes. Since such genes are often carried on plasmids,[15] it may also be possible to transfer these novel catabolic functions from one strain to another in the laboratory.

Nutrients may also trigger dormant microbes. The ability of spore-forming bacteria (e.g., *Bacillus* spp.) to remain metabolically dormant for long periods enhances their survival on the leaf surface, allowing them to survive drying, temperature extremes, and temporary nutrient deficiencies. Several strains have shown promise as potential biocontrol agents.[2,33,35] However, if these bacteria are to inhibit fungi or other bacteria through antibiosis, nutrient competition, or direct parasitism, they may need to be in a vegetative state at the appropriate time. Various chemical compounds, including inosine, adenosine, L-alanine, and tyrosine, serve as triggering agents for spore germination.[10] Hypothetically, addition of these nutrients to foliage under disease-favorable conditions could be used to "turn on" a dormant antagonist.

A wide variety of techniques are available to alter the genetic characteristics of bacteria and potentially improve their field survival and/or biocontrol efficacy. Biotechnology is an inclusive term to describe methods to manipulate microorganisms in order to obtain useful products (or processes). As it relates to a technology of epiphytic bacterial population management, biotechnology represents a spectrum of techniques and methods ranging from the traditional (microbial fermentation, phenotype selection and enrichment, formulation) to the novel (genetic engineering).

The selection of desirable phenotypes has been the basis for biological control research, but the phenotypic characteristic of inhibitory ability in vitro has tended to dominate the selection process. Spurr[34] and Andrews[1] reviewed the techniques developed to assay for potential foliar biocontrol agents. Both authors stated the need for bioassays to be performed under conditions more closely approximating those encountered in the field, but development of a methodology to select for survival in the natural environment has received relatively little attention. This is a particularly important problem, in part because we do not know the extent to which bacteria grown in the laboratory compare with their ancestors that were isolated from plant surfaces. After multiple generations in laboratory culture, strong selection pressures may result in the evolution of bacteria that are much better adapted to life in a culture flask than on the leaf surface. For example, microbes found in nature characteristically possess an extracellular glycocalyx and other structural features that function as attachment devices, permeability barriers, ion exchange resins, or protection against osmotic stress.[8,12,27,30,36] Loss of these structures, including pili and flagella, may occur in the nutrient rich, high temperature conditions that prevail in laboratory cultures.[27,32]

Unfortunately, the loss of structural features that impart survival ability to bacteria on leaf surfaces may not be a handicap under conditions used to maintain and produce large numbers of biocontrol agents. We do not yet know if high survival ability in the field is inconsistent with the need to efficiently produce large numbers of bacteria for field application. Insufficient research effort has been directed towards selecting for characteristics that enhance survival of biocontrol agents. However, a variety of techniques are available that were developed by microbial ecologists[10] and the fermentation industry[16] to select for survival and multiplication of bacteria under given environmental conditions, including temperature, osmotic pressure, radiant flux, pH, or nutritional level.

The ability to make predictable and relatively permanent alterations in the genome of a biocontrol agent offers potential for the management of epiphytic populations.[23,24] The expression of novel phenotypes in bacteria as a result of recombinant DNA techniques may lead to development of enhanced competitive ability (e.g., the ability to utilize new substrates), enhanced antagonistic ability (antibiotic production), specific markers for tracking populations, and other valuable characteristics.[3,4,15]

However, potential problems associated with the use of recombinant DNA methods in biocontrol include the possibility of recombination, both interspecific and intergeneric, occuring in nature. At one extreme, a bacterial community in a given habitat can be thought of as a single genetic pool that is "partitioned, perhaps weakly, by interconnecting units described at the genus and species level".[31] Genetic recombination processes, including conjugation, transformation, and transduction, may serve to break down those partitions. Under certain conditions, interspecific genetic transfer occurs readily in vitro, but the frequency of occurrence in the wild is unknown[31] and difficult to demonstrate.[13]

As a result of the rapid gains in the field of molecular genetics, our capability to create recombinant organisms has progressed far beyond our ability to predict and monitor their fate in the environment. It is hoped that enthusiasm for the promise of genetic engineering will not obscure the fact that we have just begun to tap the potential of more traditional techniques, including selection and formulation. Nonetheless, it has become apparent that recombinant DNA research will play a major role in future biocontrol research and will probably receive a very large share of the research funding. Plant disease biocontrol may benefit substantially from advances in genetic engineering, but by advancing our knowledge of microbial interactions in nature, and through development of techniques to track microorganisms released into the environment, biocontrol research will more than pay its own way.

Some of the problems that biocontrol agents encounter in a hostile environment may be overcome through appropriate formulation of the bacterial product. Pelletized sodium alginate formulations have been successfully used for application of mycoherbicides,[37] and should be suitable for bacterial biocontrol agents as well. Techniques of pelletization or microencapsulation may allow the timed release of bacteria or their metabolic products, as well as providing protection from dessication or ultraviolet radiation. Wettable powders containing lyophilized cells of *Pseudomonas cepacia* or spores of *Bacillus spp.* were stored at room temperature and had a shelf life of several months. They were easily applied to peanut foliage using conventional spray equipment for control of *Cercospora* leafspot.[20]

Incorporation of a nutrient source into pellets or micro-capsules might also help to circumvent nutrient competition from indigenous microbes. Thus, formulations have the potential to make-up for certain natural "deficiencies" of biocontrol agents.

III. MANAGEMENT TOOLS FOR EVALUATING FOLIAR BIOCONTROL AGENTS

Microbial behavior affecting biocontrol should be studied under natural conditions or those that mimic natural conditions. Results obtained from studies with bacteria in Petri dishes or chemostats are not always applicable to bacterial populations in the wild. Little is known about the metabolism of putative antagonists on the leaf surface, including whether antibiotic compounds that may be necessary for biocontrol are even produced there. The spatial and temporal heterogeneity of natural environments can permit the coexistence of species, one or both of which may be antagonistic in vitro, due to the wide range of niches that may be present in a small area. Thus, an environmental system is often more than simply the sum of its constituent parts.

In recent years, increased emphasis on the need to study organisms in their natural systems has led to two research approaches that have received limited attention in biocontrol studies, but hold the promise of being effective management tools. These are the use of microcosms as physical analogues of the natural environment, and the use of systems analysis and modeling techniques to create "conceptual analogues" of the field situation. Both approaches involve the creation of models, in response to the need to bridge the gap between relatively simple (and perhaps irrelevant) laboratory studies and the complexity of comprehensive field studies. These methods offer the researcher an opportunity to simulate a large array of possible environmental conditions, and to reduce the cost and logistical problems associated with field trials.

Pritchard and Bourquin[29] defined a microcosm study as "an attempt to bring an intact, minimally disturbed piece of an ecosystem into the laboratory for study in its natural state . . . within definable physical and chemical boundaries under a controlled set of experimental conditions (i.e., lighting, humidity, aeration, mixing, temperature)." Model systems to investigate the spatial and temporal dynamics of organisms in microbial films also fall into this category.[38] Much of the methodology of microcosm studies has evolved from attempts to monitor the fate of pollutants and their effects in the environment, with relatively less direct consideration of microbial interactions.[11,14,29,39] In biocontrol research, of course, we are also concerned with the fate and effects of agents (albeit biological ones) released into the environment.

Systems analysis and modeling techniques, including computer simulation, can be used to predict and explore the performance of a biocontrol system under a variety of environmental conditions. Model validation, comparing predictions of antagonist population levels or disease severity at different points in time with actual observations, provides an evaluation of how well the workings of the system are understood, as well as a basis for generating and testing new hypotheses. Survival predictions may then be used to estimate the effective dose of the agent present over time.

Knudsen and Spurr[19] presented the concept of an expert systems model that couples a computer simulation of peanut *Cercospora* leafspot development with population models for applied antagonistic bacteria. A data base of survival and efficacy parameters for different antagonist strains is accessed by the computer model, which can then be used to optimize spray timing and dosage, and to predict results of field trials. Figure 1 compares the output of submodels to predict *Pseudomonas cepacia* and *Bacillus thuringiensis* survival on peanut foliage, with actual field observations. These organisms were chosen to serve as "role models" in basic studies of foliar biocontrol. *P. cepacia* is a Gram negative, nonspore-forming rod, while *B. thuringiensis* is a Gram positive, spore-forming rod. Bacteria were

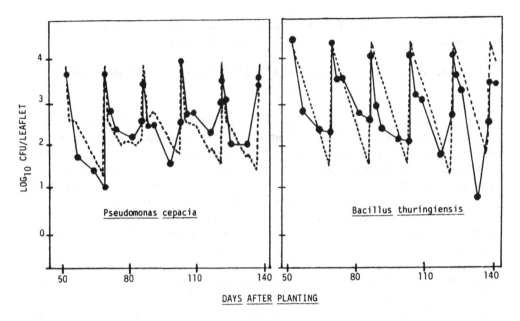

FIGURE 1.   Observed (solid line) and predicted (dashed line) population levels of *Pseudomonas cepacia* and *Bacillus thuringiensis* on the foliage of peanut cv. Florigiant. Bacteria were applied as aqueous formulations of lyophilized powders at 2 week intervals.

applied as aqueous suspensions of freeze-dried vegetative cells (*P. cepacia*) or spores (*B. thuringiensis*). The model for *P. cepacia* assumed that the mortality of the vegetative cells was very high (95%) on the day they were applied, but that thereafter cells died at a slower exponential rate during periods of relative humidity less than 95%, and increased slowly under conditions of high relative humidity. The model for *B. thuringiensis* assumed a simple exponential decline over time, and did not distinguish between endospores and vegetative cells. Parameters for the models were derived from published observations[18] and from unpublished observations made under controlled environmental conditions. Model predictions, especially for *P. cepacia*, were a reasonably good fit to field observations. This predictive ability, coupled with dose-response information, provides a basis to evaluate attempts to improve field performance. Such efforts might include modifying the application rate, increasing spray frequency, or adding nutrients to the spray formulation.

Predictions of the combined effects of survival and efficacy on disease control are shown in Figure 2. The response surface was generated by the *Cercospora* leafspot simulation model, substituting different combinations of control agent parameters in successive simulation runs, with the same weather data inputs. Predicted end-of-season disease severity following biweekly applications of *P. cepacia*, *B. thuringiensis*, or chlorothalonil fungicide are shown. It can be seen that there is a trade-off between persistence and efficacy, and that there are diminishing returns for improvements in either characteristic after some point. However, for the bacterial strains shown, any improvements in either persistence or efficacy would result in significantly greater disease control. Figure 3 shows disease progress curves from field trials in which the peanut cv. Florigiant was treated with *P. cepacia*, *B. thuringiensis*, chlorothalonil fungicide, or left untreated. Both biocontrol formulations significantly (P<0.05) reduced the area under the disease progress curve, although only *P. cepacia* significantly reduced disease severity at the end of the season.

Brand and Pinnock[7] and Pinnock et al.[28] used population models to predict the persistence of *B. thuringiensis* spores on foliage, and to relate these population levels to the mortality of insects feeding on the foliage. When the *Bacillus spp.* are used as plant disease biocontrol

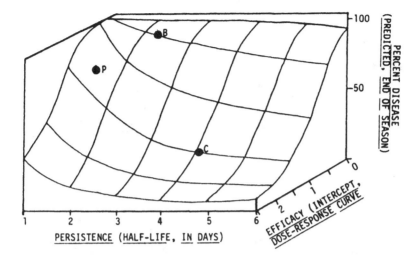

FIGURE 2. Response surface generated by a simulation model: disease severity as a function of control agent persistence and efficacy. Points indicated represent estimated parameter combinations for the bacterial biocontrol agents *Pseudomonas cepacia* (P) and *Bacillus thuringiensis* (B), and the chemical fungicide chlorothalonil (C).

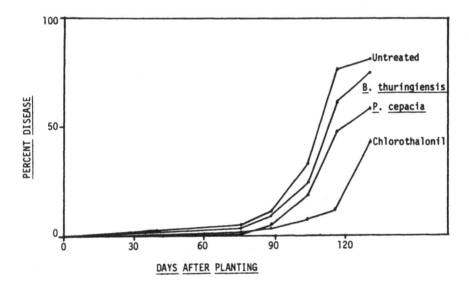

FIGURE 3. Disease progress curves for Cercospora leafspot on peanut cv. Florigiant treated with the bacterial biocontrol agents *Pseudomonas cepacia* or *Bacillus thuringiensis*, the chemical fungicide chlorothalonil, or untreated.

agents, however, it may be necessary to make predictions that distinguish between the vegetative and spore stages. These bacteria have a complex life cycle involving sporulation, dormancy, spore germination, outgrowth, and vegetative growth. The constitutive and environmental factors governing this life cycle are not fully known, but presumably include nutrients, temperature, solar radiation, and moisture. Figure 4 shows results of experiments to monitor not only total populations, but also life stages of *B. thuringiensis* on tobacco leaf disks. Aqueous suspensions of two commercial formulations were sprayed onto 1 cm leaf disks and incubated at 20°C, under constant 100% relative humidity. Periodically, disk surfaces were scraped with a scalpel, and the removed material was observed using phase-

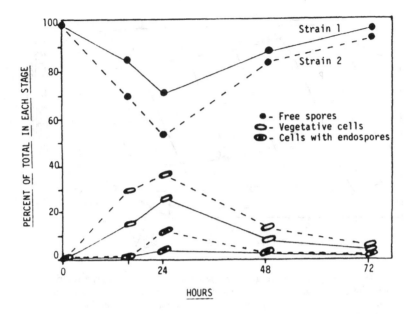

FIGURE 4.   Changes in life stages of two *Bacillus thuringiensis* strains on peanut leaf disks, over time. Disks were kept in an environment of 100% relative humidity, and sampled periodically by scraping the surfaces of sample disks, and observing using phase contrast microscopy.

contrast microscopy. Vegetative cells, cells containing refractile endospores, and free spores were counted and expressed as a percentage of the total population. Spore germination and subsequent resporulation occurred within 24 hr, and the time course of events was similar for the two strains. However, transitions between life stages were not synchronous within populations; indeed, a majority of spores apparently never germinated despite a high viability as determined by plating on nutrient agar. Thus, refinements of simple exponential decay models for *Bacillus* persistence would be necessary in order to predict population dynamics of vegetative cells.

Systems modeling techniques are especially well suited for use in conjunction with microcosm studies (indeed, both approaches generate "simulation models"). Because a microcosm is a self-contained system, the task of establishing boundaries for a system model is greatly simplified, and inputs to the system can be precisely controlled and monitored by the researcher. In other respects, the microcosm will represent the approximate level of complexity of the ecosystem it represents.[29] Thus, microcosm studies are useful for the development and initial validation of simulation models. Together, these tools help make biological control more predictable, and provide a step-wise approach to rational management of biocontrol agent populations.

IV. CONCLUSION

Foliar disease biocontrol is still evolving towards a practical plant disease control strategy, with a concomitant evolution of the attitudes and expectations of researchers. Many early hopes for rapid development of effective control measures, although unfulfilled, have led nonetheless to an increased understanding of plant surface microbiology, microbial ecology, and plant disease processes. With this greater appreciation for the complexity of leaf surface biocontrol comes the realization that we could still be a long way from practical control measures. Researchers need to accept that biocontrol is a long-term prospect, and abandon the silver bullet approach once and for all. We have suggested several avenues of research, each related to the management of bacterial populations on plant surfaces, that we feel are necessary to develop a comprehensive biocontrol strategy. The cornerstone of this management approach is the ability to make quantitative predictions about the performance of foliar biocontrol agents. Within this framework it is possible to evaluate the effects of modifying strains of antagonistic bacteria, their formulation, and their application. Biocontrol with antagonistic bacteria must become consistent and predictable, as well as effective, if it is to compete with the use of chemical fungicides to control foliar plant diseases.

REFERENCES

1. **Andrews, J. H.,** *Biological Control on the Phylloplane,* Windels, C. E. and Lindow, S. E., Eds., American Phytopathological Society, St. Paul, Minn., 1985, 31.
2. **Baker, C. J., Stavely, J. R., Thomas, C. A., Sasser, M., and MacFall, J. S.,** Inhibitory effect of *Bacillus subtilis* on *Uromyces phaseoli* and on development of rust pustules on bean leaves, *Phytopathology,* 73, 1148, 1983.
3. **Barkay, T., Fouts, D. L., and Olson, Betty, H.,** Preparation of a DNA gene probe for detection of mercury resistance genes in gram negative bacterial communities, *Appl. Environ. Microbiol.,* 49, 686, 1985.
4. **Beringer, J. E. and Hirsch, P. R.,** The role of plasmids in microbial ecology, in *Current Perspectives in Microbial Ecology,* Klug, M. J. and Preece, T. F., Eds., American Society for Microbiology, Washington, D.C., 1984, 63.
5. **Blakeman, J. P. and Brodie, I. D. S.,** Inhibition of pathogens by epiphytic bacteria on aerial plant surfaces, in *Microbiology of Aerial Plant Surfaces,* Dickinson, C. H. and Preece, T. F., Eds., Academic Press, New York, 1976, 529.
6. **Blakeman, J. P. and Fokkema, N. J.,** Potential for biological control of plant diseases on the phylloplane, *Annu. Rev. Phytopathol.,* 20, 167, 1982.
7. **Brand, R. J. and Pinnock, D. E.,** Application of biostatistical modelling to forecasting the results of microbial control trials, in *Microbial Control of Pests and Plant Diseases 1970—1980,* Burges, H. D., Ed., Academic Press, London, 1981, 667.
8. **Costerton, J. W. and Cheng, K.,** Microbe-microbe interactions at surfaces, in *Experimental Microbial Ecology,* Burns, R. G. and Slater, J. H., Eds., Blackwell Scientific, Oxford, 1982, 275.
9. **Dickinson, C. H.,** The phylloplane and other aerial plant surfaces, in *Experimental Microbial Ecology,* Burns, R. G. and Slater, J. H., Eds., Blackwell Scientific, Oxford, 1982, 412.
10. **Doetsch, R. N. and Cook, T. M.,** *Introduction to Bacteria and their Ecobiology,* University Park Press, Baltimore, Md., 1973, 371.
11. **Draggan, S.,** The role of microcosms in ecological research, *Int. J. Environ. Stud.,* 13, 83, 1979.
12. **Ferris, F. G. and Beveridge, T. J.,** Functions of bacterial cell surface structures, *BioScience,* 35, 172, 1985.
13. **Freter, R.,** Factors affecting conjugal plasmid transfer in natural bacterial communities, in *Current Perspectives in Microbial Ecology,* Klug, M. J. and Reddy, C. A., Eds., American Society for Microbiology, Washington, D.C., 1984, 105.
14. **Giesy, J. P., Ed.,** Microcosms in Ecological Research Symposium, Savannah River Ecology Laboratory, Augusta, Ga., U.S. Department of Energy Symposium Series 52 (Conf-781101), NTIS, Springfield, Va., 1980, 1110.
15. **Hall, B. G.,** Adaptation by acquisition of novel enzyme activities in the laboratory, in *Current Perspectives in Microbial Ecology,* Klug, M. J. and Reddy, C. A., Eds., American Society for Microbiology, Washington, D.C., 1984, 79.
16. **Higgins, I. J., Best, D. J., Mackinnon, G., and Warner, P. J.,** Recruitment of novel reactions: examples and strategies, in *Current Perspectives in Microbial Ecology,* Klug, M. J. and Reddy, C. A., Eds., American Society for Microbiology, Washington, D.C., 1984, 629.
17. **Jensen, V.,** The bacterial flora of beech leaves, in *Ecology of Leaf Surface Micro-organisms,* Dickinson, C. H. and Preece, T. F., Eds., Academic Press, London, 1971, 463.
18. **Knudsen, G. R. and Hudler, G. W.,** Use of a computer simulation model to evaluate a plant disease biocontrol agent, *Ecological Modeling,* 35, 45, 1987.

19. **Knudsen, G. R. and Spurr, H. W., Jr.,** Computer simulation as a tool to evaluate foliar biocontrol strategies, *Phytopathology,* 74, 1343, 1985.
20. **Knudsen, G. R. and Spurr, H. W., Jr.,** Field persistence and efficacy of five bacterial preparations for control of peanut leaf spot, *Plant Dis.,* 71, 442, 1987.
21. **Leben, C.,** Microorganisms on cucumber seedlings, *Phytopathology,* 51, 553, 1961.
22. **Leben, C.,** Epiphytic micro-organisms in relation to plant disease, *Annu. Rev. Phytopathol.,* 3, 209, 1965.
23. **Lindemann, J.,** Genetic manipulation of microorganisms for biological control, in *Biological Control on the Phylloplane,* Windels, C. E. and Lindow, S. E., Ed., American Phytopathological Society, St. Paul, Minn., 1985, 116.
24. **Lindow, S. E.,** Integrated control and role of antibiosis in biological control of fireblight and frost injury, in *Biological Control on the Phylloplane,* Windels, C. E. and Lindow, S. E., Eds., American Phytopathological Society, St. Paul, Minn., 1985, 83.
25. **Lynch, J. M., Fletcher, M., and Latham, M. J.,** Biological interactions, in *Microbial Ecology: A Conceptual Approach,* Lynch, J. M. and Poole, N. J., Eds., John Wiley & Sons, New York, 1979, 171.
26. **Morris, C. E. and Rouse, D. I.,** Role of nutrients in regulating epiphytic bacterial populations, in *Biological Control on the Phylloplane,* Windels, C. E. and Lindow, S. E., Eds., American Phytopathological Society, St. Paul, Minn., 1985, 63.
27. **Nikaido, H. and Vaara, M.,** Molecular basis of bacterial outer membrane permeability, *Microbiol. Rev.,* 49, 1, 1985.
28. **Pinnock, D. E., Brand, R. J., Milstead, J. E., Kirby, M. E., and Coe, N. F.,** Development of a model for prediction of target insect mortality following field application of a *Bacillus thuringiensis* formulation, *J. Invertebr. Pathol.,* 31, 31, 1978.
29. **Pritchard, P. H. and Bourquin, A. W.,** The use of microcosms for evaluation of interaction between pollutants and microorganisms, in *Advances in Microbial Ecology,* Vol. 7, Marshall, K. C., Ed., Plenum Press, New York, 1983, 133.
30. **Rose, A. H.,** Osmotic stress and microbial survival, in *The Survival of Vegetative Microbes,* Gray, T. R. G. and Postgate, J. R., Eds., Cambridge University Press, London, 1976, 155.
31. **Slater, J. H.,** Genetic interactions in microbial communities, in *Current Perspectives in Microbial Ecology,* Klug, M. J. and Reddy, C. A., Eds., American Society for Microbiology, Washington, D.C., 1984, 94.
32. **Sleyter, U. B. and Messner, P.,** Crystalline surface layers on bacteria, *Annu. Rev. Microbiol.,* 37, 311, 1983.
33. **Spurr, H. W., Jr.,** Experiments on foliar disease control using bacterial antagonists, in *Microbial Ecology of the Phylloplane,* Blakeman, J. P., Ed., Academic Press, London, 1981, 369.
34. **Spurr, H. W., Jr.,** Bioassays — Critical to biocontrol of plant disease, *J. Agric. Entomol.,* 98, 117, 1985.
35. **Spurr, H. W., Jr. and Knudsen, G. R.,** Biological control of leaf diseases with bacteria, in *Biological Control on the Phylloplane,* Windels, C. E. and Lindow, S. E., Eds., American Phytopathological Society, St. Paul, Minn., 1985, 45.
36. **Tempest, D. W. and Neijssel, O. M.,** Eco-physiological aspects of microbial growth in aerobic nutrient-limited environments, in *Advances in Microbial Ecology,* Vol. 2, Alexander, M., Ed., Plenum Press, New York, 1978, 105.
37. **Walker, H. L. and Connick, W. J., Jr.,** Sodium alginate for production and formulation of mycoherbicides, *Weed Science,* 31, 333, 1983.
38. **Wimpenny, J. W. T., Lovitt, R. W., and Coombs, J. P.,** Laboratory model systems for the investigation of spatially and temporally organised microbial ecosystems, in *Microbes in their Natural Environments,* Slater, J. H., Whittenbury, R., and Wimpenny, J. W. T., Eds., Cambridge University Press, London, 1983, 67.
39. **Witt, J. M. and Gillette, J. W.,** *Terrestrial Microcosms and Environmental Chemistry,* Proc. Symp., Corvallis, Ore., 1977.

Chapter 7

THE BACULOVIRUSES

B. M. Arif and P. Jamieson

TABLE OF CONTENTS

I. INTRODUCTION

The search for a safe and viable alternative to chemicals in the control of forest and agricultural pests has resulted in an increasing interest in biological agents such as viruses, bacteria, and fungi. Chemicals are becoming ecologically unacceptable because, (1) there are short and long term side effects on nontarget organisms, and (2) the development of resistance by insect pests leads to the application of higher doses of chemicals and a consequential increase in environmental pollution. In some cases insects have developed resistance to a wide range of chemical insecticides. An example of this is the cotton bollworm.[27] Insect viruses, particularly baculoviruses, appear to be potentially useful because they can be employed much like chemicals, yet without the environmental problems associated with chemical insecticides. Baculoviruses are desirable for a number of other reasons: (1) they are generally host specific, in most cases they infect only members of the host's family, (2) structurally, they do not resemble known vertebrate or plant viruses, and (3) they have been shown to be safe to nontarget organisms.

The safety of baculoviruses is a particularly important attribute, which has been demonstrated with a number of baculoviruses at the whole animal level. At the cellular level, there have been very few reports to show possible replication or gene expression in cells of nontarget organisms. It was reported that some virus replication was seen in a Chinese hamster ovary cell line infected with *Autographa californica* nuclear polyhedrosis virus (AcMNPV).[71] This observation was based on the production of particles that banded at a density of 1.24 to 1.25 g/mℓ in sucrose, which was similar to the banding of AcMNPV nonoccluded virus. The particles were infectious to a *Trichoplusia ni* cell line and were precipitated by anti AcMNPV antiserum. The authors mention that the Chinese hamster cell lines were maintained at 28°C instead of the normal 35°C and the cells did not replicate at the lower temperature. These results suggest virus replication may take place in a heterologous system under certain laboratory conditions. In another study, AcMNPV was observed to enter Hela and poikilothermic fat-headed minnow cell lines.[34] No replication was observed but infectious virus persisted for at least 6 weeks at 28°C.

Tjia et al.[114] investigated the effect of AcMNPV or its DNA on several mammalian cell lines. Cells were inoculated at multiplicities ranging from 0.1 to 100 PFU/cell and subsequently investigated for evidence of virus production or replication and for persistence of viral DNA. They found that AcMNPV did not multiply in Hela cells, human embryonic kidney cells, CV-1 monkey cells, the B3 subline of BHK21 cells, or in Muntiacus muntjak fibroblasts. Hybridization of cloned AcMNPV DNA EcoRI fragments with cellular extracts showed that the viral DNA was detectable up to 24 hr postinfection but no viral DNA replication or expression as mRNA was observed. AcMNPV DNA then disappeared after serial passage of the cells. These data strongly support the safety of baculoviruses to vertebrate systems.

Baculoviruses of several insect species have been registered, or are being considered for registration, in Canada and the U.S. They include NPVs that infect *Neodiprion lecontei* and *Neodiprion sertifer* (red-headed pine sawfly and European pine sawfly), *Heliothis zea* (cotton bollworm), *Autographa californica* (alfalfa looper), *Orgyia pseudotsugata*, and *Orgyia leucostigma* (Douglas-fir tussock moth and white-marked tussock moth), *Lymantria dispar* (gypsy moth), and a granulosis virus (GV) of *Cydia pomonella* (codling moth).

With the advent of invertebrate cell lines that supported the replication of certain NPVs, rapid progress was made in elucidating the molecular biology of baculoviruses, especially their biochemical structure and gene expression. AcMNPV became the model virus in these studies because it replicated so well in cell lines from *T. ni* and *S. frugiperda*. Several laboratories around the world showed AcMNPV to be an excellent vector for the expression of foreign genes. It was used to express α- and β-interferons, β-galactosidase interleukin-

2, and *myc*-oncogene.[64,77,82,100,101] Such sophisticated technology could be used in the future to improve baculoviruses as forest and agricultural pest control agents. The rapid advancements in baculovirology have been very interesting and promise to be even more exciting in the future.

Several reviews have so far been written on various aspects of baculoviruses and other occluded viruses.[30,34,35,38,48,52,73,74] For a review on the use of viruses in pest control, the reader is advised to read the excellent treatise by Entwistle and Evans.[27]

II. THE VIRUS

A. Nomenclature

Baculoviruses have so far been named after the Latin binomial of the insect from which they were isolated. For example, *Autographa californica* NPV and *Choristoneura fumiferana* NPV. The drawbacks of this system, indicated previously,[30] lead to confusion if more than one NPV or GV were isolated from the same host. Also, a baculovirus may have a relatively wide host range and may be isolated from more than one insect. Certain biochemical techniques, such as analysis by restriction endonuclease enzymes, could be used to solve such problems. These analyses can distinguish isolates from the same insect species or affirm the identity of a virus from different insects. Clearly the taxonomic problem will have to be resolved and we may eventually have to use biochemical as well as biological criteria.

B. Classification

The family Baculoviridae contains one genus, the baculovirus, which is divided into four subgroups, A to D, based mainly on the packaging of the nucleocapsid within a common viral membrane.[70] Subgroup A is the nuclear polyhedrosis viruses, which contain a single or many nucleocapsids enveloped within a common unit membrane (SNPVs or MNPVs). All members of this subgroup have many virions per inclusion body. Subgroup B consists of the granulosis viruses. The virus contains one nucleocapsid per membrane and only one virion is occluded in an inclusion body. Subgroup C contains baculoviruses which consist of one nucleocapsid within a viral membrane but no inclusion bodies are formed. The type species is the *Oryctes* virus from the Rhinoceros beetle. Subgroup D contains certain members of the polyDNA-viruses. The genome of these viruses consists of a population of double stranded, circular DNA with different size classes of molecules. Each circle of DNA exhibits some sequence homology to the others.[105]

C. Morphology

The virions are bacilliform in shape measuring 40 to 110 nm × 200 to 400 nm. The nucleocapsids are electron dense cylindrical forms 35 to 40 nm × 200 to 350 nm in size. An example of an MNPV was isolated from the eastern spruce budworm, *Choristoneura fumiferana* (Figure 1).

III. VIRUS REPLICATION

A. In Larvae

Thorough reviews on the pathogenesis of baculoviruses in larvae have been written earlier.[30,34,118] Basically, natural infection of larvae occurs by ingestion of viral inclusion bodies and dissolution of the polyhedral protein in the alkaline environment of the gut. Released virions fuse with microvilli on the columnar epithelial cells resulting in nucleocapsids in the cytoplasm.[34,36] Occasionally, enveloped nucleocapsids in cytoplasmic vacuoles have also been seen indicating that viropexis may also take place.[1] Genomic nucleoprotein complex appears to be released from nucleocapsids of GVs and NPVs at different sites within the

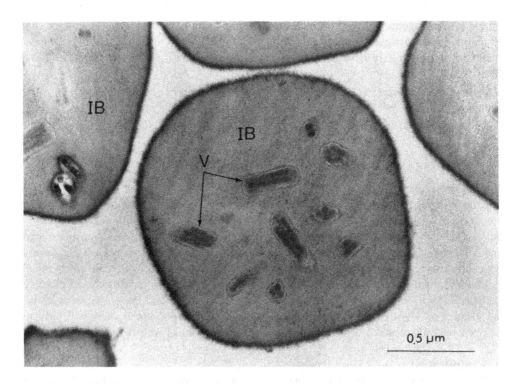

FIGURE 1.   Section through inclusion bodies of *Choristoneura fumiferana* MNPV.

cell.[34] In the case of GVs, nucleocapsids were observed adjacent to nuclear pores where presumably DNA:protein complex was extruded into the nucleus to initiate replication. The presence of nucleocapsids of AcMNPV at different stages of uncoating within the nuclei has been documented.[36] It appears, therefore, that the site of uncoating may be a basic difference between NPVs and GVs.[34]

The presence of infectious virus in the hemolymph was detected 30 min after AcMNPV polyhedra were fed to larvae,[36] indicating that NPV can invade the insect gut and release virus to the hemolymph by a mechanism that is not yet clear. In both circumstances, the virus usually initiates successful infection with swelling of the nuclei, formation of a virogenic stroma, and synthesis and subsequent envelopment of nucleocapsids. In vivo and in vitro studies have shown that envelopment of nucleocapsids may take place at various parts of the cell: (1) in the nucleus within *de novo* synthesized membranes, (2) at the nuclear membrane, and (3) at the plasma membrane.[34,41,63,85,113] It was concluded from these phenomena that replication of baculoviruses leads to the synthesis of heterogeneous nonoccluded virus (NOV) populations composed of nucleocapsids enveloped by membranes of diverse origin.[30] In the case of baculoviruses, as it is with other occluded insect viruses, two forms of virus are synthesized: (1) enveloped nucleocapsids which become occluded within the nucleus in occlusion bodies, and (2) enveloped nucleocapsids which bud from infected cells and cause secondary infection in other larval tissues.[1] The nonoccluded virus has a peplomer structure that is not present in the occluded form.[111] Also the nonoccluded virus was found to be approximately 2000 times more infectious than virus liberated by alkali from polyhedra.[126]

B. In Tissue Culture

Studies on the replication of baculoviruses in tissue culture cells have shown that they follow a similar pattern to that observed in vivo. Studies made with tissue culture systems are of course more convenient in that the investigator has the ability to infect every cell at the same time and the sequence of viral replication events can be monitored closely and more precisely than in vivo. Faulkner,[30] however, cautions that tissue cultures represent only select cells that have been maintained on defined media for a large number of generations and are not under the constraints of cells in the natural host.

The nonoccluded virus attaches, enters the cell, and becomes uncoated as has been described for the in vivo system. Virogenic stroma begins to appear in patches within the nucleus of the infected cells and the nucleus becomes enlarged. Short membrane-like structures begin to appear in the nuclei, which may arise by *de novo* morphogenesis since they were not seen to have contact with the intranuclear membrane.[55,63,106] Nucleocapsids were then observed to become associated with these membranes. This clearly appeared to be the envelopment of mature nucleocapsids. At approximately the same time occlusion bodies begin to develop. The developing occlusion bodies were seen in association with fibrous material.[18,63] It was suggested that the fibrous material condensed to form the polyhedral membrane. Under normal conditions, only fully enveloped nucleocapsids (singly or multiply) were occluded suggesting that a specific mechanism exists to allow the occlusion of enveloped nucleocapsids. This mechanism may be some type of interaction between the viral membrane and the polyhedral protein.

Serial passage of the virus in tissue culture cells eventually leads to the formation of progressively fewer occlusion bodies. In fact, two distinct forms of the virus were seen, one is associated with the formation of many polyhedra (MP) and the other forms few polyhedra (FP) within the nucleus.[31,40,84] The FP mutant appears to be considerably less infectious than the MP phenotype.[31] Studies on the structure of the viral genome from FP phenotypes have yielded good evidence on how this mutant arises (see the section on the structure of the genome).

IV. VIRION PROTEINS

A. Structural Proteins

1. The Inclusion Body

Inclusion body (IB) protein from different viruses is apparently very similar in terms of physical and chemical properties but differs serologically, in amino acid sequence and in tryptic peptide products.[88] The inclusion body protein, termed polyhedrin for NPVs and granulin for GVs, was discovered to surround the enveloped occluded viruses.[6,78] Some researchers have observed that a membrane rich in carbohydrate surrounds the IBs.[37,46,76] A thorough review of inclusion body proteins is provided by Vlak and Rohrmann.[121]

Initially, it was determined from amino acid composition that the polyhedrins of *Bombyx mori* and *Galleria mellonella* NPVs (BmMNPV and GmMNPV) have a molecular weight of 28,000 daltons (28 kDaltons).[56] Investigations which followed produced a molecular weight range of 25 to 33 kDaltons for several polyhedrins and granulins.[23,98,107,109] Protein[57] and DNA sequencing[43] allowed accurate estimations of 28 to 29 kDaltons.

N-terminal analysis comparisons between NPV and GV isolates indicated that the matrix protein consists of only one polypeptide which differs between granulin and polyhedrin.[107,108] Both polyhedrins and granulins have been shown to be rich in aspartic and glutamic acid residues[19,107] and also in strongly basic amino acids (lysine, argenine). Peptide maps of polyhedrins from numerous NPVs were essentially reproducible, with some minor differences in numbers of acidic and basic polypeptides.[56,108] Two studies on a number of polyhedrins from singly and multiply embedded NPVs found many similarities in the tryptic peptide patterns, but each produced a unique pattern.[67,107] Comparison of polyhedrin and granulin peptide maps gave similar results.[89] Passage of AcMNPV through alternate host systems, both in vivo and in vitro, caused no changes in primary structure as judged by tryptic digests and amino acid analysis.[69] The tyrosine residues (pKa 9.5 to 10.5) were clustered in two areas of the NH$_2$-terminal region and may be related to the alkali solubility of the polyhedrin. DNA sequencing has been used to derive amino acid sequences of polyhedrins and granulins.[17,91] The results indicated that they were closely related, but distinct. Certain regions of the proteins were discovered to be highly conserved although there was only about 50% amino acid sequence homology.

In vitro labeling of the polyhedrins of AcMNPV and *Rachiplusia ou* NPV (RoMNPV) with ^{32}P was detected.[25,68,108] The polyhedrins of *Trichoplusia ni* NPV (TnMNPV) and *Orgyia pseudotsugata* NPV (OpMNPV) were also discovered to be weakly phosphorylated[29,89] as were the granulins from *Trichoplusia ni* and *Plodia interpunctella*.[108,117] No evidence of glycosylation of polyhedrin or granulin was found.[107,108,117] However, Eppstein and Thoma[29] did observe that TnMNPV polyhedrin contained 3% carbohydrate.

Yamafuji et al.[130] were the first to demonstrate that alkaline protease activity was present in the inclusion bodies of a NPV. Protease activity has been found associated with the insect-derived inclusions of numerous NPVs and GVs but not in polyhedra from tissue culture cells.[28,108,115,128,132] The protease causes degradation of the inclusion body matrix protein. Wood[128] noted that the alkaline protease was required for the complete dissociation of polyhedrin from the viral envelope during alkaline dissolution of the inclusion bodies and for the subsequent removal of the envelope from the nucleocapsids by NP40.

2. Envelope, Nucleocapsid, and Capsid Proteins

Investigators have discovered between 11 and 25 polypeptides ranging in size from 9 to 160 kDalton by Coomassie brilliant blue staining or autoradiography. However, Vlak[119] was able to detect 35 AcMNPV polypeptides by SDS-discontinuous polyacrylamide gel electrophoresis (PAGE) followed by autoradiography. The polypeptide patterns obtained by electrophoresis appear to be specific for the isolate[19,39,109] although many baculoviruses contain major polypeptides in the same molecular weight range.[19,109] Two-dimensional gel electrophoresis[80] allowed the resolution of 81 acidic polypeptides from the enveloped virions of AcMNPV and *Porthetria dispar* NPV (PdMNPV).[93] Over 100 polypeptides were detected by nonequilibrium pH gradient electrophoresis (NEPHGE)[79] in the first dimension to allow both acidic and basic proteins to enter the gel.[94]

Major quantitative and qualitative differences have been observed between the polypeptide patterns of singly and multiply embedded isolates.[19,109] Smith and Summers[95] found that the occluded and nonoccluded forms of AcMNPV also differed in their polypeptide profiles by PAGE. This is in agreement with the results of others.[25,119] In addition, some polypeptides from the two phenotypes which comigrate in gels have been found to differ quantitatively or to be phosphorylated to a different extent.[68,109] In studies using reciprocal immunoblotting (Western blotting) some major structural proteins of the two forms reciprocally cross-reacted while one major structural protein from each form did not.[124] Also, antiserum against the occluded form of the virus recognized virus-induced cell surface antigens, indicating that there are antigenic similarities between the occluded form and the envelope proteins of the nonoccluded form.

Upon removal of the viral envelope of a number of NPVs and GVs using NP40 or Triton X-100 it was found that most of the high molecular weight polypeptides were derived from the envelope, as well as the majority of the remaining polypeptides.[39,109] Similarly, Yamamoto and Tanada[131] found that two strains of *Pseudaletia unipuncta* GV had six to nine polypeptides associated with the viral envelope while only three were part of the nucleocapsid fraction. However, only 5 of 15 structural proteins of *Plodia interpunctella* GV were identified as envelope polypeptides.[117] These results were confirmed using solid phase-bound lactoperoxidase and ^{125}I which radiolabel polypeptides on the external surface of the viral envelope. A study involving *Pieris brassicae* GV showed that 8 of 12 polypeptides were nucleocapsid-associated.[7] Removal of the envelopes of *Oryctes rhinoceros* baculovirus, a subgroup C virion, resulted in the disappearance of 14 out of a total of 27 protein bands from the gel pattern, suggesting that the remainder are associated with the nucleocapsid.[22] Major differences in electrophoretic mobility were observed among the envelope proteins of the GVs of *Plodia interpunctella, Trichoplusia ni, Spodoptera frugiperda*, and *Pieris brassicae*.[7,109,117] The variance in the results observed between nucleocapsid and envelope fractions may be the result of unequal success in removing all of the envelope components.

Comparative SDS-PAGE between the nucleocapsids of two subgroup C baculoviruses isolated from *Oryctes rhinoceros* and *Spodoptera littoralis* revealed that only two polypeptides comigrated.[81] This seems to be true for baculoviruses in general. A higher molecular weight protein (31 to 34 kDaltons) was discovered to be the predominant polypeptide in empty capsids of *Trichoplusia ni* and *Plodia interpunctella* GVs.[109,116] A protein of similar molecular weight is present in several other baculoviruses, possibly functioning as the major structural protein of the capsids. The other protein found in large amounts has a low molecular weight (12 to 14 kDaltons), is extremely basic, and also argenine-rich.[116] Bud and Kelly[10] determined that this polypeptide is not a histone. PAGE showed that it was not present in the polypeptide profiles of GV capsids and was therefore probably a core protein of the nucleocapsids. Electron microscopic observations indicated that it was part of a viral DNA-protein complex.[116] Several other NPV and GV nucleocapsids have been found to contain a major polypeptide of similar molecular weight.[19,39,81] Recently, similar basic, argenine-rich, lysine-poor, DNA-binding proteins have been found in the core of the capsids of *Heliothis zea* NPV, *Trichoplusia ni* NPV, *Spodoptera litura* GV, and *Oryctes rhinoceros* nonoccluded baculovirus,[14,49] as well as a minor nucleocapsid polypeptide in AcMNPV.[94] It was theorized that this 12 kDalton protein aids in condensation and protection of the genome both prior to and during encapsidation and is similar to the DNA binding protein discovered in vaccinia virus.[45]

3. Phospho- and Glycoproteins

Maruniak and Summers[68] found that AcMNPV-NOV possessed at least 9 phosphoproteins while the occluded virus (OV) contained about 14 phosphoproteins. Qualitative differences were also apparent. Dobos and Cochran[25] were able to label ten polypeptides of AcMNPV-OV with ^{32}P. Two envelope and three nucleocapsid proteins of *Plodia interpunctella* GV were found to be significantly phosphorylated.[117] Stiles and Wood[103] identified 11 structural glycoproteins in AcMNPV-NOV but only 1 in the OV form of the virion, while Dobos and Cochran[25] could label only 6 proteins using N-acetyl-D-glucosamine and mannose as substrates. However, use of a more sensitive in vitro lectin binding procedure revealed seven glycoproteins in this virus, three in the nucleocapsid fraction, and four associated with the viral envelope.[51] The glycoproteins of AcMNPV-NOV were necessary for infection of *Trichoplusia ni* cells, but not for release of enveloped, progeny virus, possibly because the glycoproteins are necessary for the envelopment of nucleocapsids.[51,104] The fact that the NOV form is much more infective in an in vitro system could be explained by the difference in numbers of glycosylated polypeptides between the NOV and OV forms.[92]

B. In Vitro Protein Synthesis

Baculoviruses from subgroup B have not been successfully propagated in an in vitro system,[33,118] while no research has been done on characterizing the polypeptides produced in tissue culture cells by subgroup D baculoviruses. Most of the work in this area of research was done with subgroup A baculoviruses.

It was observed that 30 to 35 different polypeptides appeared sequentially in AcMNPV-infected cells within 24 hr post-infection (pi).[15,25,50,68,128] This suggested a temporal regulation of virus-directed protein synthesis. The first virus-specific polypeptides appeared in infected cells by 3 hr pi. The number and amount of infected cell specific polypeptides (ICSPs) increased until approximately 20 hr pi when almost all polypeptide synthesis began to decrease.[15,68,129] One study has been carried out on the in vitro replication of a subgroup C baculovirus, *Oryctes rhinoceros*, by pulse labeling with [35S] methionine.[22] Only 8 of the 27 virus structural polypeptides could be detected. Another nonoccluded baculovirus, Hz-1, was found to produce 37 virus-induced intracellular polypeptides of which 28 were structural proteins.[11]

Three or four phases of AcMNPV polypeptide synthesis in an in vitro system have been proposed, although there was no general agreement on the duration of these phases.[25,50,68,128,129] No effects on late protein synthesis were observed after blocking viral DNA synthesis with cytosine arabinoside.[25] However, Kelly and Lescott,[50] using the same DNA synthesis inhibitor, as well as bromodeoxyuridine, found that only the two early phases taking place before onset of viral DNA synthesis occurred, while the late phases of viral polypeptide synthesis were severely inhibited. This indicates that the late phases are dependent on viral DNA synthesis. Wang et al.[127] supported these findings after performing similar experiments with various viral-specific DNA polymerase inhibitors. The functioning of the polypeptides of one phase was shown to be necessary to allow the synthesis of the polypeptides of the following phase. If the block was removed the next phase of protein synthesis proceeded, indicating a cascade induction of proteins occurs, similar to that which happens in herpes and iridoviruses.[26,42]

It was generally found that 4 to 6 early polypeptides were synthesized[15,102,129] although two studies found 1 and 12 early proteins, respectively.[25,50] No evidence was found that any early polypeptide became a major component of the virus.[102] However, they may play an important regulatory role in the infection process. The 28 kDalton polyhedral protein was first synthesized at 12 hr pi but only predominated at late stages of infection, at which time it was the only ICSP being synthesized other than a major 10 kDalton polypeptide.[15,102,112,123] At 12 hr pi NOV production and viral DNA synthesis began to decline.[26] The 10 kDalton polypeptide was found to be abundantly expressed only late in infection.[86,87,102,123] It appears to be glycosylated[51] and was very weakly labeled with [35S] methionine. Although it comigrated with a minor NOV structural polypeptide,[2,123] this did not prove that it is a structural protein.[59,86,102] No function has been determined for this protein, although Rohel[86] suggested that it may be necessary for morphogenesis of inclusion bodies.

Late in infection extra- and intracellular morphogenesis were observed to be at a maximum.[125] Carstens et al.[15] theorized that polypeptides synthesized after this time might play a role in the viral occlusion process rather than in synthesis of new virus particles. Many of the polypeptides synthesized at later stages in the infection of tissue culture cells were found to comigrate with viral structural polypeptides.[15,102,123] However, this is not enough to confirm their identity.[123] Tryptic peptide map comparisons of a few of these viral-specific polypeptides and virus structural polypeptides confirmed the relationships in the samples compared.[25] In addition, AcMNPV-RNA isolated from infected cells late (21 hr pi) in infection could be translated into proteins which had the same molecular weights as those of virion polypeptides and were immunoprecipitated with antiserum prepared against virion structural proteins.[102,123] No late protein synthesis was observed in *Oryctes rhinoceros*-infected cells and no inclusion body protein synthesis occurred, which differs from the results obtained with subgroup A baculoviruses.[22]

No significant posttranslational cleavage of viral-specific polypeptides in AcMNPV-infected cells was detected[25,50,129] except by Carstens et al.[15] However, a number of AcMNPV-specific polypeptides were found to be phosphorylated and/or glycosylated.[15,25,51,68,103]

Some reports of weak phosphorylation of the polyhedral protein exist.[25,68,108] Nine viral-induced glycoproteins were detected in *Spodoptera frudiperda* cells infected with Ac-MNPV[103] representing proteins initially synthesized as early as 2 hr pi to as late as 14 hr pi. The glycoproteins detected represented both structural and nonstructural polypeptides.

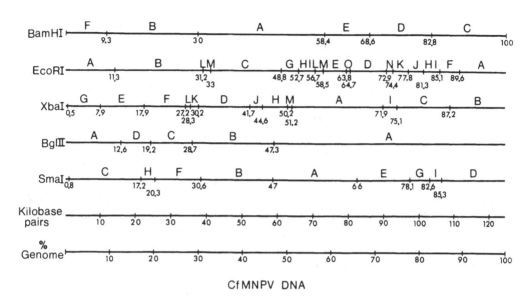

FIGURE 2.   Restriction maps of the genome of CfMNPV.

A number of glycoproteins could be detected on the surface of the viral envelopes by lactoperoxidase-catalyzed surface radioiodination. There is disagreement concerning the number of glycoproteins associated with AcMNPV-OV. Stiles and Wood[103] could detect only one glycosylated polypeptide while others observed 14 and 6 OV glycoproteins, respectively.[25,65]

V. STRUCTURE OF THE GENOME

The genome, which constitutes approximately 13% of the virus particle by weight, is double-stranded, closed circular, and supercoiled DNA ranging from 90 to 200 kilobase pairs (kb) in size. Several laboratories have shown that the viral genome was infectious to tissue cultures of insect cells.[13,16] It was found to contain reiterated sequences, which in some cases represented the incorporation of host cellular DNA and led to the formation of defective particles such as the FP mutants. Repeated DNA sequences of viral origin were also detected in the genomes of many baculoviruses. Mobile elements from the host DNA were also found to become incorporated by transposition into the viral DNA. These aspects, and others, of the viral genome will be reviewed in this section.

A. Mapping By Restriction Endonuclease Enzymes (RENs)

RENs have been an excellent tool in probing and understanding genomic structures. Indeed, they opened the door to molecular cloning and recombinant DNA research. The genomes of many baculoviruses have been analyzed by RENs and several maps of restriction sites have been generated.[5,8,21,60,62,75,95-97,120] An example of the restriction maps is shown in Figure 2. The smallest restriction fragment that contains the polyhedrin gene (BamHI-F) was located on the "left end" of the linearized maps as previously recommended.[122] Several baculovirus isolates were shown to be naturally occurring variants, the genomes of which contained differences in the restriction endonuclease patterns. The genomic changes have been indicated on the maps of several variants.[8,9,24,60,66,75,95,96] For example, the restriction sites in the genome of the multiply embedded virus *Spodoptera frugiperda* NPV (SfMNPV) and some of its variants were ordered with respect to seven RENs.[66] Four regions of the genome were found to undergo changes and give rise to viral variants characterized by insertions or deletions of DNA and cleavage site changes. Similar results were reported on differences in the genome of SfMNPV variants from four different laboratories.[54] An SfMNPV isolate from Ohio, U.S. had an extra HindIII cleavage site in the fragment HindIII-A generating the fragments HindIII-C and -D. The same isolate had additional EcoRI and BamHI cleavage sites as compared to the genome of SfMNPV-2 reported by Maruniak and co-workers.[66]

Mapping the DNA with RENs has also been useful in identifying recombinants from two viruses such as those reported between AcMNPV and *Rachiplusia ou* NPV (RoMNPV).[97] The genomes of the recombinants contained EcoRI, BamHI, and XbaI fragments with mobilities which were either (1) the same as RoMNPV DNA fragments, (2) the same as AcMNPV DNA fragments, or (3) different from RoMNPV and AcMNPV fragments. Interestingly, the virions of two recombinants, AR7 and AR8, were occluded in AcMNPV polyhedrin as shown by peptide mapping. Analysis of the viral structural protein by PAGE showed that the recombinants contained structural proteins from both parental types.[97] The distribution of recombination sites on the genomes of these two viruses does not appear to be random. Seven recombinants from AcMNPV and RoMNPV were generated in tissue culture cells and the crossover locations were mapped by RENs. In six of these recombinants, the crossover sites were located in the EcoRI-Q and -B fragments of RoMNPV and the EcoRI-I and -B fragments of AcMNPV DNAs.[110]

B. DNA Sequence Homologies Among Different Viruses

Reassociation kinetics studies on the genomes of four *Spodoptera* NPVs revealed that the DNAs were 15 to 75% homologous.[47] A more comprehensive homology analysis was carried out on DNAs of representatives from subgroups A, B, and C baculoviruses. The DNA homologies were estimated by hybridization under conditions of different stringencies.[99] There were varying degrees of homology among the representative viral DNAs and, as expected, apparent homologies increased as the stringency conditions of hybridization were decreased. At melting temperature (Tm) $-22°C$ the apparent homology ranged from 0.1 to 3% and increased 3 to 10 times at Tm $-44°C$. Some regions on the AcMNPV genomes were found to be highly conserved since they hybridized to members of subgroups A and B baculoviruses. The functional significance of the conserved regions is not yet understood but some may represent important segments of the genomes such as the origin of DNA replication.[99] The fragment that contained the polyhedrin gene in AcMNPV DNA hybridized to DNAs from all subgroups. The fragment EcoRI-P of AcMNPV hybridized to three NPVs and two GVs suggesting that it is conserved in these viruses. This fragment is known to contain the p10 gene which codes for the synthesis of a 10 kDalton protein late in the infectious cycle.[53,59]

Detailed DNA sequence homology analysis between the genomes of CfMNPV and AcMNPV was done using Southern blot hybridization.[4] The particular strain of AcMNPV used in these studies was described previously.[62] The two genomes exhibited a considerable degree of homology in that, under low stringency conditions (Tm $-44°C$), CfMNPV DNA hybridized to all AcMNPV EcoRI fragments except EcoRI-R. Cloned DNA fragments from both genomes facilitated the study and showed that the EcoRI fragments of CfMNPV DNA hybridized to the AcMNPV EcoRI fragments in approximately the same order as their locations on the physical maps.[4] This indicates that the gene order in both viruses may be the same. Nonhybridizing fragments between AcMNPV and OpMNPV were also reported and may represent hypervariable regions of DNA or of independent origin.[61]

Homology studies were also performed by hybridization of DNAs from *O. pseudotsugata* NPVs to AcMNPV DNA under conditions of varying stringency.[90] Homology of 13 to 25% was detected between OpMNPV and AcMNPV. Cloned fragments of OpMNPV DNA were then hybridized to Southern transfers of restricted AcMNPV DNA. From these studies the

HindIII map of OpMNPV was aligned with the HindIII and PstI restriction maps of AcMNPV.[61] From the cross-hybridizing regions of the two genomes it was seen that they were arranged in a predominantly colinear manner.

C. Homologous DNA Sequences in the Genome

Intragenomic DNA homologies in baculoviruses were first demonstrated to occur in four NPVs from the *Spodoptera spp.*[47] In reassociation kinetics experiments, the genomes of these viruses were shown to contain a small fraction that reannealed quickly, which is indicative of reiterated DNA. The availability of cloned DNA fragments from the genomes of several baculoviruses helped in probing the DNA for repeated sequences. In the genome of AcMNPV, hybridization was observed among the fragments HindIII-Q, -L, -F, and -A/ B suggesting that a DNA sequence(s) is repeated in these locations in the genome.[20] Plasmids containing HindIII-L and -Q were used to locate the repeated DNA on the physical map. Five locations were identified to contain homologous repeats (HR) at the following junctions: HR_1, EcoRI-B-EcoRI-I; HR_2, EcoRI-A-EcoRI-J; HR_3, EcoRI-C-EcoRI-G; HR_4, EcoRI-Q-EcoRI-L, and HR_5, EcoRI-S-EcoRI-X. The regions that contained the homologous repeats also contained a number of small EcoRI fragments ranging in size from 73 to 225 base pairs (bp). The presence and distribution of the repeated DNA was found to be conserved among several AcMNPV strains, as were the patterns and stoichiometric quantities of the small EcoRI fragments, indicating that these five regions of repeated DNA could be important in gene expression and replication of DNA.[20] Recently, the DNA genome of *Spodoptera exempta* NPV (SeMNPV) was analyzed in detail by RENs and shown to be a variant of the AcMNPV genome.[9] This virus was also found to contain homologous DNA repeats inter-dispersed at five locations.

Repeated DNA sequences were also shown to develop in the genomes of baculoviruses after serial passage in tissue culture cells for many generations. AcMNPV was passaged for 30 generations and seven plaque purified isolates of the many polyhedra phenotype were found to have an altered EcoRI pattern. They were grouped into three distinct types, termed B5, B6, and C6, according to the additional EcoRI fragments.[12] The B5 isolate had acquired three more 1.97 kb fragments which comigrated with EcoRI-P. The B6 isolate had four extra 1.55 kb fragments. The C6 isolate had extra fragments of three distinct sizes; a 3.46 kb band present in 2 molar amount, a 1.55 kb band present in 4 molar amount, and a 1.47 kb band present in 3 molar amount. These additional EcoRI fragments, which were shown to be viral DNA, appeared to have originated from locations on the genome close to where the insertions were mapped.

The genome of CfMNPV was also shown to contain repeated DNA sequences.[3,58] A "ladder" of fragments was seen in autoradiograms after HindIII restricted CfMNPV DNA was probed with nick-translated EcoRI-E or HindIII-R. The bottom three steps of the ladder were the HindIII fragments O, PQ, and R, while the other steps of the ladder (five) were submolar in amount in that they could not be seen in ethidium bromide stained gels. All the ladder fragments, except HindIII-R, were located in the EcoRI-B part of the genome (Figure 2). The CfMNPV used in these experiments was originally plaque purified and then propagated in larvae for 2 to 3 generations. It appeared, therefore, that the submolar fragments represented newly generated virus populations; each was distinguished by an additional insertion in the EcoRI-B segments of the DNA. The fragment HindIII-R has been sequenced and revealed the presence of five repeats of 61 bp in the 5′ half of the fragment.[133] Two were repeated in tandem followed by the first 19 bp of the third repeat. The third repeat contained unique insertions of 35 bp and 85 bp. The fourth and fifth were repeated in tandem and had close homology to the first two repeats.

Many other baculoviruses were shown to have homologous repeats in their DNA. The nonoccluded baculovirus Hz-1V contained a rapidly reannealing fraction that represented

13% of the total genome.[44] A domain of approximately 11% of the genome of Hz-1V is composed of 69% G + C residues which might have contributed to the observed rapid reannealing. The DNA of OpMNPV was as shown to contain reiterated sequences by hybridization. However, their sizes have not yet been determined.[61]

The function of the repeated DNA sequences in baculoviruses is not yet understood. Several observed genomic alterations appear to occur within or near the location of repeated DNA. It was noted[20] that the DNA of the plaque morphology mutant FP-D described by Potter and Miller[83] was altered near the HR_5 sequence of AcMNPV. The serial passage mutants of AcMNPV reported by Burand and Summers[12] appeared to have been duplications within Hr_2 and Hr_3. An insertion present in many AcMNPV variants and in *Galleria mellonella* NPV (GmMNPV) was close to Hr_5. It appears, therefore that the repeated DNA play a role in the biology of the virus, albeit not a well-defined role.

D. Insertion of Host DNA Into the Viral Genome

There are several reports in the literature on the acquisition of host DNA sequences by the genomes of baculoviruses. This phenomenon is, of course, well documented in the case of viruses of vertebrates. An FP mutant of AcMNPV, generated after 25 passages in a *T. ni* cell line, was found to contain new HindIII fragments (a, b, and c) and had lost the fragment HindIII-K. The mutant was designated FP-D.[72] The fragment HindIII-c was submolar in amount and remained so after several plaque purifications. FP-D was in fact found to be a mixture of two viruses, FP-DL, which was identical to FP-D, and FP-DS. The latter had lost the fragments HindIII-a, -b, and -K, but HindIII-c became molar in amount. Physical mapping experiments showed that HindIII-a and -b constituted the fragment HindIII-K. The fragment HindIII-c was actually HindIII-K plus a 270 bp insertion. FP-DL contained a large insertion of 7.3 kb and direct terminal repeats of 270 bp. This structure, designated TED, is similar to *copia* transposable elements.[72] Indeed, differences in copy number and distribution of TED homologous sequences present in DNA isolated from *T. ni* cell line compared to DNA isolated from larvae suggested that it was a movable element.

Two closely related viruses, AcMNPV and GmMNPV, were also shown to acquire host DNA after successive passages in a cell line of *T. ni* (TN-368). After three passages, FP mutants were isolated and analyzed by REN. It was shown that most of the mutations had occurred by insertions of host DNA in the HindIII-I fragment of AcMNPV and in the fragment HindIII-J of GmMNPV.[32] All host DNA insertions were located within 50 bp between 35.0 and 37.7 map units of the AcMNPV genome. The size of the host DNA in these mutants was shown to vary from 0.8 kb to 2.8 kb. The cellular DNA in one mutant, GmFP1 from GmMNPV, appeared to hybridize to repeated cellular DNA. A second mutant, GmFP2, appeared to hybridize to moderately repetitive *T. ni* DNA, while GmFP3 hybridized to unique cellular sequences.[32]

Cells infected with the FP mutants did not synthesize a 25 kDalton protein normally present in MP infected cells. It has been previously reported that a 25 kDalton protein is coded for by DNA present in the HindIII-J of the AcMNPV genome.[102,123] Whether this 25 kDalton protein is the same as the missing one from FP infected cells remains to be seen. However, the insertion of cell DNA into the baculovirus genome appeared to be responsible for the FP phenotype and the lack of the 25 kDalton protein from infected cells may be a characteristic feature of these FP mutants.[32] It should also be noted that four other FP mutants of AcMNPV, generated by serial passage in cells, were analyzed for the presence of host DNA. None was detected indicating that FP phenotypes may arise by more than one type of mutation.[83] More recently, five locations in the genome of SeMNPV, a variant of AcMNPV, appeared to be associated with insertions of cellular DNA.[9]

E. Baculoviruses As Vectors

One of the most interesting aspects of research on baculoviruses has been their utilization as vectors for the expression of foreign genes. The unique features that make baculoviruses particularly attractive as recombinant DNA vectors were outlined by Miller.[73,74] Briefly, these features are (1) the virus contains a large (100 to 250 kb) double-stranded closed, circular supercoiled DNA genome, (2) the nucleocapsid is rod-shaped, thus having the capacity to carry large passenger DNA, (3) the existence of very strong promotors (e.g., genes of the polyhedrin) that control the synthesis of a product not essential for virus replication, and (4) expression of the polyhedrin gene is temporally controlled allowing the expression of passenger genes after the synthesis of infectious virus.

Three independent laboratories used either the baculoviruses AcMNPV or *Bombyx mori* NPV (BmMNPV) as a vector for foreign DNA. Basically, the experiments involved the synthesis of a plasmid construct containing the promotor region of the polyhedrin gene upstream from a passenger gene. Permissive cells were cotransfected with this construct and with baculovirus DNA. Recombinant progeny particles that did not contain the complete polyhedrin gene were selected and were tested for their ability to synthesize the foreign product.

In this manner the human β-interferon gene was cloned in AcMNPV.[101] Approximately 10 μg of interferon were secreted by 10^6 *Spodoptera frugiperda* cells. The authors suggested that the high level of interferon production was due partly to the influence of the strong polyhedrin gene promotor and to the lack of antiviral activity on the AcMNPV itself. In contrast, 100 units/mℓ of interferon reduce the yield of SV40 by approximately 80%. In another laboratory, alpha interferon was cloned in an expression vector from BmMNPV.[64] In this case the titer of interferon increased to a maximum of approximately 5×10^6 units/mℓ in infected cell medium 4 days pi. When silkworm larvae were infected with the expression vector the kinetics of release of α-interferon into the hemolymph was equivalent to that in infected *Bombyx mori* tissue culture cells, but the relative activity was about 4×10^7 units/mℓ (in comparison, the highest reported level of α-interferon was 4×10^5 units/mℓ). There appeared to be no difference in the interferon secreted from insect and mammalian cells. Immune blot analysis showed that labeled anti-alpha interferon antibodies reacted exclusively with the interferon band produced by insect cells. A very important finding was the observation that the β-interferon was glycosylated by an N-linked glycan, the signal polypeptide was removed and the protein was secreted efficiently into the medium in the same manner as in mammalian cells.[64,101]

In another series of experiments the N-terminal region of the polyhedrin gene of AcMNPV was fused to the β-galactosidase gene of *E. coli* then inserted into the viral genome by cotransfection experiment. In the presence of X-gal, blue recombinant plaques were picked and analyzed for the synthesis of β-galactosidase.[82] β-galactosidase activity was observed 18 hr pi and increased dramatically through 48 hr pi. At 48 hr pi the observed specific activity was 8000 nmol of *o*-nitrophenyl-β-D-galactopyranoside (ONPG) cleaved per minute per milligram of protein which represented an approximately 900-fold increase in β-galactosidase activity. When insect cells, infected with AcMNPV carrying the foreign gene, were pulse-labeled with ^{35}S-methionine and analyzed by SDS-PAGE, a new protein of ca. 120K, and some other proteins not seen in cells infected with wild-type virus, were observed. The 120 kDalton protein was the proper size for a polyhedrin and β-galactosidase fusion polypeptide. The antigenic nature of the 120 kDalton protein was investigated by immune precipitation with anti-polyhedrin antibody. It precipitated, indicating that it was indeed a polyhedrin fusion product. The use of a β-galactosidase gene in these experiments was helpful in picking recombinant particles since these particles were blue, thus saving time in screening.

Recently, the human c-*myc* oncogene and human interleuken-2- were inserted into the

AcMNPV genome. In the first case the human c-*myc* gene was introduced into a transfer vector containing the strong polyhedrin promotor. *Spodoptera frugiperda* cells were cotransfected with the construct and with AcMNPV DNA. Recombinant virus containing the polyhedrin promotor and the c-*myc* gene were isolated.[77] *Spodoptera frugiperda* cells, infected with the recombinant virus and 40 hr pi, were examined for the presence of c-*myc* protein in the nucleus by immunological methods. A predominant protein of approximately 60 to 64 kDaltons was detected in cells infected with the recombinant virus, but not in cells infected with AcMNPV. A comparison of the c-*myc* protein made in *Escherichia coli*, *Saccharomyces cervisae*, and in insect cells showed that the product from each cell type revealed differences in size and relative abundance. *E. coli* cells synthesized a major 60 kDalton c-*myc* protein and a minor 58 kDalton protein. The minor protein may have been a degradation product. *S. cerevisiae* cells produced a 62 kDalton major component and a 60 kDalton minor component. Insect cells synthesized a 61 kDalton and a 65 kDalton protein in equal amounts.[77] The authors suggested that the 60 kDalton c-*myc* protein in *E. coli* represented full length and a modified product, while the larger species may represent modified forms of the protein. Indeed, the product in insect cells was shown to be phosphorylated.

Similarly, human interleuken-2 cDNA was inserted into the polyhedrin gene and was efficiently expressed under the control of the strong promotor.[100] The N-terminal amino acid sequence of interleuken-2 derived from insect cells was identical to that of the natural interleuken-2. As was found with β-interferon from insect cells, protein-processing mechanisms within insect cells recognized the preinterleuken-2 signal polypeptide and cleaved it at the correct position. The biological activity of insect derived interleuken-2 was shown by its ability to stimulate the growth of an interleuken-2 dependent T-cell line.[100]

It is very clear that our understanding of the nature and the function of the viral genome has led to significant advances in using baculoviruses as expression vectors. It should also be possible to alter the genome in order to produce viruses with other desired properties. For example, the identification of the genes that control the virulence of the virus could eventually lead to the engineering of a virus several-fold more virulent than the naturally existing one. Kirschbaum[52] points to the fact that an intact polyhedrin gene is essential for the transmission of the virus in the field and, therefore, a gene lethal or disruptive to insects could be fused either to a high level constitutive promoter or promoters controlling the synthesis of viral products early in infection. The engineered virus could contain the lethal gene instead of a presumably nonessential viral gene or be spliced into it. Also, the host range within insect families can be broadened once the genes that control this phenomenon are known and characterized. It is obvious that the consequences of research in these areas could have profitable returns in terms of controlling forest and agricultural insect pests. It was indicated that the size and the rod-shaped nature of the virus would theoretically permit the incorporation of 100 kb of passenger DNA into the genome and still produce a viable virus.[74] This is an enormous amount of genetic information which could, in the future, make baculoviruses profitable pharmaceutically as well as in the formation of insecticides with desired properties.

VI. CONCLUDING REMARKS

Significant advances have been made in understanding the biology and molecular biology of baculoviruses, particularly the viral structural proteins and virus-specific proteins synthesized in infected cells and the structure and expression of the DNA. Detailed treatises on baculovirus gene expression will soon appear as reviews in the literature. These elegant studies led to the efficient utilization of baculoviruses as expression vectors of a number of eukaryotic and prokaryotic genes. It should be possible in the future to construct baculovirus insecticides with genes tailored to manage specific forest and agricultural pests. To that end the safety of baculoviruses to nontarget species is an added advantage in disseminating them in the environment.

ACKNOWLEDGMENTS

We would like to thank Professor John Spencer (Queen's University, Kingston) for providing unpublished data on the sequence of the fragment HindIII-R of CfMNPV DNA, Mrs. Karen Jamieson for editorial comments, and Miss Donna Weeks for typing the review.

REFERENCES

1. **Adams, J. R., Goodwin, R. H., and Wilcox, T. A.,** Electron microscope investigation on invasion and replication of insect baculoviruses *in vivo* and *in vitro, Biol. Cellulaire,* 28, 261, 1977.
2. **Adang, M. J. and Miller, L. K.,** Molecular cloning of DNA complementary to mRNA of the baculovirus *Autographa californica* nuclear polyhedrosis virus: location and gene products of RNA transcripts found late in infection, *J. Virol.,* 44, 782, 1982.
3. **Arif, B. M. and Doerfler, W.,** Identification and localization of reiterated sequences in the *Choristoneura fumiferana* MNPV genome, *EMBO J.,* 3, 525, 1984.
4. **Arif, B. M., Tjia, S. T., and Doerfler, W.,** DNA homologies between the genomes of *Choristoneura fumiferana* and *Autographa californica* nuclear polyhedrosis viruses, *Virus Res.,* 2, 85, 1985.
5. **Arif, B. M., Kuzio, J., Faulkner, P., and Doerfler, W.,** The genome of *Choristoneura fumiferana* nuclear polyhedrosis virus: molecular cloning and mapping of the EcoRI, BamHI, SmaI, XbaI and BglII restriction sites, *Virus Res.,* 1, 605, 1984.
6. **Bergold, G. H.,** The molecular structure of some insect virus inclusion bodies, *J. Ultrastruct. Res.,* 8, 360, 1962.
7. **Brown, D. A., Bud, H. M., and Kelly, D. C.,** Biophysical properties of the structural components of a granulosis virus isolated from the cabbage white butterfly *(Pieris brassicae), Virology,* 81, 317, 1977.
8. **Brown, S. E., Maruniak, J. E., and Knudson, D. L.,** Physical maps of SeMNPV baculovirus DNA: an AcMNPV genomic variant, *Virology,* 136, 235, 1984.
9. **Brown, S. E., Maruniak, J. E., and Knudson, D. L.,** Baculovirus (MNPV) genomic variants: characterization of *Spodoptera exempta* MNPV DNAs and comparison with other *Autographa californica* MNPV DNAs, *J. Gen. Virol.,* 66, 2431, 1985.
10. **Bud, H. M. and Kelly, D. C.,** An electron microscope study of partially lysed baculovirus nucleocapsids: the intranucleocapsid packaging of viral DNA, *J. Ultrastruct. Res.,* 73, 361, 1980.
11. **Burand, J. P., Stiles, B., and Wood, H. A.,** Structural and intracellular proteins of the nonoccluded baculovirus HZ-1, *J. Virol.,* 46, 137, 1983.
12. **Burand, J. P. and Summers, M. D.,** Alterations of *Autographa californica* nuclear polyhedrosis virus DNA upon serial passage in cell culture, *Virology,* 119, 223, 1982.
13. **Burand, J. P., Summers, M. D., and Smith, G. E.,** Transfection with baculovirus DNA, *Virology,* 101, 286, 1980.
14. **Burley, S. K., Miller, A., Harrap, K. A., and Kelly, D. C.,** Structure of the *Baculovirus* nucleocapsid, *Virology,* 120, 433, 1982.
15. **Carstens, E. B., Tjia, S. T., and Doerfler, W.,** Infection of *Spodoptera frugiperda* cells with *Autographa californica* nuclear polyhedrosis virus. I. Synthesis of intracellular proteins after virus infection, *Virology,* 99, 386, 1979.
16. **Carstens, E. B., Tjia, S. T., and Doerfler, W.,** Infectious DNA from *Autographa californica* nuclear polyhedrosis virus, *Virology,* 101, 311, 1980.
17. **Chakerian, R., Rohrmann, G. F., Nesson, M. H., Leisy, O. J., and Beaudreau, G. S.,** The nucleotide sequence of the *Pieris brassicae* granulosis virus granulin gene, *J. Gen. Virol.,* 66, 1263, 1985.
18. **Chung, K. L., Brown, M., and Faulkner, P.,** Studies on the morphogenesis of polyhedral inclusion bodies of a baculovirus *Autographa californica* NPV, *J. Gen. Virol.,* 46, 335, 1980.
19. **Cibulsky, R. J., Harper, J. D., and Gadauskas, R. T.,** Biochemical comparison of polyhedral protein from five nuclear polyhedrosis viruses infecting *Plusiine* larvae (Lepidoptera: Noctuidae), *J. Invertebr. Pathol.,* 29, 182, 1972.
20. **Cochran, M. A. and Faulkner, P.,** Location of homologous DNA sequences interdispersed at five regions in the baculovirus AcMNPV genome, *J. Virol.,* 45, 961, 1983.

21. **Cochran, M. A., Carstens, E. B., Eaton, B. T., and Faulkner, P.**, Molecular cloning and physical mapping of restriction endonuclease fragments of *Autographa californica* nuclear polyhedrosis virus, *J. Virol.*, 41, 940, 1982.

22. **Crawford, A. M. and Sheehan, C.**, Replication of *Oryctes* baculovirus in cell culture: viral morphogenesis, infectivity and protein synthesis, *J. Gen. Virol.*, 66, 529, 1985.

23. **Croizier, G. and Croizier, L.**, Evaluation du poids moléculaire de la protéine des corps d'inclusion de divers baculovirus d'insectes, *Arch. Virol.*, 55, 247, 1977.

24. **Crook, N. E., Spencer, R. A., Payne, C. C., and Leisy, D. J.**, Variation in *Cydia pomonella* granulosis virus isolates and physical maps of the DNA from three variants, *J. Gen. Virol.*, 66, 2423, 1985.

25. **Dobos, P. and Cochran, M. A.**, Protein synthesis in cells infected by *Autographa californica* nuclear polyhedrosis virus (Ac-NPV): the effect of cytosine arabinoside, *Virology*, 103, 446, 1980.

26. **Elliott, R. M. and Kelly, D. C.**, Frog virus 3 replication: induction and intracellular distribution of polypeptides in infected cells, *J. Virol.*, 33, 28, 1980.

27. **Entwistle, P. F. and Evans, H. F.**, Viral Control, in *Comprehensive Insect Physiology Biochemistry and Pharmacology*, Vol. 12, Kerkut, G. A. and Gilbert, L. I., Eds., Pergamon Press, Elmsford, N.Y., 1985, 347.

28. **Eppstein, D. A. and Thoma, J. A.**, Alkaline protease associated with the matrix protein of a virus infecting the cabbage looper, *Biochem. Biophys. Res. Commun.*, 62, 478, 1975.

29. **Eppstein, D. A. and Thoma, J. A.**, Characterization and serology of the matrix protein from a nuclear polyhedrosis virus of *Trichoplusia ni* before and after degradation by an endogenous proteinase, *Biochem. J.*, 167, 321, 1977.

30. **Faulkner, P.**, Baculovirus, in *Pathogenesis of Invertebrate Microbial Disease*, Davidson, E. W., Ed., Allenheld, Totowa, N. J., 1981, 3.

31. **Fraser, M. J. and Hink, W. F.**, The isolation and characterization of the MP and FP plaque variants of *Galleria mellonella* nuclear polyhedrosis virus, *Virology*, 117, 336, 1982.

32. **Fraser, M. J., Smith, G. E., and Summers, M. D.**, Acquisition of host cell DNA sequences by baculoviruses: relationships between host DNA insertions and FP mutants of *Autographa californica* and *Galleria mellonella* nuclear polyhedrosis viruses, *J. Virol.*, 47, 287, 1983.

33. **Granados, R. R.**, Infection and replication of insect pathogenic viruses in tissue culture, *Adv. Virus Res.*, 20, 189, 1976.

34. **Granados, R. R.**, Infectivity and mode of action of baculoviruses, *Biotechnol. Bioeng.*, 22, 1377, 1980.

35. **Granados, R. R.**, Entomopoxviruses infections in insects, in *Pathogenesis of Invertebrate Microbial Diseases*, Davidson, E. W., Ed., Allenheld, Totowa, N.J., 1981, 101.

36. **Granados, R. R. and Lawler, K. A.**, *In vivo* pathway of *Autographa californica* baculovirus invasion and infection, *Virology*, 108, 297, 1981.

37. **Harrap, K. A.**, The structure of nuclear polyhedrosis viruses. I. The inclusion body, *Virology*, 50, 114, 1972.

38. **Harrap, K. A. and Payne, C. C.**, The structural properties and identification of insect viruses, *Adv. Virus Res.*, 25, 273, 1979.

39. **Harrap, K. A., Payne, C. C., and Robertson, J. S.**, The properties of three baculoviruses from closely related hosts, *Virology*, 79, 14, 1977.

40. **Hink, W. F. and Vail, P. V.**, A plaque assay for titration of alfalfa looper nuclear polyhedrosis virus in the cabbage looper (TN-368) cell line, *J. Invertebr. Pathol.*, 22, 168, 1973.

41. **Hirumi, H., Hirumi, K., and McIntosh, A. H.**, Morphogenesis of a nuclear polyhedrosis virus of the alfalfa looper in a continuous cabbage looper cell line, *Ann. N.Y. Acad. Sci.*, 266, 392, 1975.

42. **Honess, R. W. and Roizman, B.**, Regulation of herpes virus macromolecular synthesis. I. Cascade regulation of synthesis of three groups of viral proteins, *J. Virol.*, 14, 8, 1974.

43. **Hooft van Iddekinge, B. J. L., Smith, G. E., and Summers, M. D.**, Nucleotide sequence of the polyhedrin gene of *Autographa californica* nuclear polyhedrosis virus, *Virology*, 131, 561, 1984.

44. **Huang, Y. S., Hedberg, M., and Kawanishi, C. Y.**, Characterization of the DNA of a nonoccluded baculovirus, Hz-1V, *J. Virol.*, 43, 175, 1982.

45. **Kao, S.-Y., Ressner, E., Kates, J., and Bauer, W. R.**, Purification and characterization of a superhelix binding protein from vaccinia virus, *Virology*, 111, 500, 1981.

46. **Kawanishi, C. Y., Egawa, K., and Summers, M. D.**, Solubilization of *Trichoplusia ni* granulosis virus proteinic crystal. II. Ultrastructure, *J. Invertebr. Pathol.*, 20, 95, 1972.

47. **Kelly, D. C.**, The DNA contained by nuclear polyhedrosis viruses isolated from four *Spodoptera* sp. (Lepidoptera, Noctuidae): genome size and homology assessed by DNA reassociation kinetics, *Virology*, 76, 468, 1977.

48. **Kelly, D. C.**, Baculovirus replication, *J. Gen. Virol.*, 63, 1, 1982.

49. **Kelly, D. C., Brown, D. A., Ayres, M. D., Allen, C. J., and Walker, I. O.**, Properties of the major nucleocapsid protein of *Heliothis zea* singly enveloped nuclear polyhedrosis virus, *J. Gen. Virol.*, 64, 399, 1983.

50. **Kelly, D. C. and Lescott, T.,** Baculovirus replication: protein synthesis in *Spodoptera frugiperda* cells infected with *Trichoplusia ni* nuclear polyhedrosis virus, *Microbiologica,* 4, 35, 1981.

51. **Kelly, D. C. and Lescott, T.,** Baculovirus replication: glycosylation of polypeptides synthesized in *Trichoplusia ni* nuclear polyhedrosis virus-infected cells and the effect of tunicamycin, *J. Gen. Virol.,* 64, 1915, 1983.

52. **Kirschbaum, J. B.,** Potential implication of genetic engineering and other biotechnologies to insect control, *Annu. Rev. Entomol.,* 30, 51, 1985.

53. **Knebel, D., Lübbert, H., and Doerfler, W.,** The promotor of the late P-10 gene in the insect nuclear polyhedrosis virus *Autographa californica* activated by viral gene products and sensitivity to DNA methylation, *EMBO J.,* 4, 1301, 1985.

54. **Knell, J. D. and Summers, M. D.,** Investigations of genetic heterogeneity in wild isolates of *Spodoptera frugiperda* nuclear polyhedrosis virus by restriction endonuclease analysis of plaque-purified variants, *Virology,* 112, 190, 1981.

55. **Knudson, D. L. and Harrap, K. A.,** Replication of a nuclear polyhedrosis virus in a continuous cell culture of *Spodoptera frugiperda*: microscopy study of the sequence of events of the virus infection, *J. Virol.,* 17, 254, 1976.

56. **Kozlov, E. A., Siderova, N. M., and Serebryani, S. B.,** Proteolytic cleavage of polyhedral protein during dissolution of inclusion bodies of the nuclear polyhedrosis viruses of *Bombyx mori* and *Galleria mellonella* under alkaline conditions, *J. Invertebr. Pathol.,* 25, 97, 1975.

57. **Kozlov, E. A., Levitina, T. L., Gusak, N. M., and Serebryani, S. B.,** Comparison of the amino acid sequence of inclusion body proteins of nuclear polyhedrosis viruses *Bombyx mori, Porthetria dispar* and *Galleria mellonella, Biiorg. Khim.,* 7, 1008, 1981.

58. **Kuzio, J. and Faulkner, P.,** Regions of repeated DNA in the genome of *Choristoneura fumiferana* nuclear polyhedrosis virus, *Virology,* 139, 185, 1984.

59. **Kuzio, J., Rohel, D. Z., Curry, C. J., Krebs, A., Carstens, E. B., and Faulkner, P.,** Nucleotide sequence of the p10 polypeptide gene of *Autographa californica* nuclear polyhedrosis virus, *Virology,* 139, 414, 1984.

60. **Lee, H. H. and Miller, L. K.,** Isolation of genotypic variants of *Autographa californica* nuclear polyhedrosis virus, *J. Virol.,* 27, 754, 1978.

61. **Leisy, D. J., Rohrmann, G. F., and Beaudreau, G. S.,** Conservation of genome organization in two multicapsid nuclear polyhedrosis viruses, *J. Virol.,* 52, 699, 1984.

62. **Lübbert, H., Kruczek, I., Tjia, S., and Doerfler, W.,** The cloned EcoRI fragments of *Autographa californica* nuclear polyhedrosis virus DNA, *Gene,* 16, 343, 1981.

63. **MacKinnon, E. A., Hendersen, J. F., Stoltz, D. B., and Faulkner, P.,** Morphogenesis of a nuclear polyhedrosis virus under conditions of prolonged passage *in vitro, J. Ultrastruct. Res.,* 9, 419, 1974.

64. **Maeda, S., Kawai, T., Obinata, M., Fujiwara, H., Horiuchi, T., Saeki, Y., Sato, Y., and Furusawa, M.,** Production of human α-interferon in silkworm using a baculovirus vector, *Nature,* 315, 592, 1985.

65. **Maruniak, J. E.,** Biochemical Characterization of Baculovirus Structural and Infected TN-368 Cell Polypeptides, Glycoproteins, and Phosphoproteins, Ph.D. thesis, University of Texas, Austin, Texas, 1979.

66. **Maruniak, J. E., Brown, S. E., and Knudson, D. L.,** Physical maps of SfMNPV baculovirus DNA and its genomic variants, *Virology,* 136, 221, 1984.

67. **Maruniak, J. E. and Summers, M. D.,** Comparative peptide mapping of baculovirus polyhedrins, *J. Invertebr. Pathol.,* 32, 196, 1978.

68. **Maruniak, J. E. and Summers, M. D.,** *Autographa californica* nuclear polyhedrosis virus-phosphoproteins and synthesis of intracellular proteins after virus infection, *Virology,* 109, 25, 1981.

69. **Maruniak, J. E., Summers, M. D., Falcon, L. A., and Smith, G. E.,** *Autographa californica* nuclear polyhedrosis virus structural proteins compared from *in vivo* and *in vitro* sources, *Intervirology,* 4, 82, 1979.

70. **Mathews, R. E. F.,** Classification and nomenclature of viruses, *Intervirology,* 12, 170, 1979.

71. **McIntosh, A. H. and Shamy, R.,** Biological studies of a baculovirus in a mammalian cell line, *Intervirology,* 13, 331, 1980.

72. **Miller, D. W. and Miller, L. K.,** A virus mutant with an insertion of a *copia*-like transposable element, *Nature,* 299, 562, 1982.

73. **Miller, L. K.,** A virus vector for genetic engineering, in *Genetic Engineering in the Plant Sciences,* Panopoulos, N. J., Ed., Praeger Scientific, New York, 1981, 203.

74. **Miller, L. K.,** Exploring the gene organization of baculoviruses, in *Methods in Virology,* Vol. 7, Maramorosck, K. and Koprowski, H., Eds., Academic Press, New York, 1984, 227.

75. **Miller, L. K. and Dawes, K. P.,** Physical map of the DNA genome of *Autographa californica* nuclear polyhedrosis virus, *J. Virol.,* 29, 1044, 1979.

76. **Minion, P. C., Coons, L. B., and Broome, J. B.,** Characterization of the polyhedral envelope of the nuclear polyhedrosis virus of *Heliothis virescens, J. Invertebr. Pathol.,* 34, 303, 1979.

77. **Miyamoto, C., Smith, G. E., Farrell-Towt, J., Chizzonite, R., Summers, M. D., and Ju, G.,** Production of human c-*myc* protein in insect cells infected with a baculovirus expression vector, *Mol. Cell. Biol.,* 5, 2869, 1985.

78. **Morgan, C., Moore, D. H., and Rose, H. M.,** The macromolecular paracrystalline lattice of insect virus polyhedral bodies demonstrated in ultrathin sections examined in electron microscope, *J. Biophys. Biochem. Cytol.,* 1, 187, 1955.

79. **O'Farrell, P. C., Goodman, H. M., and O'Farrell, P. H.,** High resolution two-dimensional electrophoresis of basic as well as acidic proteins, *Cell,* 12, 1133, 1977.

80. **O'Farrell, P. H.,** High resolution two-dimensional electrophoresis of proteins, *J. Biol. Chem.,* 250, 4007, 1975.

81. **Payne, C. C., Compson, D., and De Looze, S. M.,** Properties of the nucleocapsids of a virus isolated from *Oryctes rhinoceros, Virology,* 77, 269, 1977.

82. **Pennock, G. D., Shoemaker, C., and Miller, L.,** Strong and regulated expression of *Escherichia coli* β-galactosidase in infected cells with a baculovirus vector, *Mol. Cell. Biol.,* 4, 399, 1984.

83. **Potter, K. N. and Miller, L. K.,** Correlating genetic mutations of a baculovirus with the physical map of the DNA genome, in *Animal Virus Genetics,* Vol. 18, Fields, B. N., Jaenish, R., and Fox, C. F., Eds., Academic Press, New York, 1980, 71.

84. **Potter, K. N., Faulkner, P., and MacKinnon, E. A.,** Strain selection during serial passage of *Trichoplusia ni* nuclear polyhedrosis virus, *J. Virol.,* 18, 1040, 1976.

85. **Raghow, R. and Grace, T. D. C.,** Studies on a nuclear polyhedrosis virus in *Bombyx mori* cells *in vitro.* I. Multiplication kinetics and ultrastructural studies, *J. Ultrastruct. Res.,* 47, 384, 1974.

86. **Rohel, D. Z., Cochran, M. A., and Faulkner, P.,** Characterization of two abundant mRNAs of *Autographa californica* nuclear polyhedrosis virus present late in infection, *Virology,* 124, 357, 1983.

87. **Rohel, D. A. and Faulkner, P.,** Time course analysis and mapping of *Autographa californica* nuclear polyhedrosis virus transcripts, *J. Virol.,* 50, 739, 1984.

88. **Rohrmann, G. F., Bailey, T. J., Becker, R. R., and Beaudreau, G. S.,** Comparison of the structure of C- and N-polyhedrins from two occluded viruses pathogenic for *Orgyia pseudotsugata., J. Virol.,* 34, 360, 1980.

89. **Rohrmann, G. F., Bailey, T. J., Brimhall, B., Becker, R. R., and Beaudreau, G. S.,** Tryptic peptide analysis and NH₂-terminal amino acid sequences of polyhedrins of two baculoviruses from *Orgyia pseudotsugata, Proc. Natl. Acad. Sci. USA,* 76, 4976, 1979.

90. **Rohrmann, G. F., Martingnoni, M. E., and Beaudreau, G. S.,** DNA sequence homology between *Autographa californica* and *Orgyia pseudotsugata* nuclear polyhedrosis viruses, *J. Gen. Virol.,* 62, 137, 1982.

91. **Rohrmann, G. F., Pearson, M. N., Bailey, T. J., Becker, R. R., and Beaudreau, G. S.,** N-terminal polyhedrin sequences and occluded baculovirus evolution, *J. Mol. Evol.,* 17, 329, 1981.

92. **Russell, D. L. and Consigli, R. A.,** Glycosylation of purified enveloped nucleocapsids of the granulosis virus infecting *Plodia interpunctella* as determined by lectin binding, *Virus Res.,* 4, 83, 1985.

93. **Singh, S. P., Gudauskas, R. T., and Harper, J. D.,** High resolution two-dimensional gel electrophoresis of structural proteins of baculoviruses of *Autographa californica* and *Porthetria dispar, Virology,* 125, 370, 1983.

94. **Singh, S. P., Gudauskas, R. T., Harper, J. D., and Edwards, J.,** Two-dimensional gel electrophoresis of basic proteins of *Autographa californica* nuclear polyhedrosis virus, *J. Invertebr. Pathol.,* 45, 249, 1985.

95. **Smith, G. E. and Summers, M. D.,** Analysis of baculovirus genomes with restriction endonucleases, *Virology,* 89, 517, 1978.

96. **Smith, G. E. and Summers, M. D.,** Restriction maps of five *Autographa californica* MNPV variants, *Trichoplusia ni* MNPV, and *Galleria mellonella* MNPV DNAs with endonucleases SmaI, KpnI, BamHI, SacI, XhaI, and EcoRI, *J. Virol.,* 30, 828, 1979.

97. **Smith, G. E. and Summers, M. D.,** Restriction map of *Rachiplusia ou* and *Rachiplusia ou-Autographa californica* baculovirus recombinants, *J. Virol.,* 33, 311, 1980.

98. **Smith, G. E. and Summers, M. D.,** Application of a novel radioimmunoassay to identify baculovirus structural proteins that share interspecies antigenic determinants, *J. Virol.,* 39, 125, 1981.

99. **Smith, G. E. and Summers, M. D.,** DNA homology among Subgroup A, B, and C baculoviruses, *Virology,* 123, 393, 1982.

100. **Smith, G. E., Ju, G., Ericson, B. L., Moschera, J., Lahm, H-W., Chizzonite, R., and Summers, M. D.,** Modification and secretion of human interlukin 2 produced in insect cells by a baculovirus expression vector, *Proc. Nat. Acad. Sci. USA,* 82, 8404, 1985.

101. **Smith, G. E., Summers, M. D., and Fraser, M. J.,** Production of human beta interferon in insect cells infected with a baculovirus expression vector, *Mol. Cell. Biol.,* 3, 2156, 1983.

102. **Smith, G. E., Vlak, J. M., and Summers, M. D.,** *In vitro* translation of *Autographa californica* nuclear polyhedrosis virus early and late messenger RNA species, *J. Virol.,* 44, 199, 1982.

103. **Stiles, B. and Wood, H. A.**, A study of the glycoproteins of *Autographa californica* nuclear polyhedrosis virus (AcNPV), *Virology*, 131, 230, 1983.

104. **Stiles, B., Wood, H. A., and Hughes, P. R.**, Effect of tunicamycin on the infectivity of *Autographa californica* nuclear polyhedrosis virus, *J. Invertebr. Pathol.*, 41, 405, 1983.

105. **Stoltz, D. B., Krell, P., Summers, M. D., and Vinson, S. B.**, Polydnaviridae — A proposed family of insect viruses with segmented, double-stranded, circular DNA genomes, *Intervirology*, 21, 1, 1984.

106. **Stoltz, D. B., Pavan, C., and da Cunha, A. B.**, Nuclear polyhedrosis virus: a possible example of *de novo* intranuclear membrane morphogenesis, *J. Gen. Virol.*, 19, 145, 1973.

107. **Summers, M. D. and Smith, G. E.**, Comparative studies of baculovirus granulins and polyhedrins, *Intervirology*, 6, 168, 1975.

108. **Summers, M. D. and Smith, G. E.**, *Trichoplusia ni* granulosis virus granulin: a phenol-soluble, phosphorylated protein, *J. Virol.*, 16, 1108, 1975.

109. **Summers, M. D. and Smith, G. E.**, Baculovirus structural polypeptides, *Virology*, 84, 390, 1978.

110. **Summers, M. D., Smith, G. E., Knell, J. D., and Burand, J. P.**, Physical maps of *Autographa californica* and *Rachiplusia ou* nuclear polyhedrosis virus recombinants, *J. Virol.*, 34, 693, 1980.

111. **Summers, M. D. and Volkman, L. E.**, Comparison of biophysical and morphological properties of occluded and non-occluded baculoviruses from *in vivo* host systems, *J. Virol.*, 17, 962, 1976.

112. **Summers, M. D., Volkman, L. E., and Hsieh, C-H.**, Immunoperoxidase detection of baculovirus antigens in insect cells, *J. Gen. Virol.*, 40, 545, 1978.

113. **Tanada, Y. and Hess, R. T.**, Development of a nuclear polyhedrosis virus in midgut cells and penetration of the virus into the hemocoel of the armyworm, *J. Invertebr. Pathol.*, 23, 325, 1976.

114. **Tjia, S. T., Meyer, G. A., and Doerfler, W.**, *Autographa californica* nuclear polyhedrosis virus (AcNPV) DNA does not persist in mass cultures of mammalian cells, *Virology*, 125, 107, 1983.

115. **Tweeten, K. A., Bulla, L. A., Jr., and Consigli, R. A.**, Characterization of an alkaline protease associated with a granulosis virus of *Plodia interpunctella*, *J. Virol.*, 26, 702, 1978.

116. **Tweeten, K. A., Bulla, L. A., and Consigli, R. A.**, Characterization of an extremely basic protein derived from granulosis virus nucleocapsids, *J. Virol.*, 33, 866, 1980.

117. **Tweeten, K. A., Bulla, L. A., and Consigli, R. A.**, Structural polypeptides of the granulosis virus of *Plodia interpunctella*, *J. Virol.*, 33, 877, 1980.

118. **Tweeten, K. A., Bulla, L. A., and Consigli, R. A.**, Applied and molecular aspects of insect granulosis viruses, *Microbiol. Rev.*, 45, 397, 1981.

119. **Vlak, J. M.**, The proteins of nonoccluded *Autographa californica* nuclear polyhedrosis virus produced in an established cell line of *Spodoptera frugiperda*, *J. Invertebr. Pathol.*, 34, 110, 1979.

120. **Vlak, J. M.**, Mapping of BamHI and SmaI DNA restriction sites on the genome of the nuclear polyhedrosis virus of the alfalfa looper, *Autographa californica*, *J. Invertebr. Pathol.*, 36, 409, 1980.

121. **Vlak, J. M. and Rohrmann, G. F.**, The nature of polyhedrin, in *Viral Insecticides for Biological Control*, Maramorosch, K. and Sherman, K. E., Eds., Academic Press, New York, 1985, 489.

122. **Vlak, J. M. and Smith, G. E.**, Orientation of the genome of *Autographa californica* nuclear polyhedrosis virus: a proposal, *J. Virol.*, 41, 1118, 1982.

123. **Vlak, J. M., Smith, G. E., and Summers, M. D.**, Hybridization selection and *in vitro* translation of *Autographa californica* nuclear polyhedrosis virus mRNA, *J. Virol.*, 40, 762, 1981.

124. **Volkman, L. E.**, Occluded and budded *Autographa californica* nuclear polyhedrosis virus: immunological relatedness of structural proteins, *J. Virol.*, 46, 221, 1983.

125. **Volkman, L. E. and Summers, M. D.**, *Autographa californica* nuclear polyhedrosis virus: Comparative infectivity of the occluded, alkali-liberated and non-occluded forms, *J. Invertebr. Pathol.*, 30, 102, 1977.

126. **Volkman, L. E., Summers, M. D., and Hsieh, C. H.**, Occluded and nonoccluded nuclear polyhedrosis virus grown in *Trichoplusia ni*: comparative neutralization, comparative infectivity, and *in vitro* growth studies, *J. Virol.*, 19, 820, 1976.

127. **Wang, X., Lescott, T., De Clercq, E., and Kelly, D. C.**, Baculovirus replication: inhibition of *Trichoplusia ni* multiple nuclear polyhedrosis virus by [E]-5-(2-bromovinyl)-2- deoxyuridine, *J. Gen. Virol.*, 64, 1221, 1982.

128. **Wood, H. A.**, Isolation and replication of an occlusion body-deficient mutant of the *Autographa californica* nuclear polyhedrosis virus, *Virology*, 105, 338, 1980.

129. **Wood, H. A.**, *Autographa californica* nuclear polyhedrosis virus induced proteins in tissue culture, *Virology*, 102, 21, 1980.

130. **Yamafuji, K., Yoshibata, P., and Hirayama, K.**, Protease and deoxyribonuclease in viral polyhedral crystals, *Enzymologia*, 19, 53, 1958.

131. **Yamamoto, T. and Tanada, Y.**, Comparative analysis of the enveloped virions of two granulosis viruses of the armyworm, *Pseudaletia unipuncta*, *Virology*, 94, 71, 1979.

132. **Zummer, M. and Faulkner, P.**, Absence of protease in baculovirus polyhedral bodies propagated *in vitro*, *J. Invertebr. Pathol.*, 33, 383, 1979.

133. **Spencer, J.**, Queen's University, Kingston, personal communication.

Chapter 8

BIOLOGICAL CONTROL OF NEMATODES

Geeta Saxena and K. G. Mukerji

TABLE OF CONTENTS

I. INTRODUCTION

Nematodes occupy many ecological niches and include marine, fresh water, and soil forms, as well as predators and parasites of vertebrates, invertebrates, and plants. They are serious pests of many cultivated crops around the world, which is a matter of grave concern. They serve as food for many organisms including bacteria, fungi, insects, and others. The soil is an extremely complex environment where many modifying influences and varying parameters are operating. Before making attempts to suppress undesirable elements, it is essential to document their interrelationship. This is especially the case where specific microorganisms, considered to have control potential, are introduced into the soil. The fungal antagonists of nematodes consist of a great variety of organisms, which include nemato-phagous fungi, both predatory and endoparasitic, parasites of nematode eggs, and parasites of nematode cysts and fungi, which produce metabolites toxic to nematodes. It is remarkable that fungi belonging to widely divergent orders and families occur in each of these groups. Predaceous, parasitic, and biochemical relationships with nematodes have evolved among all major groups of soil fungi from lower to higher classes. Considering the long co-evolution of nematodes and fungi, which obviously occurred in the close confines of the soil habitat, it is evident that a great variety of associations have developed between the two groups. Observations made, and experiments conducted to date, indicate that certain bacteria such as *Bacillus penetrans* and certain fungi may be the most suitable organisms for man to manipulate in controlling plant-parasitic nematodes. There is sufficient information on these fungi to make at least preliminary interpretations of their value and potential role in biological control. There is not much information regarding the role of bacteria and viruses as biocontrol agents. Most of the papers published over the past 50 years on the control of nematodes have involved nematode-destroying fungi.

Interactions between fungal antagonists and nematodes in agricultural soils have been known for many years.[2,14,15,56,57] Nematophagous fungi are common and abundant in various soils, and they undoubtedly play a role in maintaining the balance of microbial life, although their total contribution to soil biology is not fully understood. Fungi capable of destroying nematodes consist of a large variety of organisms that fall into major groups based on their mode of attack, predatory and endoparasitic. The specialized trapping devices developed by predators include adhesive or nonadhesive structures (Figures 1 to 12). Adhesive organs of capture are hyphae, branches, knobs, and nets. Nonadhesive devices of capture are either constricting or nonconstricting rings. Endoparasites exist in the environment principally in the form of spores. Members of lower fungi produce evacuation tubes externally for the release of zoospores from the sporangia, and these flagellate spores either directly adhere or produce adhesive buds for attachment to a nematode cuticle. In Deuteromycetes, conidia are either ingested by the nematode, or they infect by adhering to the cuticle. A few fungi parasitize the eggs and cysts of nematodes rather than the larvae or adults. Colonization of reproductive structures of plant-parasitic nematodes, especially cysts and eggs by fungi, has been known for some years.

Nematodes belonging to the family Heteroderidae are vulnerable to attack by fungi at sedentary stages of their life cycles, either within root tissue or more frequently when exposed in the rhizosphere or in the soil. In the genera *Heterodera* and *Globodera*, females breaking through the root cortex become increasingly susceptible to attack. In the rhizosphere, where fungal growth is stimulated and enhanced by root exudates, the likelihood of a fungal invasion is increased. Cysts and eggs released into the soil, where they may remain for a lengthy period of time, are additionally vulnerable.

A renewed search for the mechanisms involved in natural disease control has been stimulated by a recognition of the widespread occurrence of suppressive soils where populations of plant-parasitic nematodes are reduced.[9,73] As the phenomena which bring about suppres-

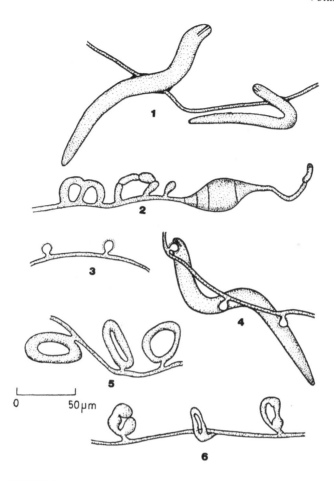

FIGURE 1 to 6.   Adhesive hyphae, branches, knobs, and constricting rings. (1) Two nematodes attached to the hypha of *Stylopage leiohypha* by means of adhesive at the point of contact. (2) Hyphae emerging from the conidium of *Monacrosporium cionopagum* after its germination, producing adhesive branches, which are joining to form loops. (3) Adhesive knobs of *M. parvicollis*. (4) Nematode captured by adhesive knobs of *M. parvicollis*. (5) Open rings produced on mycelial hypha of *Arthrobotrys dactyloides*. (6) Closed and open constricting rings of *A. brochopaga*.

sion are elucidated, the ability to enhance the biocontrol of nematodes by manipulation of the soil environment should become increasingly possible.

II. VIRUSES

There are few reports of nematode diseases caused by viruses. Loewenberg et al.[51] demonstrated that an agent which passed through a bacteriological filter caused sluggishness in *Meloidogyne incognita* larvae and prevented them from forming galls. The infective agent was thought to be a virus. However, virus particles were not observed in diseased nematodes.

Hollis[31] first observed the phenomenon of the congregation of nematodes in masses, a phenomenon that is known to occur in at least 13 species of plant-parasitic nematodes.[62] Virus-like particles were observed in the hypodermis, muscle layers, digestive and reproductive systems, and on the surface of the cuticle of *Tylenchorhynchus martini*, and were isolated from homogenates of swarming nematodes. However, there is lack of conclusive

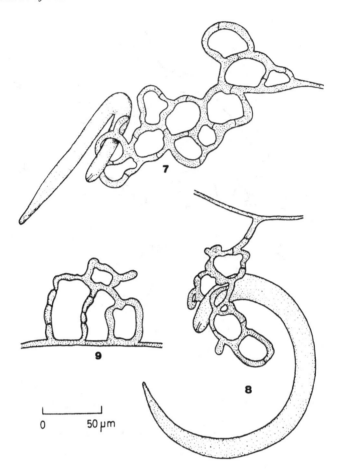

FIGURE 7 to 9.   Adhesive nets. (7) Freshly captured nematode in the adhesive network of *A. conoides*. (8) Portion of the hypha of *A. superba*, with nematode entangled in its adhesive network. (9) Hypha of *M. salinum* with adhesive network.

evidence that virus-like particles associated with swarming nematodes cause a disease that results in the slow death of nematodes. Because viruses attack many groups of lower invertebrates and microorganisms, viral diseases of nematodes certainly occur. Technical difficulties probably explain the reason for the existence of only a few documented cases. Inactive nematodes are not isolated by existing methods of nematode recovery. It is also difficult to differentiate nematodes which have a viral disease, from nematodes that are immobile due to other reasons.

III. BACTERIA

Several bacterial diseases of nematodes have been reported, and there are records of the presence of bacteria within the bodies of nematodes. *Bacillus penetrans* has been found infecting a large number of nematodes, and it has been the subject of intensive study in recent years as a promising biocontrol agent of nematodes. *B. penetrans* is the most specific obligate parasite of nematodes yet discovered, and it has a life cycle remarkably well adapted to the parasitism of certain plant-parasitic nematodes. Its spores attach to, and penetrate, the nematode cuticles, perhaps by enzymatic activity.[59] In root-knot nematodes, germination of the spore occurs about 8 days after an encumbered nematode enters a root and initiates

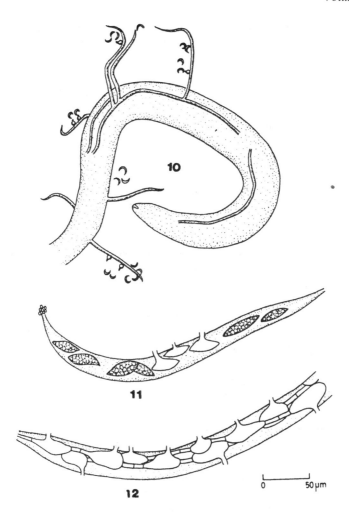

FIGURE 10 to 12.   Endoparasites. (10) Conidiophores of *Harposporium lilliputanum* emerging from an infected nematode, with phialides and conidia. (11) Nematode body filled with empty and mature sporaniga of *Myzocytium papillatum*, zoospores at its mouth. (12) Linear series of empty sporangia of *Caternaria anguillulae* connected with each other by isthmuses and evacuation tubes produced by them outside the nematode body.

feeding in the host.[76] The entire female body is invaded, and the nematode may be killed or reach maturity without producing eggs.[56] The spores are released upon decomposition of the dead nematode, and they remain free in the soil until contacted by another nematode. Spores may live in the soil for at least 6 months and they are not affected by drying or by normal field temperatures. This bacterium appears to be perfectly synchronized with the root-knot nematode's development and physiology.

B. penetrans has been found to exert effective control of root-knot nematodes. Most of the biological control research has been on one or more species of the root-knot nematodes, the *Meloidogyne* spp.[54]

In a greenhouse experiment, air-dried soil infested with spores of *B. penetrans* was planted with tomato seedlings, to which 10,000 root-knot nematode larvae were added. After 70 days plants in the air-dried spore infested soil were found healthier than plants in soil free of spores. Roots of these plants had considerably less nematode galling as compared with other treatments. Higher dry weight of above ground parts was also found.[55]

In an experiment, Mankau[54] used the following treatments:

(1)   Air-dried soil containing spores was placed in holes 3 in. wide and 6 in. deep;
(2)   Seedlings were grown in spore-infested soil and then transplanted into microplots; and
(3)   Encumbered with spores, 240,000 larvae were added to plots at a depth of 4 in.

Eleven months after cropping, 98% of the larvae emerging from the first treatment were heavily encumbered with spores. In the second treatment only 53% carried spores, and there were only a few spores/larva. In the third treatment, only 7% were lightly infected. The results suggested that small amounts of the spore-infested soils were a suitable means for effectively introducing the endoparasite into noninfested field plots.

Mankau[55] noted that spores of *B. penetrans* originating from the *Meloidogyne* sp. reduced the recovery of *Pratylenchus scribneri* from soil planted with beans. Observations by Mankau[56] indicate that populations of *B. penetrans* do not increase rapidly in field soil. Mankau and Prasad[61] determined the compatibility of seven nematicides with *B. penetrans*. Only nemagon was found to be slightly toxic, others had no noticeable effect.

Greenhouse experiments have shown that a population of *Meloidogyne incognita* can be destroyed by *B. penetrans* in a few generations of the host.[88] In Florida, when soil in outdoor plots were infested with *Meloidogyne incognita* and *B. penetrans* and planted with successive host crops, less damage was caused by the nematode to plants in plots containing *B. penetrans* than to plants in those plots without *B. penetrans*.[78]

The ability of *B. penetrans* to kill root-knot nematodes, as well as several other species of nematodes, indicates its use for the biological control of a major crop pest. The resistance of these long-lived spores to heat and desiccation, and their favorable compatibility with nematicides, are characteristics well suited for their eventual practical use in field soils.[56]

The major problem in using *B. penetrans* as a biocontrol agent is that we are unable to culture the bacterium in vitro. The use of a bacterium on a large scale will probably depend on the commercial, in vitro cultivation of the organism. Research into the survival and dispersal of *B. penetrans* in field soil is needed. Determination of the host ranges among nematode species and the distribution of biotypes are also important. The promising characteristics of *B. penetrans* justify the efforts needed for their use as a potential biocontrol agent.

IV. NEMATOPHAGOUS FUNGI

Nematophagous fungi are common in all types of soils. Most of them are not specific, and any particular fungus may attack a wide range of nematode species. Therefore, they have been considered as agents for biological control since the 1930s. Linford[46] and Linford et al.[50] started work in this area in Hawaii. They added chopped, green, pineapple tops to soil in pots infested with nematodes and estimated the activity of the predatory fungi as well as the nematode populations. The resultant effect was a marked increase in the number of free-living nematodes in the soil. This in turn stimulated the development of predatory species. They rapidly destroyed nematodes, and there was a reduction in the nematode population to below the original level in the pots. This gave a measure of disease control caused by the nematodes. In subsequent experiments, Linford and Yap[48,49] added five species of nematophagous fungi to soil in which pineapple plants were grown. These species showed only a slight beneficial effect. The history of attempts to use predatory fungi for the control of plant-parasitic nematodes has been summarized by Duddington.[17]

Dobbs and Hinson[13] established that spores of many fungi fail to germinate when placed in natural soils considered favorable for their germination. This is due to soil mycostasis. Duddington et al.[18-20] used *Arthrobotrys oligospora* and *Dactylaria thaumasia* in a number of early experiments and found them to be very sensitive to fungistasis, and few of their conidia germinated. Mycelia added to soil do not persist even in the presence of suitable food supplies.[9] Conidia produced on diseased nematodes of the *Nematoctonus* spp. were less sensitive to fungistasis and lysis when added to soil, than conidia added from pure cultures.[27] Such applications are impractical because of the difficulties in raising large numbers of nematodes. Endoparasites with sticky spores may be less sensitive to fungistasis and better agents for biological control, as they do not germinate until they touch a host. Fungal resting structures, particularly those which germinate after stimulation from a host, could also overcome fungistasis in soil.[77] In southern California, Mankau[52] studied soil fungistasis with respect to nematophagous fungi. He concluded that there was a water-diffusible substance in all of the soils tested which inhibited germination. He suggested that the ability of these fungi to grow through soil in a saprophytic phase may be limited. He found that in highly antagonistic soils conidia generally formed traps immediately on germination, after which they were entirely dependent on the nematodes for survival. It is possible that mycostasis inhibits normal spore germination. On the other hand, nematodes present produce morphogenic factors which stimulate immediate trap formation. Sensitivity to mycostasis by nematophagous fungi has been demonstrated by Cooke and Satchuthananthavale[10] in 15 predatory species. They found that all showed a sensitivity to mycostasis except one, and there was a variation in the degree of inhibition. Antagonistic factors in the soil may have inhibitory effects on germinated spores; therefore, successful germination of conidia in soil may not lead to successful establishment of the fungus. Their view is that attempts to establish these fungi in agricultural soil, for the purpose of the biological control of nematodes, are unlikely to meet with success.

Olthof and Estey[69] found enhanced predaceous activity of *Arthrobotrys oligospora* on a medium supplemented with dextrose and ammonium nitrate. Similarly, in sterilized soil amended with dextrose and ammonium nitrate, a reduction in root-knot of tomato caused by *Meloidogyne hapla* was observed. It can be assumed that in natural soils, green manure may provide sufficient nutrients to stimulate the activity of nematophagous fungi irrespective of the direct effects of the nematodes themselves. The best solution to control nematode pests is to manipulate microbial populations by green manuring, which will initiate a succession of events favorable for the growth of nematode-destroying fungi. When nutrients become available, as in green manuring, there is a major microbial shift and a subsequent triggering of spore germination.

Cooke[9] reviewed the ecological relationships of fungal predators and identified some of the problems associated with using them as biocontrol agents. Before predation can occur, mycelial growth and trap formation must occur. Both these processes require energy, which can be supplied by a readily available nutrient source. Consequently, the addition of organic amendments to soil is usually followed by a short period in which the activity that nematode-trapping fungi increases. He suggested that nematode capture may have been a means of escaping competition during phases of intense microbial activity in microhabitats with readily available organic substrates. He pointed out that although the addition of organic amendments to soil results in an increase in the population of free-living nematodes, predation is apparently not related to the density of the nematode population. In fact, increasing the amount of amendment may result in a reduction of the predaceous activity of a fungus. This occurs because of an intensification in the activity of the soil microbes which compete with the predaceous fungi for nutrients. Complex interactions between these fungi, other soil microbes, and decomposing organic substances determine the final level of each in the soil.

The nematophagous fungi have an extraordinary capacity to destroy nematodes in laboratory experiments. With increasing knowledge of these fungi in their natural environment, they should be regarded as potential biological control agents.[32] Progress in the use of parasites or predators against nematode pests is hindered by a lack of knowledge of their basic biology, and by the absence of methods to estimate nematode kills and to demonstrate their effects in soil. Most of the methods for the extraction of nematodes depend on their mobility, and since parasitized nematodes have reduced activity, many individuals fail to be recovered. However, elutriation techniques have increased the recovery of fungal parasites from eggs, and the number of eggs containing parasitic fungi can be reliably determined. Populations of nematode parasites in soil are difficult to estimate. Semiquantitative methods for the recovery of nematode-trapping fungi are available,[55] but obligate parasites that have yet to be cultured on artificial media can be recovered only from diseased nematodes. The effects of parasites and predators on nematode numbers may be measured by suppressing or stimulating the activity of such organisms already present in the soil. The addition of organic matter may stimulate the growth of predaceous fungi,[16] but it may also stimulate plant growth and offset nematode damage.[36] During decomposition, organic matter may produce nematicidal breakdown products.[60] There are numerous situations where nematodes cannot be effectively controlled, particularly in the tropics. Various aspects of the biological control of plant-parasitic nematodes have been reviewed a number of times by several workers.[8,16,22,40,54,56,74,92] Much of the research was descriptive, and few attempts have been made to quantify the effects of other organisms on nematode numbers.

Despite the generally equivocal performance of nematophagous fungi in applied biological control in the past, recently there have been a few examples with apparently successful results. *Paecilomyces lilacinus*, infecting eggs of *Meloidogyne incognita acrita*, was discovered in Peru, and when placed in soil containing nematode-infected plants, 70 to 90% of the eggs of *M. incognita acrita* and *Globodera pallida* were infected after 1 month.[23,33] In field trials against *M. incognita*, parasitizing potato, 86% of the egg masses were infected and 55% of the eggs in those masses were destroyed.[34] In a 3-year trial with *P. lilacinus*, added only in the first year, crop damage by *M. incognita* was reduced each year. The root-galling index on the original crop was severe, but was reduced to moderate amounts on successive crops.[35] *P. lilacinus* is a typical soil-borne fungus and seems to be relatively common and ubiquitous in the tropics and subtropics. Godoy et al.[30] indicated, by their greenhouse experiment, that this fungus was effective in reducing *Meloidogyne arenaria* infestations. This was the first report of its implication in the study of phytonematode disease in vivo in the U.S. It has also been shown to have some antagonistic activity, and thus competitive capacity, against bacteria and fungi.[5] This fungus has been found with high frequency in soils treated with fungicides such as benomyl, captan, and PCNB.[91] Morgan-Jones et al.[67] found, by ultrastructural studies, that *P. lilacinus* is capable of parasitizing and thus destroying both the eggs and larvae of *Meloidogyne*. These facts suggest that *P. lilacinus* is a strong competitor, capable of successfully establishing itself when artificially infested in natural soil. Its ability to compete effectively, colonize natural soil, and to exercise a degree of control against *Meloidogyne*, suggest that it may be used as a good biological control agent. It has a number of advantages as a biocontrol agent. Its optimum growth temperature is 26 to 30°C. It readily produces abundant inoculum in the form of long chains of conidia. It has been reported to degrade chitin.[68] This characteristic is of significance since the egg shell of *Meloidogyne incognita* has been reported to be made up mostly of protein and chitin.[3] It has the greatest potential of all the parasites of eggs and cysts of phytonematodes, for application as a biocontrol agent in subtropical and tropical agricultural soils.

Dactylella oviparasitica was found parasitizing eggs of the *Meloidogyne* spp. in a peach orchard.[79,82,83] Further research showed that the economic control of the *Meloidogyne* spp. by *D. oviparasitica* did not occur on grape and tomato plants.[85] They were controlled on peaches due to the restricted capacity of the *Meloidogyne* females to produce eggs on Lovell rootstock. The fungus can survive saprophytically on dead roots,[80] and can parasitize other nematode eggs, thus it may be useful in certain limited conditions. Stirling and Mankau[84] detected chitinase production in cultures of *D. oviparasitica*. They suggested a possible involvement of the enzyme activities in the process of parasitism.

Nematode-trapping fungi have been reported to kill the eggs and larvae of some nematodes. In one in vitro experiment, a strain of *Arthrobotrys oligospora* trapped up to 90% of the larvae of *Haemonchus contortus*. Trapping efficiency is both temperature and density dependent.[90]

Two fungal parasites, *Nematophthora gynophila* and *Verticillium chlamydosporium*, which attack females and eggs of the cereal cyst nematode *Heterodera avenae*, have reduced populations to nondamaging levels in a wide range of soils in Europe.[37-40,43] *V. chlamydosporium* is reported as an effective parasite of the maturing females and eggs of nematodes. It has been implicated, on a number of occasions, in the pathology of both cyst and root-knot nematodes. Willcox and Tribe[94] were the first to document the association between *V. chlamydosporium* and nematodes. Kerry[37] commonly found this fungus in the females and cysts of *Heterodera avenae* where continuous cereal cultivation was practiced. The fungus was found to be capable of killing all eggs within young cysts. It was isolated from eggs and from infected second stage larvae, suggesting invasion at a relatively early stage in embryonic development. In pot experiments, Kerry and Crump[42] added eggs to soils known to contain this fungus. Recovery of cysts from host plants after a period of time showed the fungus infecting eggs of *Globodera rostochiensis* and six species of *Heterodera*. Tribe[87] showed this fungus to be implicated in the pathology of cysts and eggs of *H. avenae* and *H. schachtii* in Belgium, Czechoslovakia, Italy, Netherlands, Poland, Sweden, and England. Stirling[81] recorded the presence of this fungus in the brown cysts of *H. avenae* in Australia. It had an adverse effect on *H. avenae* multiplication in fields and played a significant role in controlling cyst nematode populations in Europe.[41,44,45] Under intense cereal cultivation, *H. avenae* populations had been found to decline.[24] Tribe[86] categorized *V. chlamydosporium* as a major egg-pathogen and described some of its characteristics. High frequency of *V. chlamydosporium* was noted by him in German *H. schachtii* cysts, where beet monoculture had been practiced. Sufficient fungal inoculum was built up to exercise significant and detectable control in crop monoculture and as a result its attendant phytonematode populations. In crop-plant monocultures, not only does pathogenic fungal biomass build up, but selection pressures may favor a particular fungal species such as *V. chlamydosporium*. *V. chamydosporium* was found to be a parasite of *Meloidogyne arenaria* females and eggs in Alabama peanut monoculture-field soils.[30,65] In a greenhouse experiment, the fungus was found to have a considerable suppressive effect on nematodes. Gintis et al.[26] encountered this fungus in low frequency as chlamydospores in cysts of *Heterodera glycines* from Alabama soybean-field soils. It can certainly be considered a fungus which may be operative as a biological control agent of nematodes in certain soils. Although its capacity to penetrate females and eggs and to increase its biomass within these structures is well documented, the precise mode of action of its parasitism has not been elucidated. Likewise, little is known of its physiology. Two possible types of activity, operating separately or in combination, are thought to be involved in the parasitism of cysts and eggs by fungi. It seems likely that diffusable fungal toxins can deleteriously affect eggs and even lead to premature death of larvae. Exoenzymes, which might affect the permeability of egg shells, could operate in tandem with mycotoxins to predispose eggs to potential invasion. One or more of these factors might render the females and hatched larvae vulnerable to infection.[66] The success of parasitic fungi in controlling nematodes like *H. avenae* suggests that more research is warranted, and the biological agents could be of use in control programs in the future.

Tribe[86,87] reviewed the involvement of fungi in cyst-nematode pathology and the possible role of fungi as pathogens of these organisms. Much, however, remains to be learned about the fate and ecology of cysts remaining in soil over an extended period of time and of the factors controlling the levels of egg viability. The role of fungi in the progressive degradation of nematodes and their cysts in the soil is likewise an area where research is needed in order to fully understand fungal-nematode relationships and the possible significance of fungi indicating nematode population dynamics. Gintis et al.[25] demonstrated that at least some fungi are capable of invading freshly exposed, young cysts at about the same time. Godoy et al.[29] investigated the in vitro capacity of 13 fungal species, isolated from the cysts and eggs of *Heterodera glycines*, to parasitize eggs of that nematode as well as those of *Meloidogyne arenaria*. Of these, only *Verticillium lamellicola* and *V. leptobactrum* succeeded in parasitizing the eggs of these nematodes with high frequency. Gintis et al.[26] revealed an increase in colonization with the age of *H. glycines*, both in the percentage of cysts invaded and in the diversity of fungal species. Although some fungi occurred in more than one stage of the developing nematode, some change in the composition of the mycoflora was discernible from one developmental stage to another. Factors which might influence fungal associations with specific stages of the nematode cyst may include the natural habitat of the fungus and the fungal antagonisms. The very low incidence of colonization of the earliest stage of *H. glycines* females indicates that the root provides a degree of protection. Studies on the female parasites of *H. avenae* have shown that juveniles within roots are not parasitized unless nematodes develop semiendoparasitically. Both enzymatic activity and/or toxic metabolite biosynthesis are thought to be involved in cyst-nematode pathology. Fungi capable of invading cysts and rapidly penetrating and destroying egg contents in significant numbers probably invade eggs primarily through enzymatic degradation of chitinous egg-shell membranes.[3]

There are few reports published about fungi associated with root-knot nematode genus *Meloidogyne*. Initially Linford and Oliveira[47] concentrated on nematode-trapping fungi. Egg masses are clearly susceptible to invasion, particularly those exposed to the rhizosphere, since root exudates and nutrients released as a result of cortex disruption may enhance fungal growth activity and negate natural soil fungistasis. Dunn et al.[21] demonstrated by scanning electron microscopy the capacity of some isolates to coloinze nematode eggs in vitro. Culbreath et al.[11] described an agar disc method which permits the isolation of fungi associated with colonized eggs and an estimation of egg colonization in different soils. It is useful as a means of determining which nematode-colonizing fungi are present in a particular soil. A decrease in the number of species implicated in colonization in response to various amendments to the soil can also be detected. This method documents the presence of fungi capable of colonizing eggs and gives some estimate of what might be occurring in a given soil where natural egg masses are exposed to fungal antagonists.

Mankau and Das,[58] Miller et al.,[64] and Mian et al.[63] studied the effect of chitin amendments in soil on nematode populations. When chitin is added to soil it is depolymerized through the action of chitinases to yield N-acetylglucosamine. Ammonia is liberated from it or from one of its derivatives.[1] Ammonia is nematicidal to several species of ectoparasitic and endoparasitic nematodes;[71,89] however, it nematicidal activity in the field is short lived.[71] The action of chitin against nematodes has been attributed mainly to the release of ammonia[64] and to a lesser extent to stimulation of microorganisms resulting in an unfavorable environment for nematodes. Godoy et al.[30] studied the effect of chitin added to soil for the control of *Meloidogyne arenaria* and its effect on the soil microflora. They found that the addition of chitin to soil reduced nematode populations. Godoy and Rodriguez-Kabana[28] devised an enzymatic technique for the isolation of *Meloidogyne* females for biological control studies. This permits isolation of the females and prevents contamination by slow growing fungi. It is well established that the addition of some organic amendments to soil can result in a degree of nematode control.[63] This can be attributed to increased soil microbial activity, especially nematode-destroying fungi. Rodriguez-Kabana et al.[72] added crustacean chitin to soil at rates of 0.5 to 4.0% (w/w) which resulted in the control of *Heterodera glycines* in the roots of soybeans. Results of their study indicated that chitin amendments to soil can control the soybean-cyst nematode and other endoparasitic and ectoparasitic nematodes. These results confirm previous studies in which amending the soil with the polysaccharides was shown to control root-knot nematodes. Additions of 1% of chitin to soil are necessary to obtain consistent results. Culbreath et al.[12] added hemicellulosic waste material to soil at six levels (0 to 2.0% w/w), alone and in combination with two levels (0 and 2.0% w/w) of crustacean chitin to control *Meloidogyne arenaria*. Control of root-knot nematodes was maintained in all soils treated with chitin regardless of the amount of hemicellulose added. This indicates that the factors involved in the nematicidal activities of chitin were not affected by the addition of hemicellulose. Extended control observed with amendment in their experiment suggests that in addition to the nematicidal activity of ammonia, other microbial or chemical antagonism of nematodes is promoted by the addition of chitin.

Selection for fungal species capable of parasitizing nematode eggs by the addition of chitin amendments may be responsible for the control of the nematodes in such treated soils. The development of special soil mycoflora in response to the addition of chitin has been reported. An increase in number and activity of a specialized mycoflora, rather than an increase in general fungal activity, is likely responsible for the extended control of plant-parasitic nematodes observed in soil amended with chitin.

No organism has been used commercially to control nematode populations but there are instances of a few nematophagous fungi. By testing a group of these fungi for compatibility with *Agaricus bisporus*, Cayrol et al.[7] developed the use of a commercially prepared isolate of *Arthrobotrys* to protect commercial mushrooms from attack by the destructive mycophagous nematode *Ditylenchus myceliophagus*. They chose *A. robusta* var *antipolis*, a rapidly growing species, for development as a biological control agent (Royal 300). Trials with the fungus, seeded simultaneously with *A. bisporus* into mushroom compost, increased harvests of mushroom by over 28% and reduced *Ditylenchus* populations in the compost by about 40%. The results justified the commercial use of the fungus for nematode control in mushroom culture. Cayrol and Frankowski[6] developed another isolate of *Arthrobotrys* against root-knot nematodes in tomato. They grew *A. irregularis* commercially (Royal 350) and tried it in fields of several vegetable growers, at a rate of 140 g/m^2. This rate resulted in good protection of tomatos against *Meloidogyne* and satisfactory colonization of the soil by the fungus. They indicated that in the case of heavy populations of *Meloidogyne*, this fungus would be more appropriate as a secondary control measure after an initial treatment with a nematicide. Rhoades[70] compared the efficacy of commercially prepared *A. amerospora* and fenamiphos for controlling plant-parasitic nematodes. Fenamiphos significantly reduced nematode population whereas *A. amerospora* failed to do so. Barron[2] and Sayre[75] reported that in general, little or no beneficial results were obtained by adding only predatory fungi to the soil. If, however, such fungi are added with significant amounts of organic matter, beneficial effects may be obtained.

V. CONCLUSION

Nematodes, destroying and causing diseases of plants, are of great concern since they lead to the loss of several economically important crop plants, and thus indirectly harm man by affecting his food supply. Chemicals used to reduce nematode numbers in the soil are expensive and often required high application rates. In experimental plots, addition of formalin as a nematicide (38% formaldehyde) increased *Heterodera avenae* populations.[93] Formalin suppressed the activity of nematode parasitic fungi.[43] Ethylene dibromide (EDB) and dibromochloropropane (DBCP) did not reduce the activity of nematode-destroying fungi, whereas a dichloropropane-dichloropropene (DD) mixture did.[53] The production of DBCP is associated with health hazards. Nevertheless, economic as well as environmental back-lashes are leading towards ways of reducing nematicide use and the application of biological control methods on a large scale.

A few nematophagous fungi may be the most appropriate organisms for controlling plant-parasitic nematodes. Accumulated evidences based on the frequency of occurrence, microscopic observations of egg penetration, and known enzymatic capacity of the implicated organisms suggest that measure of biological control may already be in place in many cultivated soils. The details of its operation and the extent of its occurrence is likely to vary from soil to soil, but measures exist that may permit us to enhance the natural suppression of cyst-nematode populations.

Pests are rarely eradicated by natural enemies, but biological control usually results in a dynamic equilibrium between the nematode population and the parasite and predator. The equilibrium population density of the nematode pest should be below the economic threshold for damage to the crop. Much information is needed on their ecological requirements and on the limitations imposed by the soil environment and soil microorganisms.

It is important to have an idea of the characteristics of organisms used as biocontrol agents including competitive ability, growth requirements, enzymatic capacity, and capability of biosynthesis of toxic metabolites. The potential biocontrol organisms must be chosen with care based on a knowledge of host-parasite relationships, ecological requirements, mass-rearing, storing, shipping, and release considerations.[78] Joint efforts will be required to realize the commercialization of most, if not all, of the biological control organisms. It is expected that one day biological control organisms will assume a significant role in the continual struggle to protect plants from pests and disease-producing organisms.

REFERENCES

1. **Alexander, M.,** *Introduction to Soil Microbiology,* 2nd Ed., John Wiley & Sons, New York, 1977.
2. **Barron, G. L.,** The nematode-destroying fungi, in *Topics in Mycobiology,* Vol. 1, Canadian Biological Publications, Guelph, Ontario, Canada, 1977, 140.
3. **Bird, A. F. and McClure, M. A.,** The tylenchid (Nematoda) egg shell: structure, composition and permeability. *Parastiology,* 72, 19, 1976.
4. **Boosalis, M. G. and Mankau, R.,** Parasitism and predation of soil microorganisms, in *Ecology of Soil-borne Plant Pathogens,* Baker, K. F. and Snyder, W. C., Eds., University of California Press, Berkeley, 1965, 374.
5. **Brian, P. W. and Hemming, H. G.,** Production of antifungal and antibacterial substances by fungi, preliminary examination of 166 strains of Fungi Imperfecti, *J. Gen. Microbiol.,* 1, 158, 1947.
6. **Cayrol, J. C. and Frankowski, J. P.,** Une methode de lutte biologique contre les nematodes a galles des racines appartenant au genre *Meloidogyne, Rev. Hortic.,* 193, 15, 1979.
7. **Cayrol, J. C., Frankowski, J. P., Laniece, A., d'Hardemare, G., and Talon, J. P.,** Contre les nematodes en champignonniere. Mise au point d'une methode de lutte biologique a l'aide d'un hyphomycete predateur: *Arthrobotrys robusta* souche *antipolis* (Royal 300), *Rev. Hortic.,* 184, 23, 1978.
8. **Christie, J. R.,** Biological control-predaceous nematodes, in *Nematology, Fundamentals and Recent Advances with Emphasis on Plant Parasitic and Soil Forms,* Sasser, J. N. and Jenkins, W. R., Eds., University of North Carolina Press, Chapel Hill, 1960, 466.
9. **Cooke, R. C.,** Relationships between nematode-destroying fungi and soil-borne phytonematodes, *Phytopathology,* 58, 999, 1968.
10. **Cooke, R. C. and Satchuthananthavale, V. E.,** Sensitivity to mycostasis of nematode-trapping Hyphomycetes, *Trans. Br. Mycol. Soc.,* 51, 555, 1968.
11. **Culbreath, A. K., Rodriguez-Kabana, R., and Morgan-Jones, G.,** An agar disc method for isolation of fungi colonizing nematode eggs, *Nematropica,* 14, 145, 1984.
12. **Culbreath, A. K., Rodriguez-Kabana, R., and Morgan-Jones, G.,** The use of hemicellulosic waste matter for reduction of the phytotoxic effects of chitin and control of root-knot nematodes, *Nematropica,* 15, 49, 1985.

13. **Dobbs, C. G. and Hinson, W. H.,** A widespread fungistasis in soil, *Nature,* 172, 197, 1953.

14. **Drechsler, C.,** Predaceous fungi, *Biol. Rev.,* 16, 265, 1941.

15. **Duddington, C. L.,** The predaceous fungi: Zoopagales and Moniliales, *Biol. Rev.,* 31, 152, 1956.

16. **Duddington, C. L.,** Biological control-predaceous fungi, in *Nematology, Fundamentals and Recent Advances with Emphasis on Plant Parasitic and Soil Forms,* Sasser, J. N. and Jenkins, W. R., Eds., University of North Carolina Press, Chapel Hill, 1960, 461.

17. **Duddington, C. L.,** Predaceous fungi and the control of eelworms, in *View Points in Biology,* Vol. 1, Duddington, C. L. and Carthy, J. D., Eds., Butterworth, London, 1962.

18. **Duddington, C. L., Jones, F. G. W., and Moriarty, F.,** The effect of predaceous fungus and organic matter upon the soil population of beet eelworm *Heterodera schachtii* Schm., *Nematologica,* 1, 344, 1956.

19. **Duddington, C. L., Jones, F. G. W., and Williams, T. D.,** An experiment on the effect of a predaceous fungus upon the soil population of potato root eelworm, *Heterodera rostochiensis* Woll., *Nematologica,* 1, 341, 1956.

20. **Duddington, C. L., Everard, C. O. R., and Duthoit, C. M. G.,** Effect of green manuring and a predaceous fungus on cereal root eelworm on oats, *Plant Pathol.,* 10, 108, 1961.

21. **Dunn, M. T., Sayre, R. M., Carrell, A., and Wergin, W. P.,** Colonization of nematode eggs by *Paecilomyces lilacinus* (Thom) Samson, as observed with scanning electron microscopy, *Scanning Electron Microsc.,* 3, 1351, 1982.

22. **Esser, R. P. and Sobers, E. K.,** Natural enemies of nematodes, *Proc. Soil Crop. Sci. Soc. FL.,* 24, 326, 1964.

23. **Franco, J., Jatala, P., and Bocangel, M.,** Efficiency of *Paecilomyces lilacinus* as a biocontrol agent of *Globodera pallida J. Nematol.,* 13, 438, 1981.

24. **Gair, R., Mathias, P. L., and Harvey, P. N.,** Studies of cereal nematode populations and cereal yield under continuous or intensive culture, *Ann. Appl. Biol.,* 63, 503, 1969.

25. **Gintis, B. O., Morgan-Jones, G., and Rodriguez-Kabana, R.,** Mycoflora of young cysts of *Heterodera glycines* in North Carolina soils, *Nematropica,* 12, 295, 1982.

26. **Gintis, B. O., Morgan-Jones, G., and Rodriguez-Kabana, R.,** Fungi associated with several developmental stages of *Heterodera glycines* from an Alabama soybean field soil, *Nematropica,* 13, 181, 1983.

27. **Giuma, A. Y. and Cooke, R. C.,** Potential of *Nematoctonus* conidia for biological control of soil-borne phytonematodes, *Soil Biol. Biochem.,* 6, 217, 1974.

28. **Godoy, G. and Rodriguez-Kabana, R.,** An enzymatic technique for obtaining *Meloidogyne* females for biological control studies, *Nematropica,* 13, 75, 1983.

29. **Godoy, G., Rodriguez-Kabana, R., and Morgan-Jones, G.,** Parasitism of eggs of *Heterodera glycines* and *Meloidogyne arenaria* by fungi isolated from cysts of *H. glycines, Nematropica,* 12, 111, 1982.

30. **Godoy, G., Rodriguez-Kabana, R., and Morgan-Jones, G.,** Fungal parasites of *Meloidogyne arenaria* eggs in an Alabama soil. A mycological survey and greenhouse studies, *Nematropica,* 13, 201, 1983.

31. **Hollis, J. P.,** Induced swarming of a nematode as a means of isolation, *Nature,* 182, 956, 1958.

32. **Jansson, H. B.,** Nematophagous fungi and biological control of plant-parasitic nematodes, *Vaxtskyddsnotiser,* 44, 146, 1980.

33. **Jatala, P., Kaltenbach, R., and Bocangel, M.,** Biological control of *Meloidogyne incognita acrita* and *Globodera pallida* on potatoes, *J. Nematol.,* 11, 303, 1979.

34. **Jatala, P., Kaltenbach, R., Bocangel, M., Devaux, A. J., and Campos, R.,** Field application of *Paecilomyces lilacinus* for controlling *Meloidogyne incognita* on potatoes, *J. Nematol.,* 12, 226, 1980.

35. **Jatala, P., Salas, R., Kaltenbach, R., and Bocangel, M.,** Multiple application and long-term effect of *Paecilomyces lilacinus* in controlling *Meloidogyne incognita* under field conditions, *J. Nematol.,* 13, 445, 1981.

36. **Jones, F. G. W.,** Aspects of plant nematology in Great Britain, *Agric. Rev.,* 3, 8, 1958.

37. **Kerry, B. R.,** Fungi and the decrease of cereal cyst-nematode populations in cereal monoculture, *EPPO Bull.,* 5, 353, 1975.

38. **Kerry, B. R.,** Natural control of the cereal cyst nematode by parasitic fungi, *Agric. Res. Counc. Res. Rev.,* 4, 17, 1978.

39. **Kerry, B. R.,** Biocontrol: Fungal parasites of female cystnematodes, *J. Nematol.,* 12, 253, 1979.

40. **Kerry, B. R.,** Progress in the use of biological agents for control of nematodes, in *Biological Control in Crop Production, BARC Symposium No. 5,* Papavizas, G. C., Ed., Allenheld, Totowa, N. J., 1981, 79.

41. **Kerry, B. R.,** The decline of *Heterodera avenae* populations, *EPPO Bull.,* 12, 491, 1982.

42. **Kerry, B. R. and Crump, D. H.,** Observations on fungal parasites of females and eggs of the cereal cyst nematode, *Heterodera avenae* and other cyst nematodes, *Nematologica,* 23, 193, 1977.

43. **Kerry, B. R., Crump, D. H., and Mullen, L. A.,** Parasitic fungi, soil moisture and the multiplication of the cereal cyst nematode, *Heterodera avenae* and other cyst nematodes, *Nematologica,* 26, 57, 1980.

44. **Kerry, B. R., Crump, D. H., and Mullen, L. A.,** Studies of the cereal cyst-nematode, *Heterodera avenae* under continuous cereal, 1975—1978 II. Fungal parasitism of nematode females and eggs, *Ann. Appl. Biol.,* 100, 489, 1982.

45. **Kerry, B. R., Crump, D. H., and Mullen, L. A.,** Natural control of the cereal cyst-nematode *Heterodera avenae* Woll. by soil fungi at three sites, *Crop Protection*, 1, 99, 1982.
46. **Linford, M. B.,** Stimulated activity of natural enemies of nematodes, *Science*, 85, 123, 1937.
47. **Linford, M. B. and Oliveira, J. M.,** Potential agents of biological control of plant-parasitic nematodes, *Phytophathology*, 28, 14, 1938.
48. **Linford, M. B. and Yap, F.,** Root-knot injury restricted by a nematode-trapping fungus, *Phytopathology*, 28, 14, 1938.
49. **Linford, M. B. and Yap, F.,** Root-knot nematode injury restricted by a fungus, *Phytopathology*, 29, 596, 1938.
50. **Linford, M. B., Yap, F., and Oliveira, J. M.,** Reduction of soil populations of the root-knot nematode during decomposition of organic matter, *Soil Sci.*, 45, 127, 1938.
51. **Loewenberg, J. R., Sullivan, T., and Schuster, M. L.,** A virus disease of *Meloidogyne incognita*, the southern root-knot nematode, *Nature*, 184, 1896, 1959.
52. **Mankau, R.,** Soil fungistasis and nematophagous fungi, *Phytopathology*, 52, 611, 1962.
53. **Mankau, R.,** Effect of nematocides on nematode-trapping fungi associated with the citrus nematode, *Plant Dis. Rep.*, 52, 851, 1968.
54. **Mankau, R.,** Utilization of parasites and predators in nematode pest management ecology, *Proc. Tall Timbers Conf. Ecol. Anim. Control Habitat Manag.*, 4, 129, 1972.
55. **Mankau, R.,** *Bacillus penetrans* n. comb causing a virulent disease of plant-parasitic nematodes, *J. Invertebr. Pathol.*, 26, 333, 1975.
56. **Mankau, R.,** Biocontrol: Fungi as nematode control agents, *J. Nematol.*, 12, 244, 1980.
57. **Mankau, R.,** Biological control of nematode pests by natural enemies, *Annu. Rev. Phytopathol.*, 18, 415, 1980.
58. **Mankau, R. and Das, S.,** The influence of chitin amendments on *Meloidogyne incognita*, *J. Nematol.*, 1, 15, 1969.
59. **Mankau, R. and Imbriani, J. L.,** The life-cycle of an endoparasite in some Tylenchid nematodes, *Nematologica*, 21, 89, 1975.
60. **Mankau, R. and Minteer, R. J.,** Reduction of soil populations of the citrus nematode by the addition of organic materials, *Plant Dis. Rep.*, 46, 375, 1962.
61. **Mankau, R. and Prasad, N.,** Possibilities and problems in the use of a sporozoan endoparasite for biological control of plant-parasitic nematodes, *Nematropica*, 2, 7, 1972.
62. **McBride, J. M. and Hollis, J. P.,** Phenomenon of swarming in nematodes, *Nature*, 211, 545, 1966.
63. **Mian, I., Godoy, G., Shelby, R. A., Rodriguez-Kabana, R., and Morgan-Jones, G.,** Chitin amendments for control of nematodes in infested soil, *Nematropica*, 12, 71, 1982.
64. **Miller, P. M., Sands, D. C., and Rich, S.,** Effect of industrial residues, wood fibre wastes and chitin on plant-parasitic nematodes and some soil-borne diseases, *Plant Dis. Rep.*, 57, 438, 1973.
65. **Morgan-Jones, G., Godoy, G., and Rodriguez-Kabana, R.,** *Verticillium chlamydosporium*, fungal parasite of *Meloidogyne arenaria* females, *Nematropica*, 11, 115, 1981.
66. **Morgan-Jones, G., White, J. F., and Rodriguez-Kabana, R.,** Phytonematode pathology: Ultrastructural studies I. Parasitism of *Meloidogyne arenaria* eggs by *Verticillium chlamydosporium*, *Nematropica*, 13, 245, 1983.
67. **Morgan-Jones, G., White, J. E., and Rodriguez-Kabana, R.,** Phytonematode pathology: Ultrastructural studies II. Parasitism of *Meloidogyne arenaria* eggs and larvae by *Paecilomyces lilacinus*, *Nematropica*, 14, 57, 1984.
68. **Okafor, N.,** Decomposition of chitin by microorganisms isolated from a temperate and tropical soil, *Nova Hedwigia*, 13, 209, 1967.
69. **Olthof, Th. H. A. and Estey, R. H.,** Carbon and nitrogen levels of a medium in relation to growth and nematophagous activity of *Arthrobotrys oligospora* Fres., *Nature*, 209, 1158, 1966.
70. **Rhoades, H. L.,** Comparison of fenamiphos and *Arthrobotrys amerospora* for controlling plant nematodes in Central Florida, *Nematropica*, 15, 1, 1985.
71. **Rodriguez-Kabana, R., King, P. S., and Pope, M. H.,** Combinations of anhydrous ammonia and ethylene dibromide for control of nematodes parasitic of soybeans, *Nematropica*, 11, 27, 1981.
72. **Rodriguez-Kabana, R., Morgan-Jones, G., and Gintis, B. O.,** Effects of chitin amendments to soil on *Heterodera glycines*, microbial populations, and colonization of cysts by fungi, *Nematropica*, 14, 10, 1984.
73. **Rovira, A. D.,** Organisms and mechanisms involved in some soils suppressive to soil-borne plant dieseases, in *Suppressive Soils and Plant Diseases*, Schneider, R. W., Ed., American Phytopathological Society, St. Paul, Minn., 1982, 23.
74. **Sayre, R. M.,** Biotic influence in soil environment, in *Plant Parasitic Nematodes*, Zuckerman, B. M., Mai, W. F., and Rohde, R. A., Eds., Academic Press, New York, 1971, 235.
75. **Sayre, R. M.,** Promising organisms for biocontrol of nematodes, *Plant Dis.*, 64, 526, 1980.
76. **Sayre, R. M. and Wergin, W. P.,** Bacterial parasite of a plant nematode: Morphology and ultrastructure, *J. Bacteriol.*, 129, 1091, 1977.

77. **Schroth, M. N. and Hilderbrand, D. C.**, Influence of plant exudates on root infecting fungi, *Annu. Rev. Phytopathol.*, 2, 101, 1964.
78. **Smart, G. C., Jr.**, Biological control of and by nematodes, *J. Georgia Entomol. Soc.*, 19, 28, 1984.
79. **Stirling, G. R.**, The role of *Dactylella oviparasitica* and other antagonists in the biological control of root-knot nematodes (*Meloidogyne* spp.) on peach (*Prunus persica*), Ph.D. thesis 148, University of California, Riverside, 1978.
80. **Stirling, G. R.**, Techniques for detecting *Dactylella oviparasitica* and evaluating its significance in field soils, *J. Nematol.*, 11, 99, 1979.
81. **Stirling, G. R.**, Parasites and predators of cereal cyst nematode (*Heterodera avenae*) in South Australia, *Aust. Plant Pathol. Soc. 4th Nat. Conf. (Abstr.)* Perth, Australia, May 12 to 14, 1980.
82. **Stirling, G. R. and Mankau, R.**, *Dactylella oviparasitica*, a new fungal parasite of *Meloidgyne* eggs, *Mycologia*, 70, 774, 1978.
83. **Stirling, G. R. and Mankau, R.**, Parasitism of *Meloidogyne* eggs by a new fungal parasite, *J. Nematol.*, 10, 236, 1978.
84. **Stirling, G. R. and Mankau, R.**, Mode of parasitism of *Meloidogyne* and other nematode eggs by *Dactylella oviparasitica*, *J. Nematol.*, 11, 282, 1979.
85. **Stirling, G. R., McKenry, M. V., and Mankau, R.**, Biological control of root-knot nematodes (*Meloidogyne* spp.) on peach, *Phytopathology*, 69, 806, 1979.
86. **Tribe, H. T.**, Pathology of cyst-nematodes, *Biol. Rev.*, 52, 477, 1977.
87. **Tribe, H. T.**, Extent of disease in populations of *Heterodera* with special reference to *H. schachtii*, *Ann. Appl. Biol.*, 92, 61, 1979.
88. **U.S. Department of Agriculture**, Biological agents for pest control, status and prospectus, Stock No. 001-000-03756-1, Library of Congress Cat. Card No. 77-600057, Superintendant of documents, U.S. Government Printing Office, Washington, D.C., 1978.
89. **Vasalo, M. A.**, The nematicidal power of ammonia, *Nematologica*, 13, 155, 1967.
90. **Virat, M. and Peloille, M.**, *In vitro* predatory activity of a strain of *Arthrobotrys oligospora* towards an animal parasitic nematode, *Ann. Rech. Vet.*, 8, 51, 1977.
91. **Wainwright, M. and Pugh, G. J. F.**, The effects of fungicides on certain chemical and microbial properties of soils, *Soil Biol. Biochem.*, 6, 263, 1974.
92. **Webster, J. M.**, Nematodes and biological control, in *Economic Nematology*, Webster, J. M., Ed., Academic Press, London, 1972, 469.
93. **Williams, T. D.**, The effects of formalin, nabam, irrigation, and nitrogen on *Heterodera avenae* Woll., *Ophiobolus graminis* Sacc. and the growth of spring wheat, *Ann. Appl. Biol.*, 64, 325, 1969.
94. **Willcox, J. and Tribe, H. T.**, Fungal parasitism in cysts of *Heterodera*. I. Preliminary investigations., *Trans. Br. Mycol. Soc.*, 62, 585, 1974.

Chapter 9

BIOCONTROL OF WEEDS WITH MICROBES

S. Hasan

TABLE OF CONTENTS

I. INTRODUCTION

Biological control of weeds is the use of natural enemies to exert pressure on the population of their host plant to reduce it to below levels of economic importance. Where natural controls are absent, e.g., when a plant is introduced from a different geographical region without its natural enemies or they are ineffective, as is sometimes the case with endemic weeds, the weed can become so abundant that it becomes a serious problem. Biological control seeks to redress the situation, either by introducing natural enemies in the first case or by increasing their effectiveness in the second. The diversity of natural enemies may include vertebrates, arthropod herbivores, pathogens, etc. Amongst these, arthropods have received most attention and for the last several decades have been used to control weeds.[7,71] However, some workers in this field were aware of the important role of diseases in weed control.[41,73] Thus the restricted spread of the crofton weed (*Eupatorium adenophorum*) has been attributed to the combined effect of the gall-forming fly *Procecidochares utilis* and the leaf spot fungus *Cercospora eupatorii* which was accidentally introduced into Australia on the body of the fly.[33] A similar case is the control of prickly pear with the introduced moth *Cactoblastis cactorum*. The final task of the larvae was completed by pathogens, including bacteria and fungi, which infected the partially eaten plants.[32]

In recent years there has been an increasing interest in the deliberate use of all possible organisms, especially plant pathogens, as biocontrol agents. These pathogens are being used as the sole agents or as part of a complex with other biocontrol organisms.[29,39,51-54,83,85,103,107,128]

Among the various methods of weed control, including mechanical and chemical, biological control holds an important place because it is specific, economical, and in certain cases it produces long and durable weed control. In this it has some distinct advantages over chemical control. Biocontrol agents may be highly specific and therefore harmless to non-target crops, even when the latter are growing close to the target weed, they do not leave herbicide residues on plants, in the soil, or underground water, they have long-term control action and thus may not need to be reapplied, and they do not cause the environmental changes that can be produced by chemicals.

II. SUITABILITY OF A WEED FOR BIOLOGICAL CONTROL

There are several reasons why a weed may be selected for biological control — it may be difficult to control by conventional methods, may have developed resistance or tolerance to the chemical herbicides, or the use of herbicides may not be allowed, be undesirable, or be uneconomical. However, it is advisable to carefully analyze the weed problem to determine whether it is suitable for biological control.[44,95]

It is important to correctly identify the weed species, its biology, mode of dispersal, and mode of reproduction. In the case of the skeleton weed (*Chondrilla juncea*), an apomict, three clones were found to occur in Australia. Searches were therefore made in Europe for strains of the rust *Puccinia chondrillina* for each individual clone.[47,51,55] Whether a weed is a single plant species or several species together is important. European weedy blackberries in Australia are composed of several species of *Rubus*, also known as taxa of *Rubus fruticosus* aggregate.[3] It was therefore necessary to collect several strains of the rust *Phragmidium violaceum* with different degrees of pathogenicity on individual taxa and make a pool of spores from these strains, before testing against other members of Rosaceae.[13,14]

Information should be gathered on the origin of the weed — whether it is introduced or endemic. In the case of the former, the country of origin and other areas where it commonly occurs should be located.

An important point in selecting a weed for biological control is to evaluate not only the economic losses it causes, but also any of its beneficial aspects. If the weed presents some beneficial properties, such as flowers useful to the honey industry, a loss-benefit ratio will give an idea if one should proceed with biological control of the weed. This will obviously be affected by the extent of the weed's distribution and the economic importance of the industry it affects. In cases of conflict of interest the cost and feasibility of the other means of control is relevant. However, in some situations biological control may be the only solution to a weed problem.

III. STRATEGIES IN BIOLOGICAL CONTROL OF WEEDS WITH PLANT PATHOGENS

Traditionally, the "classical strategy" of the biological control of weeds has been practiced by introducing exotic arthropods from the original habitat of the weed or elsewhere to control perennial naturalized weeds commonly occupying wide areas of undisturbed land.[68] In principle, an introduction on one occasion is adequate to allow the build-up of a sufficiently large population to have a long term impact on the weed density. This stragety has more recently been applied to annual weeds[7] and extended mostly by the use of plant pathogens.[54,56,57] Thus plant pathogens have been searched for in the original habitat of the weed and introduced into the new environment, thereafter relying on their ability to self-perpetuate and reduce the host population below the economic threshold. Despite the traditional use of the classical method against weeds in large stable situations of low economic return and with single dominating weed species, the successful biological control of *Chondrilla juncea* with *Puccinia chondrillina* in wheat crops in south eastern Australia[28,61] has shown that the method is also effective in annual cropping systems.

Recently a new strategy of biological control of weeds has been developed using virulent strains of pathogens already occurring on the weed and enhancing their destructive action. Pathogens are mass produced and used against weeds in a manner similar to conventional herbicides. In this "bioherbicide or inundative strategy" both endemic and exotic pathogens can be used as "mycoherbicides," though to date mostly endemic fungi have been so employed.[54,57,103,107] According to Templeton[104] the reasons for the lack of effect of such endemic pathogens on indigenous weeds under normal conditions are multiple, including cultivation and the use of selective pesticides or management practices that disrupt or elminate stages of the pathogens' life cycle. In order to overcome these constraints the pathogens are collected from undisturbed areas and multiplied in large fermentation tanks. A suspension of spores and mycelium is applied so as to inundate the host plant with inoculum at the most propitious time for infection to eliminate the weed in its early stages of growth. This technique is applied mostly in the annual cropping systems but it has also proved to be efficient in orchards[90] and waterways.[26]

Among other strategies that can be cited is the integration of biological control into pest management systems ("integrated weed control"). It is a relatively new field compared to integrated control of arthropods. Attempts are being made to combine the destructive effect of pathogens with compatible herbicides and mechanical or cultural practices.[20,88,97]

The above mentioned strategies may be applied alone or combined together to solve a weed problem, the nature of the problem and the origin of the weed determining which of them is most suitable.

IV. WEED BIOLOGICAL CONTROL WITH EXOTIC PLANT PATHOGENS — THE CLASSICAL STRATEGY

As the success of control depends on the ability of the pathogen to self-perpetuate, disperse, and reduce the host population in the infested region, special care is taken to select agents which are strongly pathogenic to the plant and are ecoclimatically suitable for the target area. Pathogens should also be specific to the weed or at least to a small group of related plants as long as these include no economically or environmentally important species. A general account of the procedures to be considered in the use of plant pathogens for biological control of weeds has been published elsewhere.[54] However, some of the basic considerations in the selection of pathogens are outlined here.

A. Selection of Pathogens

All kinds of plant pathogens including viruses, bacteria, mycoplasmas, fungi, and nematodes, that can cause reduction in plant growth or reproduction, can be considered as biocontrol agents of weeds. So far fungi have received the most attention, probably because they are numerically more important, have a well defined taxonomy, in many cases their host specialization is well known, and at the same time more is known about the mechanism and stability of that specialization. Moreover, most viruses require insects for transmission, many bacteria are not specific, and relatively few nematodes with potential are known as yet. Additionally, in classical programs only those pathogens not already present in the target area or only represented by ineffective host specialized races would be candidates for selection.

B. Survey of Weed Diseases

A literature survey should enable one to gather all possible information on the occurrence of the weed and its diseases as well as their geographical distribution. It is necessary to determine whether certain organisms are already present and the extent of their damage to the target weed in the area where control is desired. This information helps in the selection of areas to be surveyed. During subsequent field surveys, collections are made of all pathogens causing damage to the weed, not only those already known, and if this can be done at different times of the season it will enable organisms attacking different plant stages to be collected. Priority should be given to pathogens attacking all stages of development, from very young seedlings to the mature plants. Those pathogens with high virulence and widespread occurrence are normally considered as suitable for further studies of their biological control effectiveness.

C. Field Assessment of Effectiveness

The biocontrol effectiveness of a pathogen can be assessed in its native habitat by ecological observations to evaluate the relationship between changes in the weed population and variation in attack by the pathogen, preferably throughout the weed's growing season. The destructive action of *Puccinia chondrillina* was demonstrated by this method.[64]

Climate, geography, and topography are often responsible for the different levels of the expression of virulence, disease intensity, infectivity, and dispersal of plant pathogens in the field. Therefore, assessments are best made in the native environment of a candidate pathogen, in climates analogous to those existing in the proposed area of introduction.[116] If the pathogen is found to be virulent and widespread in its native range, it can be expected to show similar properties after introduction. Such observations were made with skeleton weed rust, and it established and became widespread in Australia according to expectations.[29,64]

D. Host Specificity

Plant pathogens can only be used in the biological control of weeds if there is sufficient evidence that they will not be harmful to useful plants. Most of this evidence can be obtained by glasshouse tests for host specificity, but field observations can also provide supporting evidence. Further indications regarding host specificity of a pathogen can also be had from its taxonomic position. A literature survey will indicate if it is recorded only on the weed or on other plants, and also the level of specificity characteristic of its close relatives.

However, less information is usually available for pathogens infecting weeds than for those causing diseases in crop plants.

Detailed studies on the host range of a pathogen are carried out in the glasshouse by exposing it to a wide range of important plants under optimum conditions of inoculation and incubation.[47,55] However, it is clearly impossible to test each and every plant species, so methods have been developed to select representative species.[117,118] Amongst these, the most appropriate is the "centrifugal phylogenetic method".[118] Plants are selected on the basis of their phylogenetic relationship to the target weed. This method is most commonly used in current testing programs and includes, in order of relationship: (1) other forms of the target species, (2) other species of the same genus, (3) other members of the tribe, (4) other members of the subfamily, (5) other members of the family, and (6) other members of the order.[119] In some cases where a pathogen has never been recorded on plants outside a family, the list of test plants may include only members of that family, it not being necessary to include members of the order. Thus *Phragmidium violaceum* from blackberry was tested only on plants belonging to the Rosaceae family, as the genus *Phragmidium* has been recorded only on plants of this family.[14] On the other hand, if the biology of a pathogen is dependent on more than one plant species, as is the case in heteroecious rust fungi, the above testing scheme should be applied independently for each host.

When determining the host specificity of a fungal pathogen, a visual examination is made to assess the external symptoms produced by the pathogen. Detailed microscopic observations must also be made to determine if the plant is immune or resistant to the pathogen. In the latter case, there is some development of a mycelium within or outside the host tissue and this can vary according to the environment or physiological condition of the plant. Microscopic examination of cleared and stained leaves or stem sections showing spore germination, germ tube penetration, and further development of the mycelium leads to more accurate conclusions concerning the host specificity of a fungal pathogen.[12,55]

E. Major Successes in Classical Biological Control with Plant Pathogens
1. Skeleton Weed Rust

One of the foremost examples of the use of exotic plant pathogens in the biocontrol of weeds is the successful introduction into Australia of *Chondrilla* rust (*Puccinia chondrillina*) for the biological control of skeleton weed, *Chondrilla juncea*. The latter was a serious weed, mainly in the wheat growing areas of Australia (Figure 1).[78] The plant is native to Mediterranean Europe and the Middle East where it is not considered a weed largely due to the presence of natural enemies, in particular the rust fungus.[120] *C. juncea* is an apomict and in Australia it occurs in three morphological forms: narrow-, intermediate-, and broad-leaf.[69] The plant is also considered a weed in America[94] and Argentina.[129]

Among various insect parasites and pathogens discovered in Europe and the Middle East *P. chondrillina* was found to be the most damaging.[46,49,50,65,115] This macrocyclic and autoecious rust remains active throughout the year in the Mediterranean area, where it is common and infects all the aerial parts of *Chondrilla* plants. In autumn and spring the germinating seedlings and the rosettes regenerating from old rootstocks are heavily damaged by the brown eruptive uredinia (Figures 2A, B). There are several generations of uredinia, giving rise to abundant powdery urediniospores. These also appear on the flowering shoot in late spring and continue to multiply throughout summer, infecting flower buds as well (Figure 2C, D). Telia appear in late summer but these do not seem to play an important role in the life-cycle of the fungus in the Mediterranean climates.[46] Heavily infected plants are either killed or severely damaged, and seed production and root reserves are reduced. The rust is also responsible for the reduction of skeleton weed populations in the Mediterranean region in situations homologous to those in Australia.[64]

Several strains of *P. chondrillina* were collected during surveys in the Mediterranean

FIGURE 1.   A dense patch of natural infestation of skeleton weed in Australia (Photographed in 1966).

region, and among these a strain (IT32) from Vieste (S. Italy) was found to be strongly pathogenic to the narrow leaf form of the skeleton weed, the most common of the three Australian forms. The specificity of this strain of the rust was tested by inoculating all important crop plants grown in Australia as well as cultivated and wild composites closely related to *Chondrilla*. All these plants remained unattacked, thus demonstrating the rust's specificity.[47] The stability of the host specificity to *Chondrilla* was further demonstrated by inoculating closely related plants under varying climatic conditions.[63] These results satisfied Australian plant quarantine authorities, and the strain IT32 was multiplied under aseptic conditons and introduced into Australia.[51]

The rust was released in the field in Australia in 1971 where it established itself extremely successfully on the narrow leaf form of the skeleton weed. The spread of the disease was so rapid that in 12 generations it was found as far as 320 km from the original site, some 7 months after its release.[29] Within another few months the rust was widespread in skeleton weed infestations and thereafter steadily reduced the populations of this weed, providing

FIGURE 2.   Infection of *Puccinia chondrillina* on skeleton weed. Uredinia on seedlings (A), rosette leaf (B), stem (C), and flower buds (D).

savings of several million dollars to the farming community.[28] This action of the rust in reducing the weed density has continued, and in several areas skeleton weed is no longer a problem.[61] However, the introduced strain of the rust was virulent only on the narrow leaf form and the other two forms remained unaffected and gradually began to increase in some areas.[15] It was therefore necessary to search for strains of *P. chondrillina* virulent to these other forms.

Spore samples were collected from many sites in Europe and the Middle East and their pathogenicity studied on the Australian forms of skeleton weed. Several of the rust strains from eastern and central Mediterranean areas were found to be virulent on the intermediate form. One strain from Turkey (TU21) and another from southern Italy (IT36), after being shown to be host specific, were recently introduced into Australia and released in the field for biological control of the intermediate form.[55,58] A few strains infecting the broad leaf form were also isolated from spore samples collected in Greece and Turkey, and one of these, TU28 from central Turkey, was recently tested against several cultivated and wild composites and, like previously screened strains of *P. chondrillina,* was demonstrated to be specific to skeleton weed and safe for eventual introduction into Australia.[131] However, like most other strains collected in Turkey, though this strain infects both intermediate and broad leaf forms, it is less aggressive on the latter.[59] This strain is therefore being kept in reserve and searches to discover more virulent strains still continue, especially in Turkey, where so far most of the strains virulent to the broad leaf form have been found.

To increase the possibility of discovering suitable strains two major techniques have been adopted:

1.   As on previous occasions urediniospores of *P. chondrillina* are sampled from various sites from a large area and their pathogenicity is tested on the Australian *C. juncea.* The degree of virulence on the infected plants is noted and selected strains are compared among themselves and with those already under investigation. Also, seedlings of the Australian forms are exposed among local stands of skeleton weed in Turkey, and spores produced on these plants are similarly collected for further investigation. Electrophoretic studies of protein enzymes enable these various strains of the rust to be distinguished and classified.

2.   Skeleton weed seeds collected from wide areas, especially in Turkey, are also currently the subject of study by electrophoresis, such that their protein enzymes can be compared with those of the broad leaf form. This technique may permit the identification of localities where plants similar to the broad leaf form can be found and thus delimit the areas which are likely to yield suitable strains of the pathogen.

P. chondrillina has also been released in four states of the western U.S. for the control of skeleton weed. Seven virulent isolates of the rust were collected from different parts of the world and screened against skeleton weed from California, Oregon, Washington, and Idaho showing four different patterns of virulence on the host populations.[35] After release the rust was able to survive Washington's cold winter, it became widespread within a short time, and it produced all four spore stages of the rust fungus.[1,2]

2. Blackberry Rust

Several forms of European blackberry (*Rubus fruticosus* agg.), Rosaceae, have become serious weeds in many parts of the world which are ecoclimatically similar to their area of origin.[4] Their extensive growth forms impenetrable thickets in wastelands, forests, national parks, riverbanks, roadsides, and agricultural lands.

The macrocyclic and autoecious rust fungus *Phragmidium violaceum* was found to be highly damaging to blackberries in Europe.[83] It multiplies mainly by the urediniospores produced on leaves in yellowish-brown eruptive uredinia (Figure 3A) and overwinters on old leaves in the form of dark telia (Figure 3B). Under favorable conditions the rust attack causes severe defoliation of the brambles and an important reduction in the vigor of the canes. It also decreases both the vegetative propagation and the seed production of the weed.

P. violaceum was released in Chile in 1973 for the biological control of *Rubus constrictus* and *Rubus ulmifolius.* Within 2 years the rust became widespread and controlled the population of *R. constrictus.* The attack was less important on the *R. ulmifolius.* No plants other than these two blackberries were found to be infected by the pathogen.[85]

Encouraged by the success of this project, a program of research was undertaken to

FIGURE 3.   Characteristic symptoms of *Phragmidium violaceum* on blackberry
leaves, (A) uredinia, (B), telia.

investigate the possibility of using *P. violaceum* to control blackberries of European origin
which are serious weeds in the southeastern parts of Australia. Studies on the biology of
the rust, its effectiveness as a biological control agent, and its host specificity in relation to
various members of Rosaceae, including those native to Australia, have recently been completed.[12-14] An application will soon be made requesting permission to introduce strains of
P. violaceum into Australia.

3. Pamakani Leaf Spot

Hamakua pamakani (*Ageratina (Eupatorium) riparia*), a composite of Mexican origin,
was first observed in Hawaii in 1925, and since then it has spread rapidly and become a
serious problem. It has no forage value and as a vigorous competitor for space it eliminates
valuable forage species from range land and pastures, rendering them uneconomical.

The leaf-spot fungus *Cercosporella ageratina* was introduced into Hawaii from Jamaica
and successfully controlled pamakani weed.[109] The pathogen was demonstrated to be host
specific after inoculating 40 plant species belonging to 29 families.[130]

Several other pathogens are currently being studied with a view to using them against weeds outside their natural habitats. *Uromyces heliotropii* and the *Cercospora* spp. are under investigation in Europe for the biocontrol of the common heliotrope, *Heliotropium europaeum*, in Australia.[60,62] The rust fungi *Puccinia jaceae* and *P. carduorum*, commonly occuring in Europe, are being studied to control, respectively, diffuse knapweed, *Centaurea diffusa*, and musk thistle, *Carduus nutans*, in North America.[79,89,122]

V. DEVELOPMENT OF BIOHERBICIDES FOR WEED CONTROL

Even though bioherbicides are usually pathogens already present in the infested region, inundation of the environment with a massive dose of inoculum still requires them to be host specific, and to be effective they should still be sufficiently virulent. The general procedure for the development of bioherbicides is, therefore, broadly similar to that mentioned for the classical strategy, but a few points are worth mentioning.

A. The Pathogen

As for classical biological control, any plant pathogen can be considered as a bioherbicide. However, Deuteromycetes or Fungi Imperfecti have received major attention as fungi belonging to this group are mostly facultative and can be grown easily on artifical media. This facilitates the mass production of the infective stage of the fungus, either in the form of reproductive spores or as vegetative mycelium.

An extensive search is made for effective strains of the pathogen, preferably in the area where control is desired. These are further studied in the laboratory for their high level of pathogenicity against the host plant.

Evaluation of the field effectiveness of the pathogen is of less importance since the massive inoculum used compensates for any ecological shortcomings. However, to be economically worthwhile, a successful bioherbicide should be able to be used in a diversity of climatic conditions.

The methods used to determine host specificity are the same as for classical biocontrol. However, the fact that the pathogen is already present in the control area has allowed strains with a limited host range, but belonging to species with large host ranges, to be used as mycoherbicides.[103]

B. Notable Examples of Bioherbicides
1. Anthracnose of Northern Jointvetch

The foremost commercial production of a fungal pathogen as a mycoherbicide is the use of the endemic anthracnose disease *Colletotrichum gloeosporioides* f. sp. *aeschynomene* for the biocontrol of northern jointvetch, *Aeschynomene virginica*. *A. virginica* is a leguminous weed in rice fields, and to some extent in soybeans in Arkansas.[98] The weed is responsible for a reduction in the yield and quality of rice. *C. gloeosporioides* f. sp. *aeschynomene* occurs naturally on its leguminous host, but the infection is not very severe and the disease rarely kills the plant. Field observations indicated that the limited effect was due to the low level of inoculum in the early stages of growth, while the disease became epiphytotic in later stages, appreciably reducing the population level and/or seed production.[99]

The fungus grows rapidly on solid and liquid media and produces abundant spores. A water suspension of spores at a concentration of 2 to 15 million spores per milliliter, when sprayed on the weed at the rate of 94 to 374 ℓ/ha gave control in growth chambers, the greenhouse, and in the field. In field tests, control of northern jointvetch was achieved by spraying a spore suspension of 2 to 6 million spores per milliliter on the weed at dusk.[106] Young seedlings were rapidly killed, but control averaged 99% when plants were 66 cm tall. In other tests 99% weed kill was achieved by aerial spraying with a concentration of

2 million spores per milliliter. Death rate was progressive within 46 days, and plants escaping death were severely diseased and stunted. The potential of the fungus as a biocontrol agent was further confirmed in 1975 when 95 to 100% control of northern jointvetch plants was achieved in 17 rice fields totalling 240 ha.

The host specificity of *C. gloeosporioides* f. sp. *aeschynomene* was demonstrated by exposing 15 crop plants, including rice and soybean, and 15 common weeds to the spores.[30] All species remained unaffected except *Aeschynomene indicum*. In nature, this plant is not susceptible to the pathogen although it grows amidst the host weed in the same environment.[99] Also, inoculation tests on laboratory animals showed that the fungus was sufficiently safe for use as a biocontrol agent applied artificially in large quantities. *C. gloeosporioides* f. sp. *aeschynomene* has now been commercialized under the trade name "Collego"® as a mycoherbicide against northern jointvetch.[108]

2. Phytophthora Rot of Milkweed Vine

A strain of the stem and root rot fungus initially identified as *Phytophthora citrophthora* was isolated from dying milkweed vine, *Morrenia odorata*, a serious weed in the citrus groves of Florida.[16,17] It girdles the stem of infected plants near the soil surface, killing the seedlings and older plants. The pathogen was evaluated as a mycoherbicide. When applied to the infested fields as zoospores or chlamydospores, most of the milkweed vines were diseased and eventually killed. Detailed comparative studies later indicated that the *Phytophthora* isolate from the diseased milkweed vine was in fact *P. palmivora*.[36,37] The pathogen was found to be stable even after manipulation for several years and proved to be host specific when tested against 58 members of 12 families. A few plants other than milkweed were found to be infected, but only with a very high dose of the fungus.[90] A formulation of *P. palmivora* was registered in 1981 to control milkweed vine ("De Vine"®), and is marketed by Abbott Laboratories, U.S.[103,127]

Among other pathogens which have reached an advanced stage of development and are likely to be used as mycoherbicides are *Cercospora rodmanii* against water hyacinth, which will be described under Integrated Weed Control later in this chapter, and *Alternaria macrospora* to control spurred anoda, *Anoda cristata*. The latter is a serious weed in cotton fields in the U.S. *A. macrospora* has been evaluated as a mycoherbicide and found effective especially against young seedlings. A spore suspension sprayed in the infested fields destroyed most of the *A. cristata* plants. The fungus was also effective when a mixture of spores and mycelium was applied in the soil as a granular formulation with vermiculite.[112,114]

VI. INTEGRATED WEED CONTROL

Integration of biological control with more conventional methods of weed control practice has received little attention compared with the integrated control of arthropods. With both classical and bioherbicidal strategies, plant pathogens are subjected to a range of biotic and abiotic factors, including the impact of prevailing cultural practices and especially the use of other control measures. It is therefore necessary to take into account the whole agroecosystem for the management and control of weeds. Integrated biological control requires a detailed knowledge of the ecology of the weed, the environment where it is a problem, its effect on the agroecosystem, and the extent of control desired.[5] Information on the ecology of the control agents and their behavior will also be important. Such practices as the use of fertilizers, herbicides, and pruning may alter the action of biocontrol agents. Other authors have also discussed the possibility of integrating biological control into pest management systems.[96,97,100]

The use of plant pathogens has been combined with compatible herbicides and mechanical or cultural practices. The microbial herbicide *Phytophthora palmivora*, used for the control of *Morrenia odorata* in citrus groves of Florida, has been reported to be inhibited by some herbicide treatments such as diuron, glyphosate, and paraquat.[97] However, when a suspension of chlamydospores was applied 3 weeks after spraying glyphosate, these together were effective in controlling the weed.

Research is currently underway in the Netherlands on the possible use of a combination of the fungus *Cochliobolus lunatus (Curvularia lunata)* and a conventional herbicide against *Echinochloa cursgalli*. The pathogen causes leaf and stem necrosis, but the damage to the weed is limited even with the most virulent strains so far isolated. However, the destructive action of the pathogen is much higher when it is applied on *E. crus-galli* along with 10% of the normal dose of a chemical herbicide.[93]

Similarly, in the U.S., the possibility of using the indigenous rust fungus *Puccinia canaliculata* in conjunction with herbicides against yellow nutsedge (*Cyperus esculentus*), a serious weed in different parts of the world has been suggested.[88] The rust usually appears in late summer or autumn when the yellow nutsedge plants have produced seeds and tubers and their life cycle is complete. Efforts were made to use the pathogen in the early spring and to integrate it into the weed management system. In the presence of properly selected fungicides and appropriate timing of the treatment of the crop, the rust was able to develop epiphytotics in cultivation and inhibit nutsedge growth.

The combined effect of plant pathogens and insects for the biological control of weeds has also been suggested as a form of integrated control.[20,97] The biological control of the water hyacinth (*Eichhornia crassipes*) is a good example where the leaf spot fungus *Cercospora rodmanii*, shortly to be commercialized as a mycoherbicide, is likely to be used to control the weed in Florida, in combination with arthropod parasites and herbicides.[20,38] The pathogen was first discovered in 1973 and found to cause leaf spot, leaf necrosis (Figure 4A), and root rot in water hyacinth.[24] It was also found to be highly virulent and capable of causing declines in water hyacinth. *C. rodmanii* was demonstrated to be specific to the weed by testing 85 plant species belonging to 22 families.[26] In another test it was shown that the fungus would not be harmful to fish, a major concern for use on lakes.[25]

Further studies were made to assess the biocontrol potential of *C. rodmanii*. The fungus was grown on artificial media and a combination of spores and mycelia were applied on the edges of infested lakes, whereupon the disease readily spread over the entire mat of water hyacinth and killed individual plants, especially in low nutrient areas.[26,27] However, recently it has been shown that even in higher nutrient lakes, water hyacinth can be destroyed by *C. rodmanii* in combination with arthropod parasites and low rates of herbicides.[20] Among various arthropod parasites of water hyacinth,[6] three exotic and two native arthropods exert pressure on the weed in the south eastern U.S. These are the mottled weevil, *Neochetina eichhorniae*, the chevroned weevil, *N. bruchi*, the moth *Sameodes albiguttalis*, the mite *Orthogalumna terebrantis*, and the moth *Arzema densa*. The pathogen or the arthropods alone were not able to destroy the water hyacinth infestations (Figure 4B). However, when *C. rodmanii* was applied on the weed damaged mainly by the two weevils, 99% of the treated plants died after 6 months. Dead plants decayed and sank leaving behind open water (Figures 4C, D).

In a similar study the compatibility of two herbicides, 2,4-D amine and diquat, with *C. rodmanii*, was demonstrated.[20] Water hyacinth plants treated with 6.4 and 0.3% of the recommended dose, respectively, of 2,4-D and diquat, were highly susceptible to *C. rodmanii* attack.

Recent field studies have shown that the mycoherbicide *C. gloeosporioides* can be integrated with agricultural pesticides.[72] The pathogen was tank mixed with two herbicides and did not lose its activity or affect the herbicides. Similarly, tests were carried out in a glasshouse with several insecticides, and while some of these did reduce the activity of *C. gloeosporioides* against northern jointvetch, others did not inhibit the fungal action. It was

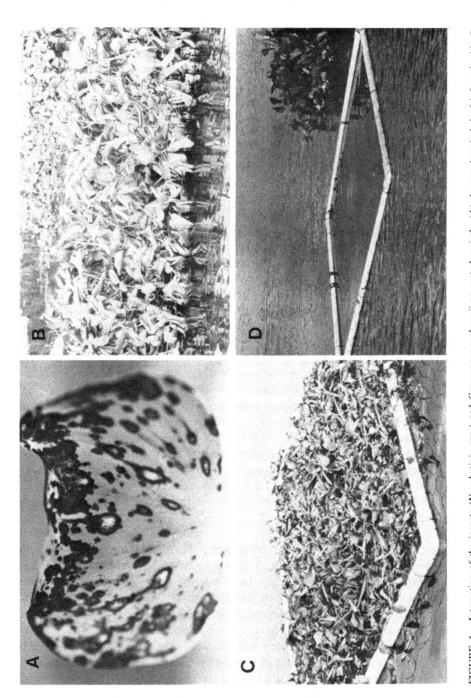

FIGURE 4. Interaction of the insects (*Neochetinia* spp.) and *Cercospora rodmanii* on waterhyacinth. (A) A waterhyacinth leaf infected with *C. rodmanii*. (B) Plants kept free of insects and the pathogen by spraying with insecticide and fungicide. (From Charudaltan, R., Phytopathogens, in *Water Hyacinth*, Gopal, B., Ed., Elsevier, Amsterdam, 1987, 471. With permission.) (C) Plants attacked by insects alone. (D) The results of attack by the insect and the pathogen. The plants are destroyed leaving an empty frame. Photos B to D taken at the end of the growing season.

also demonstrated in field tests that the pathogen could be used with pesticides when applied in sequence.[103]

VII. PRESENT STATUS OF WEED BIOCONTROL WITH PLANT PATHOGENS

Currently several projects are underway on the biocontrol of weeds with plant pathogens. Several authors have published lists which include plant pathogen projects.[19,39,71,93,104,105] An updated list is given in Table 1. As shown in the table, the projects included represent a diversity of plant species including algae, aquatic, semiaquatic, terestrial, annual, perennial, herbacious and woody plants, and poisonous and parasitic weeds. These weeds are found in different situations ranging from waterways, lakes, range land, and pastures, to cultivated crops and forests. The plant pathogens used include viruses, bacteria, fungi, and nematodes, but most of the projects utilized fungi. Programs of which the present status is unknown have not been included in the list. Usually only one reference has been cited, but this will permit the status of a project to be followed further if necessary.

VIII. CONCLUDING REMARKS

The use of plant pathogens in the biocontrol of weeds has gained increasing importance in recent years. However, it is still behind the use of the phytophagous arthropods which have been utilized for several decades. The foremost example where the use of insects was completely overshadowed by the effects of a pathogen is that of the successful introduction of *P. chondrillina* into Australia to control *C. juncea*. This success clearly demonstrated the feasibility of the classical strategy of biocontrol, so far used only by entomologists, and opened the way for the development of several other projects. The weed diseases have become a part of plant pathology which traditionally dealt with diseases of crop plants.

Classical biocontrol has the great advantage of providing continuing economic returns. After introduction of the pathogen into an extensive weed population, it is allowed to perpetuate and become widely established on its own. The experience with *Chondrilla* rust in Australia showed that only one release would have been sufficient for the epidemic spread and establishment of the pathogen within a short time throughout the skeleton weed infestation. Also, control of this latter weed in wheat crops in Australia has proved that contrary to the previously accepted generalization that the classical strategy was more suitable for introduced weeds dominating native vegetation in stable environments, it can also be effective in certain annual cropping systems.

This strategy is still most suitable for exotic weeds, but in some cases, endemic weeds which have become cosmopolitan, e.g., water hyacinth and field bind weed, may have well-adapted biocontrol agents found elsewhere in their distribution. Also, strains of endemic pathogens virulent to an endemic weed may be transferred from one part of the weed's distribution to another normally separated by natural barriers.

However, one important shortcoming of classical biocontrol is that once an agent is released in a country, it cannot be kept restricted to only those situations where control is desired. The weed is likely to be attacked wherever it is found within the area of introduction. This shortcoming can be overcome by the use of a pathogen such as a bioherbicide or mycoherbicide. This technique has been developed very recently, mainly for endemic weeds, though it can also be used for exotic species if suitable pathogens are available. As shown in Table 1, several projects are underway in this field, both with foliar sprays as well as soil applications. The two successful examples to date, *C. gloeosporioides* f. sp. *aeschynomene* and *P. palmivora* as mycoherbicides to control, respectively, northern jointvetch and milkweed vine, are being commercialized after several years of field testing. Other

Table 1
STATUS OF PROJECTS ON WEED CONTROL WITH PLANT PATHOGENS

Target weed	Pathogen	Strategy[a]	Country	Present[b] status	Ref.
Algae (Blue Green)	Viruses	bi	Various	1	18
Amaranthaceae					
Alternanthera philoxeroides	Alternaria alternantherae	bi	U.S.	1	67
Asclepiadaceae					
Morrenia odorata	Araujia mosaic virus	cl	U.S.	1	23
	Phytophthora palmivora	bi	U.S.	3	90, 103
Boraginaceae					
Heliotropium europaeum	Cercospora spp.	cl	Australia	1	53
	Uromyces heliotropii	cl	Australia	1	53
Cannabaceae					
Cannabis sativa	Fusarium oxysporum f. sp. cannabis	cl	U.S.	1	66
Chenopodiaceae					
Chenopodium album	Ascochyta caulina	bi	Netherlands	1	93
	Cercospora chenopodii	bi	Netherlands	1	93
Compositae					
Ageratina riparia (Eupatorium reparia)	Cercosporella ageratinae sp. n.	cl	Hawaii	3	109, 81
Ambrosia artemisiifolia	Albugo tragopogi	cl	Canada	2	45
Ambrosia trifida	Puccinia xanthii f. sp. Ambrosia-trifidae	cl	U.S.	1	8
Carduus nutans	Puccinia carduorum	cl	U.S.	1	89
Centaurea diffusa	Puccinia jaceae	cl	Canada	1	122
	Puccinia centaureae	cl	Canada	1	122
Centaurea repens	Paranguina picridis (Nematode)	cl	Canada	1	121
Chondrilla juncea	Puccinia chondrillina	cl	Australia	3	29, 51, 64
	Erysiphe cichoracearum	cl	Australia	1	52
	Leveillula taurica f. sp. chondrillae	cl	Australia	1	48

Table 1 (continued)
STATUS OF PROJECTS ON WEED CONTROL WITH PLANT PATHOGENS

Target weed	Pathogen	Strategy[a]	Country	Present status[b]	Ref.
Cirsium arvense	*Puccinia punctiformis*	bi	Canada	1	123
	Verticullium dahliae	cl	Denmark	1	75
	Phomopsis cirsii	bi	Denmark	1	75
	Septoria cirsii	bi	Denmark	1	75
Senecio jacobaea	*Puccinia expansa* (*P. glomerata*)	cl	Australia	1	61
Xanthium spp.	*Puccinia xanthii*	cl	Australia	1	53
	Colletotrichum xanthii	bi	Australia	1	82
Convolvulaceae					
Convolvulus arvensis	*Erysiphe convolvuli*	cl	U.S.	1	91
Cucurbitaceae					
Cucurbita texana	*Fusarium solani*	bi	U.S.	1	9
Cyperaceae					
Cyperus esculentus	*Puccinia canaliculata*	bi	U.S.	1	88
Ebenaceae					
Diospyros virginiana	*Cephalosporium diospyri*	bi	U.S.	1	126
Euphorbiaceae					
Euphorbia cyprissias	*Melampsora euphorbiae*	cl	U.S.	1	111
	Uromyces scutellatus	bi	Switzerland	1	31
		cl	U.S.	1	31
Fagaceae					
Quercus spp.	*Ceratocystis fagacearum*	bi	U.S.	1	40
Graminae					
Echinochloa crus-galli	*Curvularia lunata*	bi	Netherlands	1	92
Hydrocharitaceae					
Hydrilla verticillata	*Fusarium roseum* "Culmorum"	bi	U.S.	1	22
Hypolepidaceae					
Pteridium aquilinum	*Ascochyta pteridium*	bi	U.S.	1	124

					Morris[c]
Leguminosae					
Acacia spp.	Uromycladium tepperianum	cl	South Africa	1	
Aeschynomene virginica	Colletotrichum gloeosporioides f. sp. aeschynomene	bi	U.S.	3	107
Albizzia julibrissin	Fusarium oxysporum f. sp. perniciosum	bi	Hawaii	2	101
Cassia obtusifolia	Alternaria cassiea	bi	U.S.	1	113
Cassia surrattensis	Cephalosporium sp.	bi	Hawaii	2	110
Galega officinalis	Uromyces galegae	cl	Chile	1	84
Liliaceae					
Asphodelus fistulosus	Puccinia barbeyi	cl	Australia	1	53
Malvaceae					
Abutillon theophrasti	Fusarium lateritium	bi	U.S.	1	10
Anoda cristata	Alternaria macrospora	bi	U.S.	1	114
Sida spinosa	Colletotrichum malvarum	bi	U.S.	1	74
	Fusarium lateritium	bi	U.S.	1	10
Onagraceae					
Jussiaea decurrens	Colletotrichum gloeosporioides f. sp. jussiaeae	bi	U.S.	2	11
Orobanchaceae					
Orbanche spp.	Fusarium oxysporum var. orthoceras	bi	U.S.S.R.	2	43
Oxalidaceae					
Oxalis sp.	Puccinia oxalidis	bi	France	1	34
Polygonaceae					
Emex spp.	Cercospora tripoliana	cl	Australia	1	53
	Peronospora rumicis	cl	Australia	1	53
Rumex crispus	Uromyces rumicis	cl	U.S.	1	70
Rumex spp.	Ovularia obliqua	bi	Switzerland	1	102
Pontederiaceae					
Echhornia crassipes	Acremonium zonatum	bi	U.S.	1	77
	Cercospora rodmanii	bi	U.S.	2	26
	Myrothecium roridum	bi	Indonesia	1	76
	Uredo echhorniae	cl	U.S.	1	21
Proteaceae					
Hakea sericea	Colletotrichum gloeosporioides	bi	South Africa	1	80

Table 1 (continued)
STATUS OF PROJECTS ON WEED CONTROL WITH PLANT PATHOGENS

Target weed	Pathogen	Strategy[a]	Country	Present[b] status	Ref.
Rosaceae					
Prunus serotina	*Chondrostereum purpureum*	bi	Netherlands	1	92
Rubus spp.	*Phragmidium violaceum*	cl	Australia	1	57
		cl	Chile	3	85
Solanaceae					
Solanum elaeagnifolium	*Nothanguina phyllobia* (Nematode)	bi	U.S.	2	86
Viscaceae					
Arceuthobium spp.	*Collectotrichum gloeosporioides*	bi	U.S.	1	87, 125
	Nectria fuckeliana var. *macrospora*	bi	U.S.	1	42

[a]   The biological control strategies: classical (cl); bioherbicidal (bi).

[b]   Evaluation studies: (1) limited field trials, (2) introduction for classical biocontrol, (3) commercial production of bioherbicide.

[c]   M. J. Morris, Weed Research Subdivision, Plant Protection Research Institute, Private Bag X5017, Stellenbosch 7600, S. Africa (unpublished).

pathogens will have to undergo similarly extensive studies to reach an equivalent level of development. Since the pathogen has to be applied more than once, at particular intervals, it becomes much more of a commercial operation. However, if the weed is restricted to a small area or to only a few crops, commercialization is less attractive to industry.

Compared with chemical herbicides, bioherbicides present one great attraction, that of not being harmful to the environment, while at the same time, still seeming to be effective against annual weeds in annual crops where rapid control of the weed is desired.

Integration of mycoherbicides or other weed pathogens with effective weed and pest management systems including crop rotation, soil and water management practices, cultivation practices, chemical herbicides, etc., seems to have a great potential for solving weed problems, and it is a field which remains to be explored. However, present efforts to use mycoherbicides with compatible chemical herbicides should improve the prospects of weed control compared with using only one of them alone, and at the same time should ensure a quality environment.

ACKNOWLEDGMENTS

I thank R. Charudattan (Gainesville University, Fla.) and E. Bruzzese (Keith Turnbull Research Institute, Frankston, Australia) for contributing photographic illustrations, respectively, on water hyacinth and blackberry. I am indebted to J. M. Cullen of CSIRO's Division of Entomology for critical review of the manuscript.

REFERENCES

1. **Adams, E. B. and Line, R. F.,** Biology of *Puccinia chondrillina* in Washington, *Phytopathology,* 74, 742, 1984.
2. **Adams, E. B. and Line, R. F.,** Epidemiology and host morphology in the parasitism of rush skeletonweed by *Puccinia chondrillina, Phytopathology,* 74, 745, 1984.
3. **Amor, R. L. and Miles, B. A.,** Taxonomy and distribution of *Rubus fruticosus* L. agg. (Rosaceae) naturalised in Victoria, *Muelleria,* 3, 37, 1974.
4. **Amor, R. L. and Richardson, R. G.,** The biology of Australian weeds. II. *Rubus fruticosus* L. agg., *J. Aust. Inst. Agric. Sci.,* 46, 87, 1980.
5. **Andres, L. A.,** Integrating weed biological control agents into a pest-management program, *Weed Sci.,* 30, 25, 1982.
6. **Andres, L. A. and Bennet, F. D.,** Biological control of aquatic weeds, *Annu. Rev. Entomol.,* 20, 31, 1975.
7. **Andres, L. A., Davis, C. J., Harris, P., and Wapshere, A. J.,** Biological control of weeds, in *Theory and Practice of Biological Control,* Huffaker, C. B. and Messenger, P. S., Eds., Academic Press, New York, 1976, 481.
8. **Batra, S. W. T.,** *Puccinia xanthii* forma specialis *Ambrosia trifidae.* A microcyclic rust for the biological control of giant ragweed, *Ambrosia trifida* (Compositae), *Mycopathologia,* 73, 61, 1981.
9. **Boyette, C. D. and Templeton, G. F.,** Evaluation of a *Fusarium solani* strain for biological control of texas gourd, *Phytopathology,* 71, 862, 1981.
10. **Boyette, C. D. and Walker, H. L.,** Factors influencing biocontrol of velvet leaf (*Abutilon theophrasti*) and prickly sida (*Sida spinosa*) with *Fusarium lateritium, Weed Sci.,* 33, 209, 1985.
11. **Boyette, C. D., Templeton, G. E., and Smith, R. J., Jr.,** Control of winged waterprimrose (*Jussiaea decurrens*) and northern jointvetch (*Aeschynomene virginica*) with fungal pathogens, *Weed Sci.,* 27, 497, 1979.
12. **Bruzzese, E. and Hasan, S.,** A whole leaf clearing and staining technique for specificity studies of rust fungi, *Plant Pathol.,* 32, 335, 1983.
13. **Bruzzese, E. and Hasan, S.,** The collection and selection in Europe of isolates of *Phragmidium violaceum* (Uredinales) pathogenic to species of European blackberry naturalised in Australia, *Ann. Appl. Biol.,* 108, 527, 1986.

14. **Bruzzese, E. and Hasan, S.,** Host specificity of the rust *Phragmidium violaceum* a potential biological control agent of European blackberry, *Ann. Appl. Biol.,* 108, 585, 1986.

15. **Burdon, J. J., Marshall, D. R., and Groves, R. H.,** Aspects of weed biology important to biological control, *Proc. 5th Int. Symp. Biol. Control Weeds,* Brisbane, Australia, 1980, 21.

16. **Burnett, H. C., Tucker, D. P. H., Patterson, M. E., and Ridings, W. H.,** Biological control of milkweed vine with a race of *Phytophthora citrophthora, Proc. Fl. State Hortic. Soc.,* 86, 111, 1973.

17. **Burnett, H. C., Tucker, D. P. H., and Ridings, W. H.,** *Phytophthora* root and stem rot of milkweed vine, *Plant Dis. Rep.,* 58, 355, 1974.

18. **Cannon, R.,** Field and ecological studies on blue-green algal viruses, *Proc. Symp. Water, Qual. Manage. Biol. Control.,* Gainsville, Fla., 1975, 112.

19. **Charudattan, R.,** *Biological Control Projects in Plant Pathology, A Directory,* Institute of Food and Agricultural Science, Gainesville, Fla., 1978, 69.

20. **Charudattan, R.,** Integrated control of waterhyacinth (*Eichhornia crassipes*) with a pathogen, insects and herbicides, *Weed Sci.,* 34(Suppl. 1), 26, 1986.

21. **Charudattan, R. and Conway, K. E.,** Comparison of *Uredo eichhorniae,* the waterhyacinth rust, and *Uromyces pontederiae, Mycologia,* 67, 653, 1975.

22. **Charudattan, R. and McKinney, D. E.,** A Dutch isolate of *Fusarium roseum* Culmorum may control *Hydrilla verticillata* in Florida, *Proc. Eur. Weed Res. Soc. 5th Int. Symp. Aquatic Weeds,* Amsterdam, 1978, 219.

23. **Charudattan, R., Zettler, F. W., Cordo, H. A., and Christie, R. G.,** Partial characterisation of a potyvirus infecting the milkweed vine, *Morrenia odorata, Phytopathology,* 70, 909, 1980.

24. **Conway, K. E.,** Evaluation of *Cercospora rodmanii* as a biological control of waterhyacinths, *Phytopathology,* 66, 914, 1976.

25. **Conway, K. E. and Cullen, R. E.,** The effect of *Cercospora rodmanii,* biological control for waterhyacinth, on the fish, *Gambusia affinis, Mycopathologia,* 66, 113, 1978.

26. **Conway, K. E. and Freeman, T. E.,** The potential of *Cercospora rodmanii* as a biological control of waterhyacinths, *Proc. 4th Int. Symp. Biol. Control Weeds,* Gainesville, Fla., 1976, 207.

27. **Conway, K. E., Freeman, T. E., and Charudattan, R.,** Development of *Cercospora rodmanii* as a biological control for *Eichhornia crassipes, Proc. Eur. Weed Res. Soc. 5th Int. Symp. Aquatic Weeds,* Amsterdam, 1978, 225.

28. **Cullen, J. M.,** Evaluating the success of the program for the biological control of *Chondrilla juncea* L., *Proc. 4th Int. Symp. Biol. Control Weeds,* Gainesville, Fla., 1976, 117.

29. **Cullen, J. M., Kable, P. F., and Catt, M.,** Epidemic spread of a rust imported for biological control, *Nature,* 244, 462, 1973.

30. **Daniel, J. T., Templeton, G. E., Smith, R. J., Jr., and Fox, W. T.,** Biological control of northern jointvetch in rice with endemic fungal disease, *Weed Sci.,* 21, 303, 1973.

31. **Défago, G., Kern, H., and Sedlar, L.,** Potential control of weedy spurges by the rust *Uromyces scutellatus* s.l., *Weed Sci.,* 33, 857, 1985.

32. **Dodd, A. P.,** The biological control of prickly pear in Australia, *Monogr. Biol.,* 8, 565, 1959.

33. **Dodd, A. P.,** Biological control of *Eupatorium adenophorum* in Queensland, *Aust. J. Sci.,* 23, 356, 1961.

34. **Durrieu, G.,** *Puccinia oxalidis,* a help in the control of *Oxalis, Proc. 4th Int. Symp. Biol. Control Weeds,* Gainesville, Fla., 1976, 241.

35. **Emge, R. G., Stanley Melching, J. S., and Kingsolver, C. H.,** Epidemiology of *Puccinia chondrillina,* a rust pathogen for the biological control of rush skeleton weed in the United States, *Phytopathology,* 71, 839, 1981.

36. **Feichtenberger, E., Zentmyer, G. A., and Menge, J. A.,** Identity of *Phytophthora* isolates from milkweed vine, *(Morrenia odorata), Phytopathology,* 71, 215, 1981.

37. **Feichtenberger, E., Zentmyer, G. A., and Menge, J. A.,** Identity of *Phytophthora* isolated from milkweed vine, *Phytopathology,* 74, 50, 1984.

38. **Freeman, T. E. and Charudattan, R.,** *Cercospora rodmanii* Conway, a biocontrol agent for water hyacinth, Tech. Bull. No. 842, Agriculture Experimental Station, Institute of Food and Agricultural Science, Gainesville, Fla., 1984, 18.

39. **Freeman, T. E., Charudattan, R., and Conway, K. E.,** Status of the use of plant pathogens in the biological control of weeds, *Proc. 4th Int. Symp. Biol. Control Weeds,* Gainesville, Fla., 1976, 201.

40. **French, I. W. and Schroeder, D. B.,** The oak wilt fungus, *Ceratocystis fegacearum* (Bretz) Hunt. as a selective silvicide, *Forest Sci.,* 15, 198, 1969.

41. **Fullaway, D. T.,** Biological control of cactus in Hawaii, *J. Econ. Entomol.,* 47, 696, 1954.

42. **Funk, A., Smith, R. B., and Baranvay, J. A.,** Canker of dwarf mistletoe swellings on western hemlock caused by *Nectria fuckeliana* var. *macrospora., Can. J. For. Res.,* 3, 71, 1973.

43. **Girling, D. J., Greathead, D. J., Mohyuddin, A. I., and Sankaran, T.,** The potential for biological control in the suppression of parasitic weeds, *Biocontrol News Inform.,* (sample issue), 7, 1979.

44. **Harris, P.,** Current approaches to biological control of weeds, Techn. Comm., Commonwealth Institute of Biological Control, Silwood Park, Ascot, U.K., 4, 67, 1971.

45. **Hartman, H. and Watson, A. K.,** Damage to common ragweed *(Ambrosia artemisiifolia)* caused by the white rust fungus *(Albugo tragopogi)*, *Weed Sci.*, 28, 632, 1980.

46. **Hasan, S.,** The possible control of skeleton weed, *Chondrilla juncea* Bubak & Syd., *Proc. 1st Int. Symp. Biol. Control Weeds*, Délémont, Switzerland, 1969, 11, 1970.

47. **Hasan, S.,** Specificity and host specialisation of *Puccinia chondrillina.*, *Ann. Appl. Biol.*, 72, 257, 1972.

48. **Hasan, S.,** Behaviour of the powdery mildews of *Chondrilla juncea* L. in the Mediterranean, *Actas 3rd Congr. Un. Fitopat. Medit.*, Oeiras, Portugal, 1972, 171.

49. **Hasan, S.,** Host specialisation in *Chondrilla* fungi, *Proc. 2nd Int. Symp. Biol. Control Weeds*, Rome, 1971, 134.

50. **Hasan, S.,** The powdery mildews as potential biological control agents of skeleton weed *(Chondrilla juncea* L.), *Proc. 2nd Int. Symp. Biol. Control Weeds*, Rome, 1971, 114.

51. **Hasan, S.,** First introduction of a rust fungus in Australia for the biological control of skeleton weed, *Phytopathology*, 64, 253, 1974.

52. **Hasan, S.,** Host specialisation of a powdery mildew, *Erysiphe cichoracearum* from *Chondrilla juncea.*, *Aust. J. Agric. Res.*, 25, 459, 1974.

53. **Hasan, S.,** Current research on plant pathogens as biocontrol agents for weeds of Mediterranean origin, *Proc. EWRS Symp. The Influence of Different Factors on the Development and Control of Weeds*, Mainz, West Germany, 1979, 333.

54. **Hasan, S.,** Plant pathogens and biological control of weeds, *Rev. Plant Pathol.*, 59, 349, 1980.

55. **Hasan, S.,** A new strain of the rust fungus *Puccinia chondrillina* for biological control of skeleton weed in Australia, *Ann. Appl. Biol.*, 99, 119, 1981.

56. **Hasan, S.,** Present status and prospects of the program in Europe for the microbiological control of Australian weeds, *Proc. 5th Int. Symp. Biol. Control Weeds*, Brisbane, Australia, 1980, 333.

57. **Hasan, S.,** Biological control of weeds with plant pathogens — status and prospects, *Proc. 10th Int. Cong. Plant Prot.*, Brighton, U.K., 1983, 759.

58. **Hasan, S.,** Recherches de souches virulentes de *Puccinia chondrillina* pour la lutte biologique centre la mauvaise herbe *Chondrilla juncea*, *Proc. 26th Meet. Fr. Phytopathol. Soc.*, Avignon, France, 1984, 167.

59. **Hasan, S.,** Search in Greece and Turkey for *Puccinia chondrillina* strains suitable to Australian forms of skeleton weed, *Proc. 6th Int. Symp. Biol. Control Weeds*, Vancouver, 1984, 625. Ottawa: Agriculture Canada.

60. **Hasan, S.,** Prospects for biological control of *Heliotropium europaeum* by fungal pathogens, *Proc. 6th Int. Symp. Biol. Control Weeds*, Vancouver, 1984, 617. Ottawa: Agriculture, Canada.

61. **Hasan, S. and Cullen, J. M.,** Studies in the Mediterranean region on biological control of Australian weeds, *Proc. Eur. Weed Res. Soc. 3rd Symp. Weed Problems in the Mediterranean Area*, Oeiras, Portugal, 1984, 365.

62. **Hasan, S. and Cullen, J. M.,** *Heliotropium europaeum* and its control by natural enemies, *Proc. 7th Int. Symp. Weed Biol. Ecol. Systematics*, Paris, 1984, 117.

63. **Hasan, S. and Jenkins, P. T.,** The effect of some climatic factors on the infectivity of the skeleton weed rust, *Puccinia chondrillina*, *Plant Dis. Rep.*, 56, 858, 1972.

64. **Hasan, S. and Wapshere, A. J.,** The biology of *Puccinia chondrillina* a potential biological control agent of skeleton weed, *Ann. Appl. Biol.*, 74, 325, 1973.

65. **Hasan, S., Giannotti, J., and Vago, C.,** Virus-like particles associated with a disease of *Chondrilla juncea*, *Phytopathology*, 63, 791, 1973.

66. **Hildebrand, D. C. and McCain, A. H.,** The use of various substrates for large-scale production of *Fusarium oxysporum* f. sp. *cannabis* inoculum, *Phytopathology*, 68, 1099, 1978.

67. **Holcomb, G. E.,** *Alternaria alternantherae* from alligatorweed also is pathogenic on ornamental amaranthaceae species, *Phytopathology*, 68, 265, 1978.

68. **Huffaker, C. B.,** Fundamentals of biological weed control, in *Biological Control of Insect Pests and Weeds*, DeBach, P., Ed., Chapman and Hall, London, 1970, 631.

69. **Hull, V. J. and Groves, R. H.,** Variation in *Chondrilla juncea* L. in south-eastern Australia, *Aust. J. Bot.*, 21, 113, 1973.

70. **Inman, R. E.,** A primary evaluation of *Rumex* rust as a biological control agent of curly dock, *Phytopathology*, 61, 102, 1971.

71. **Julien, M. H.,** *Biological Control of Weeds. A World Catalogue of Agents and their Target Weeds*, Slough: Commonwealth Agricultural Bureau, Farmham Royal, Slough, U.K., 1982, 108.

72. **Klerk, R. A.,** Use of microbial herbicide for weed control in rice *(Oryza sativa)*, M.S. thesis, University of Arkansas, Fayetteville, 1983, 104.

73. **King, L. J.,** *Weeds of the World, Biology and Control*, Interscience, New York, 1966, 526.

74. **Kirkpatrick, T. L., Templeton, G. E., and TeBeest, D. O.,** Potential of *Colletotrichum malvarum* for biological control of prickly sida, *Plant Dis.*, 66, 323, 1982.

75. **Leth, V. and Hass, H.,** The potential for biological weed control in Denmark, in 1st *Danske Planteva-ernskonference, Ukrudt,* Slagelse, Denmark, Institut for Ukrudtsbekaempelse, 1984, 221.
76. **Mangoendihardjo, S., Setyawati, O., Syed, R. A., and Sosromarsona, S.,** Insects and fungi associated with some aquatic weeds in Indonesia, *Proc. 6th Asian-Pacific Weed Soc. Conf.,* Jakarta, Indonesia, 1977, 440.
77. **Martyn, R. D. and Freeman, T. E.,** Evaluation of *Acremonium zonatum* as a potential biocontrol agent of waterhyacinth, *Plant Dis. Rep.,* 62, 604, 1978.
78. **McVean, D. N.,** Ecology of *Chondrilla juncea* L. in south-eastern Australia, *J. Ecol.,* 54, 345, 1966.
79. **Mortensen, K.,** Reaction of safflower cultivars to *Puccinia jaceae,* a potential biocontrol agent for diffuse knapweed, *Proc. 6th Int. Symp. Biol. Control Weeds,* Vancouver, 1984, 447. Ottawa: Agric. Canada.
80. **Morris, M. J.,** Evaluation of field trials with *Colletotrichum gloeosporioides* for the biological control of *Hakea sericea, Phytophylactica,* 15, 13, 1983.
81. **Nakao, H. K. and Funasaki, G. Y.,** Introductions for biological control in Hawaii: 1975 and 1976, *Proc. Hawaii. Entomol. Soc.,* 13, 125, 1979.
82. **Nikandrow, A., Weidemann, C. J., and Auld, B. A.,** Mycoherbicide Co-operative Project, *Proc. 7th Aust. Weeds Conf.,* 1, 129, 1984.
83. **Oehrens, E. B. and Gonzales, S. M.,** Introduccion de *Phragmidium violaceum* (Schulz) Winter como factor de control biologico de zarzamora (*Rubus constrictus* Lef. et M. y *R. ulmifolius* Schott.), *Agro Sur.,* 2, 30, 1974.
84. **Oehrens, E. B. and Gonzales, S. M.,** Introduction de *Uromyces galegae* (Opiz) Saccardo como factor de control biologico de galega (*Galega officinalis* L.), *Agro. Sur,* 3, 87, 1975.
85. **Oehrens, E. B. and Gonzales, S. M.,** Dispersion, ciclo biologico y danos causdos por *Phragmidium violaceum* (Schulz) Winter en zarzamora (*Rubus constricuts* Lef. et M. y *R. ulmifolius* Schott.) en las zonas centro-sur y sur de Chile, *Agro Sur,* 5, 73, 1977.
86. **Orr, C. C.,** *Nothanguina phyllobia,* a nematode biocontrol of silverleaf nightshade, *Proc. 5th Int. Symp. Biol. Control Weeds,* Brisbane, Australia, 1980, 389.
87. **Parmeter, J. R., Jr., Hood, J. R., and Sharpf, R. F.,** *Colletotrichum* blight of dwarf mistletoe, *Phytopathology,* 49, 812, 1959.
88. **Phatak, S. C., Sumner, D. R., Wells, H. D., Bell, D. K., and Glaze, N. C.,** Biological control of yellow nutsedge with the indigenous rust fungus *Puccinia canaliculata, Science,* 219, 1446, 1983.
89. **Politis, D. J., Watson, A. K., and Bruckart, W. L.,** Susceptibility of musk thistle and related composites to *Puccinia carduorum, Phytopathology,* 74, 687, 1984.
90. **Ridings, W. H., Mitchell, D. J., Schoulties, C. L., and El-Gholl, N. E.,** Biological control of milkweed vine in Florida citrus groves with a pathotype of *Phytophthora citrophthora, Proc. 4th Int. Symp. Biol. Control Weeds,* Gainsville, Fla., 1976, 224.
91. **Rosenthal, S. S. and Buckingham, G. R.,** Natural enemies of *Convolvulus arvensis* in western Mediterranean Europe, *Hilgardia,* 50(2), 1, 1982.
92. **Scheepens, P. C.,** Bestrijding van onkruiden met pathogene microörganismen; perspectiven voor de Nederlands situatie, *Gewasbecherming,* 10, 113, 1979.
93. **Scheepens, P. C. and Van Zon, H. C. J.,** Microbial herbicides, in *Microbial and Viral Pesticides,* Kurstak, E., Ed., Marcel Dekker, New York, 1982, 624.
94. **Schirman, R. and Robocker, W. C.,** Rush skeleton weed — Threat to dryland agriculture, *Weeds,* 15, 310, 1967.
95. **Schroeder, D.,** Biological control of weeds, in *Recent Advances in Weed Research,* Fletcher, W. W., Ed., Commonwealth Agric. Bureau, Farmham Royal, Slough, U.K., 1983, 41.
96. **Shaw, W. C.,** Integrated weed management systems technology for pest management, *Weed Sci.,* 30(Suppl. 1), 2, 1982.
97. **Smith, R. J., Jr.,** Integration of microbial herbicides with existing pest management programs, in *Biological Control of Weeds with Plant Pathogens,* Charudattan, R. and Walker, H. L., Eds., John Wiley & Sons, New York, 1982, 189.
98. **Smith, R. J., Jr. and Shaw, W. C.,** Weeds and their control in rice production, U.S. Department of Agriculture Handbook, 292, U.S. Government Printing Office, Washington, D.C., 1986, 64.
99. **Smith, R. J., Jr., Daniel, J. T., Fox, W. T., and Templeton, G. E.,** Distribution in Arkansas of a fungus disease used for biocontrol of northern jointvetch in rice, *Plant Dis. Rep.,* 57, 695, 1973.
100. **Soerjani, M.,** Integrated control of weeds in aquatic areas, in *Integrated Control of Weeds,* Fryer, J. D. and Matsunaka, S., Eds., University of Tokyo Press, Japan, 1977, 121.
101. **Stipes, R. J. and Phipps, P. M.,** *Fusarium oxysporum* f. sp. *perniciosum* on *Fusarium*-wilted mimosa trees, *Phytopathology,* 65, 188, 1975.
102. **Strässle, A., Défago, G., Kern, H., and Sedlar, L.,** *Ovularia obliqua.* Host specifity and conidiation in axenic culture, *Biologia,* 41, 847, 1986.
103. **TeBeast, D. O. and Templeton, G. E.,** Mycoherbicide: Progress in the biological control of weeds, *Plant Dis.,* 69, 6, 1985.

104. **Templeton, G. E.**, Status of weed control with plant pathogens, in *Biological Control of Weeds with Plant Pathogens,* Charudattan, R. and Walker, H. L., Eds., John Wiley & Sons, New York, 1982, 29.

105. **Templeton, G. E. and Smith, R. J., Jr.**, Managing weeds with pathogens, in *Plant Diseases. An Advanced Treatise,* Vol. 1, Horsfall, J. G. and Cowling, E. B., Eds., Academic Press, New York, 1977, 167.

106. **Templeton, G. E., TeBeast, D. O., and Smith, R. J., Jr.**, Development of endemic fungal pathogen as a mycoherbicide for biocontrol of northern jointvetch in rice, *Proc. 4th Int. Symp. Biol. Control Weeds,* Gainsville, Fla., 1976, 214.

107. **Templeton, G. E., TeBeast, D. O., and Smith, R. J., Jr.**, Biological weed control with mycoherbicides, *Annu. Rev. Phytopathol.,* 17, 301, 1979.

108. **Templeton, G. E., TeBeast, D. O., and Smith, R. J., Jr.**, Biological weed control in rice with a strain of *Colletotrichum gloeosporioides* (Penz.) Sacc. used as a mycoherbicide, *Crop Protec.,* 3, 409, 1984.

109. **Trujillo, E. E.**, Biological control of hamakua pamakani with plant pathogens, *Proc. Am. Phytopathol. Soc.,* 3, 298, 1976.

110. **Trujillo, E. E. and Obrero, F. P.**, *Cephalosporium* wilt of *Cassia surattensis* in Hawaii, *Proc. 4th Int. Symp. Biol. Control Weeds,* Gainsville, Fla., 1976, 217.

111. **Bruckart, W. L., Turner, S. K., Stucker, E. M., Vonmoos, R., Sedlar, L., and Défago, G.**, Relative virulence of *Melampsora euphorbiae* from central Europe toward North America and European spurges, *Plant Dis.,* 70, 847, 1986.

112. **Walker, H. L.**, Spurred anoda (*Anoda cristata* (L.) Schlecht) biocontrol with a plant pathogen, *Proc. 33rd Annu. Meet. South. Weed Sci. Soc.,* 1980, 65.

113. **Walker, H. L. and Boyette, C. D.**, Biocontrol of sicklepod (*Cassia obtusifolia*) in soybeans (*Glycine max*) with *Alternaria cassiae, Weed Sci.,* 33, 212, 1985.

114. **Walker, H. L. and Sciumbato, G. L.**, Evaluation of *Alternaria macrospora* as a potential biocontrol agent for spurred anoda (*Anoda cristata*): Host range studies, *Weed Sci.,* 27, 612, 1979.

115. **Wapshere, A. J.**, Assessment of biological control potential of the organisms attacking *Chondrilla juncea* L., *Proc. 1st Int. Symp. Biol. Control Weeds,* Délémont, Switzerland, 1969, 81.

116. **Wapshere, A. J.**, Towards a science of biological control of weeds, *Proc. 3rd Int. Symp. Biol. Control Weeds,* Montpellier, France, 1973, 3.

117. **Wapshere, A. J.**, A strategy for evaluating the safety of organisms for biological weed control, *Ann. Appl. Biol.,* 77, 201, 1974.

118. **Wapshere, A. J.**, A protocol for programs for biological control of weeds, *Pest Artic. News Summ.,* 21, 295, 1975.

119. **Wapshere, A. J. and Kirk, A. A.**, The biology and host specificity of the *Echium* leaf miner *Dialectica scalariella* (Zeller) (Lepidoptera: Gracillariidae), *Bull. Entomol. Res.,* 67, 627, 1977.

120. **Wapshere, A. J., Hasan, S., Wahba, W. K., and Caresche, L.**, The ecology of *Chondrilla juncea* in the western Mediterranean, *J. Appl. Ecol.,* 11, 783, 1974.

121. **Watson, A. K.**, The biological control of Russian knapweed with a nematode, *Proc. 4th Inter. Symp. Biol. Control Weeds,* Gainsville, Fla., 1976, 221.

122. **Watson, A. K. and Alkhoury, I.**, Response of safflower cultivars to *Puccinia jaceae* collected from diffuse knapweed in eastern Europe, *Proc. 5th Int. Symp. Biol. Control Weeds,* Brisbane, Australia, 1980, 301.

123. **Watson, A. K. and Keogh, W. J.**, Mortality of Canada thistle due to *Puccinia punctiformis, Proc. 5th Int. Symp. Biol. Control Weeds,* Brisbane, Australia, 1980, 325.

124. **Webb, R. and Lindow, S. E.**, Evaluation of *Ascochyta pteridium* as a potential biological control agent of bracken fern, *Phytopathology,* 71, 911, 1981.

125. **Wicker, E. F. and Shaw, C. G.**, Fungal parasites of dwarf mistletoes, *Mycologia,* 60, 372, 1968.

126. **Wilson, C. L.**, Use of plant pathogens in weed control, *Annu. Rev. Phytopathol.,* 7, 411, 1969.

127. **Woodhead, S.**, Field efficacy of *Phytophthora palmivora* for control of milkweed vine, *Phytopathology,* 71, 913, 1981.

128. **Zettler, F. W. and Freeman, T. E.**, Plant pathogens as biocontrol of aquatic weeds, *Annu. Rev. Phytopathol.,* 10, 455, 1972.

129. **Cullen, J. M.**, Commonwealth Scientific and Industrial Research Organization, Division of Entomology, personal communication.

130. **Trujillo, E. E.**, College of Tropical Agriculture, Dept. of Plant Pathology, University of Hawaii at Manoa, personal communication.

131. **Hasan, S.**, Commonwealth Scientific and Industrial Research Organization, Division of Entomology, unpublished data.

Chapter 10

BIOLOGICAL CONTROL OF SCLEROTIAL DISEASES

Abdul Ghaffer

TABLE OF CONTENTS

I. INTRODUCTION

Plant diseases caused by sclerotial fungi are widespread and cause considerable losses to crop plants. The sclerotial propagules of disease causing pathogens are difficult to eliminate because they are well adapted to survive under adverse environmental conditions. Where resistant varieties are not available and chemical control is not effective, biological control has shown promise as a practical agricultural method for the control of sclerotial plant diseases. An attempt is made here to review the available data on biological control of fungal sclerotia, and to reveal the mechanism of the process.

The term sclerotia includes diverse fungal bodies characteristic of certain taxa such as *Sclerotium, Sclerotinia, Claviceps, Typhula, Botrytis, Colletotrichum, Gloeocercospora, Helicobasidium, Macrophomina, Mycosphaerella, Phymatotrichum, Rhizoctonia,* and *Verticillium*. With the exception of *Claviceps*, the majority of sclerotial fungi are soil-borne plant pathogens. The sclerotia produced by air borne fungal pathogens are variable in size and can attain a maximum diameter of 2 cm. They usually produce fruiting bodies such as the apothecia of the *Sclerotinia* sp. or the perithecia of the *Claviceps*. In contrast, sclerotia of root infecting fungi are smaller, up to 2 mm in diameter, somewhat regular in shape, and uniform in size such as those of *Verticillium dahliae, Sclerotium rolfsii, S. cepivorum, Phymatotrichum omnivorum.*

Sclerotia are complex vegetative structures formed by the aggregation of hyphae in some filamentous fungi. This quiescent viable state is maintained by many fungi, in the absence of a host, to overcome unfavorable conditions. Besides, the sclerotia have an inherited adaptation for rapid mycelial growth and colonization of a wide variety of substrates. Survival of these under adverse environmental conditions have been reviewed by several workers.[38,76,134,152,183] The adverse environmental conditions, as described by Sussman,[152] include dessication, starvation, and toxic chemicals present in the soil or atmosphere or that may accumulate as a result of the activities of other organisms and/or solar radiation.

Survival of sclerotia for varying periods of time is known.[38] There does not appear to be any definite relationship between longevity and the structure of sclerotia, with special reference to the presence of rind tissue. In those in which no rind is present, the sclerotia consist of a mass of loosely interwoven hyphae of barrel-shaped cells without any well marked zonation (*Rhizoctonia solani*), or all the cells of sclerotia have a thick wall (*Macrophomina phaseolina*) or consist of matrix of thin-walled hyaline cells and thick-walled heavily pigmented cells (*Verticillium dahliae*). Where a rind is present, the mature sclerotium consists of a more or less homogenous medulla (*Claviceps purpurea, Phymatotrichum omnivorum, Sclerotinia* spp.), or a cortex of pseudoparenchymatous cells and a large medulla of loosely arranged filamentous hyphae (*Botyrtis allii, B. cinerea*), or a thick cuticle, a rind two to four cells thick with a cortex of thin-walled cells and medulla of loose filamentous hyphae (*Sclerotium rolfsii*). Damage to the rind of sclerotia may lead to an increase in susceptibility to colonization by other microorganisms. The types of sclerotial structures and their morphology has been comprehensively reviewed.[27,38,183]

II. HOST-PATHOGEN-ENVIRONMENT INTERACTION

Biological control has been defined by Garrett[57] as ''any condition under which, or practice whereby, survival or activity of a pathogen is reduced through the agency of any other living organisms except man with the result that there is reduction in the incidence of disease caused by the pathogen''. Garrett's definition is fulfilled by the process by which plant residues, soil amendments, or nutrients stimulate the multiplication of organisms that are competitive, antagonistic, or parasitic to the pathogen. The concept of biological control as given by Cook and Baker[40] has been largely accepted by the plant pathologists working with soil-borne plant pathogens. It deals with the "reduction of the amount of inoculum or disease producing activity of a pathogen accomplished by or through one or more organisms other than man". Biological control may operate at any stage of the pathogen survival or disease development through antibiosis, lysis, parasitism, or competition.[131] In the scheme of interaction of the host, pathogen, and the environment,[19] the biological control agent may operate in the following two pathways:

1. Outside the host, the biocontrol agent may be antagonistic and thereby reduce the activity, efficiency, and/or inoculum density of the pathogen through antibiosis, competition, or hyperparasitism.
2. In the host tissue the biocontrol agent may initiate a resistance response in the host becoming antagonistic or transmitting factors rendering the pathogen avirulent through cross protection, production of inhibitors, hypovirulence, and/or competition for host receptor sites.

Examples of the biological control of plant diseases caused by sclerotial fungi are given in this review.

A. *Botrytis* spp.

The *Botrytis* spp. cause blossom blight, fruit rot, stem canker, leaf spot, tuber, corm, bulb, and root rot or storage rot on a variety of vegetables, ornamentals, fruits, and field crops in different parts of the world.[6] The fungus overwinters in the soil as mycelium on decaying plant debris and as sclerotia which survive for about a year.[111]

In field soil the presence of a mycoparasite, *Gliocladium roseum*, affects the survival of the sclerotia of *Botrytis allii* Munn, the causal agent of neck rot of onion.[114] *Trichoderma viride* has been isolated from decayed sclerotia of *Botrytis convoluta* Whetzel and Drayton associated with Iris rhizome rot.[110] Whereas *Acrostalagmus roseus* was able to parasitize and destroy sclerotia,[135] *T. pseudokoningii* isolated from wind-fall apples completely suppressed the rot caused by *B. cinerea*. Strains of *T. lignorum* differed in their parasitism of sclerotia of *B. cinerea*.[99] *Coniothyrium minitans*, a well established mycoparasite was shown to parasitize sclerotia of *B. cinerea*, *B. fabae* and *B. narcissicola*.[164] Application of a *C. minitans* inoculum in the form of pycnidial dust to sclerotia placed on compost results in high levels of infection by the mycoparasite. Sclerotia of *B. cinerea* were infected by *Teratosperma oligocladum* which may prove useful in the biological control of sclerotial plant pathogens.[16]

B. *Claviceps purpura* (Fr.) Tul.

Claviceps purpurea (Fr.) Tul., the ergot fungus of barley, wheat, oats, rye, and other grasses is cosmopolitan.[103] The sclerotia of the fungus are reported to survive for up to 1 year under laboratory conditions.[115] *Acrostalagmus roseus* was able to parasitize and destroy sclerotia of *C. purpurea* when they were placed on the surface of the soil.[135] *Fusarium heterosporum (Gibberella gordonia)*, a highly specialized hyperparasite, was found colonizing *C. purpurea* on *Lolium perrenne*.[81] *C. minitans* was shown to parasitize sclerotia of *C. purpurea* to varying degrees.[164]

C. *Macrophomina phaseolina* (Tassi) Goid.

Macrophomina phaseolina (Tassi) Goid., syn. *Rhizoctonia bataticola* (Taub.) Butl., produces charcoal rot, damping-off, root rot, stem rot, or pod rot on over 400 species of plants[47,139,189] in the tropical and subtropical countries of the world. In soil the fungus survives for over 10 months[65] primarily as black sclerotia which are formed in the host tissue and released into soil as tissue decays.[41] Populations as high as 1000 sclerotial propagules/g soil have been reported.[129]

In agar culture, *M. phaseolina* was inhibited by fungi, viz., *Arachniotus* sp., *Aspergillus aculeatus*,[45] *A. flavus*,[89] *A. niger*,[138,174] *Cepahalosporium humicola*,[45] *Chaetomium cupreum*, *Eurotium chevelieri*, *Fusarium moniliforme*, *Penicilium citrinum*, *P. variabile*, *Trichothecium roseum*,[45] *Stachybotrys atra*,[28,45] *Trichoderma harzianum* (Shahzad and Ghaffar, unpublished), *T. lignorum*,[45,155,174,179] *T. viride*;[59] actinomycetes, viz., *Streptomyces albus*, *S. griseus*, *S. noursei*,[62] *Streptomyces* sp.,[146] *Chainia antibiotica*;[153] and bacteria, viz., *Bacillus subtilis*,[100,154,173] *Bacillus* sp.,[146] and *Rhizobium japonicum*.[30] *Rhizopogon vinicolor*, a mycorrhizal fungus of Douglas fir (*Pseudotsuga menziesii*), also inhibited growth of *M. phaseolina* in agar culture.[190] In sterile amended and unamended soils, *M. phaseolina* was inhibited by *Aspergillus flavus*, *A. rugulosus*, *A. terreus*, *Trichoderma lignorum*, and *Bacillus cereus*.[122] Norton[122] found that growing mycelium of *T. lignorum* invaded *M. phaseolina* mycelial colony and checked its growth. Hyphae of *T. lignorum*, *and A. niger*,[174] and *T. viride*[59] parasitized and coiled around the hyphae of *M. phaseolina* resulting in coagulation of the protoplasmic contents of the hyphal cells followed by cell wall dissolution.

Whereas sclerotia of *M. phaseolina* germinate in nutrient amended soil,[12,148] the mycelium is sensitive to antagonistic microorganisms[12,45,95] and does not survive in soil for more than 7 days.[58] At 50 and 80% moisture holding capacity (MHC) mycelial growth decreased and the number of bacteria attacking hyphae and accompanying lysis of hyphal cells increased.[95] Similarly, the fungus could not be recovered from fiberglass cloth pieces buried in soil after 2 days at 0 and 100% MHC and after 5 days at 25 to 75% MHC.[58]

Successful biological control of black root rot of Slash pine (*Pinus elliotti*) seedlings caused by *M. phaseolina* was demonstrated.[44] Pine seedlings died 3 days after inoculation with *M. phaseolina*, but they did not become infected when inoculated with an unidentified basidiomycete and remained alive after 3 months. An increase in seedling stand from 57% in soil artificially infested with *M. phaseolina* to 81% in soil similarly inoculated with *M. phaseolina* and an unidentified sterile basidiomycete was observed.

Of the eight different species of *Streptomyces* tested, significant reduction in *Macrophomina* infection of cotton was obtained when *Streptomyces albus* and *S. griseus* were used without deleterious effect on cotton growth.[62] Similarly, unidentified isolates of bacteria and actinomycetes were effective in the control of *M. phaseolina* on okra, *Abelmoschus esculentus*.[67] *Aspergillus flavus* growing on peanut shells and kernels suppressed the rate and extent of kernel infection by *M. phaseolina*.[89] *Macrophomina* infection on *Dolichos biflorus* was completely checked in the presence of *T. lignorum*.[56] Survival of *M. phaseolina* in jute rhizosphere was poor in the presence of *Streptomyces fradiae*.[118]

Seedling emergence of urid bean increased when the *Arachniotus* sp. or *Aspergillus aculeatus* was applied to the seed or soil in the presence of *M. phaseolina*, controlling pre and postemergence damping-off and giving a total stand of 73 and 69%, respectively, as compared to 1 to 3% in control.[45] The biological control of *Macrophomina* on gram was obtained by coating seed with the *Bacillus* sp. and *Streptomyces* sp., resulting in increased plant growth.[146] Treatment of potato seed pieces and whole tubers with *Bacillus subtilis* reduced the frequency of charcoal rot at harvest.[154] Similarly, the bacterization of seeds and roots with *Rhizobium japonicum* reduced the charcoal rot of soybeans.[30] Culture extracts of *R. japonicum* has since been identified as rhizobitoxine.

Since disease potential can be reduced by reducing the inoculum density, organic amendments of soil with a low Carbon/Nitrogen (C/N) ratio have been found more effective than amendments with high C/N ratio in reducing inoculum density.[46,66] Numbers of sclerotia of *M. phaseolina* decreased rapidly in soils amended with leaf and stem fragments of lucerne and clover.[73,74] The reduction in inoculum densities is brought about through changes in the microbial balance of soil. Ghaffer et al.[66] showed that soil amended with either alfalfa or

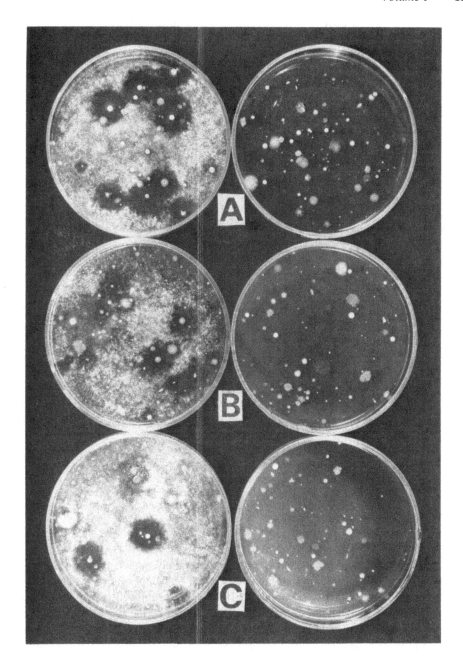

FIGURE 1.   Right row: Soil dilution plates for actinomycetes and bacteria from soil amended with alfalfa meal (A), barley straw (B), and no amendment (C). Soil samples taken 15 days after amendment. Left row: Similar dilution plates sprayed with a mycelial suspension of *Macrophomina phaseolina* showing antagonism of actinomycetes and bacteria to *M. phaseolina*. (From Ghaffar, A., Zentmyer, G. A., and Erwin, D. C., *Phytopathology*, 59, 1267, 1969. With permission.)

barley straw increased propagules of *Bacillus subtilis* and actinomycetes antagonistic to *M. phaseolina* (Figure 1 and 2), and increase in antagonists was correlated with a reduction in *Macrophomina* infection on cotton (Figure 3).

D. *Phymatotrichum omnivorum* (Shear) Duggar

Phymatotrichum omnivorum (Shear) Duggar is a serious pathogen of more than 2000

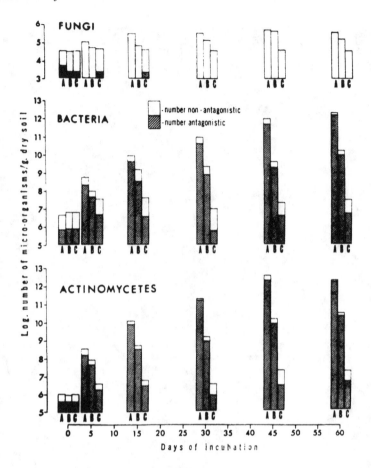

FIGURE 2.   Total population of fungi, bacteria, and actinomycetes and population of antagonists to *Macrophomina phaseolina* in soil amended with alfalfa meal (A), or barley straw (B), compared to no amendment (C). (From Ghaffar, A., Zentmyer, G. A., and Erwin, D. C., *Phytopathology*, 59, 1267, 1969. With permission.)

species of dicotyledonous plants and is not reported on any monocotyledonous plant. Phymatotrichum root rot, also known as Texas root rot or Ozonium root rot, is one of the most destructive diseases of plants like cotton, alfalfa, etc., grown in soils which are alkaline and have high summer temperatures. It is considered as one of the most difficult diseases of plants to control. There have been reports of the occurrence of Phymatotrichum root rot disease in India and Pakistan[172] but these have not been substantiated.[151] Sclerotia and strands of hyphae enable the fungus to persist up to 12 years in soil.[93]

Significant reduction in the Phymatotrichum root rot of cotton was observed by the application of green manure amendments to infested soil which supported high populations of microorganisms competitive with, or antibiotic to, *P. omnivorum*.[92] In Houston black clay the disappearance of sclerotia of *P. omnivorum* from soil amended with green manure was attributed to a 100% increase in actinomycetes, rod shaped bacteria, and fungi.[116] Mostly alfalfa,[92] various legumes, and green manure crops like wheat straw, *Vicia sativa*,[37] sweet clover (*Melilotus indica*),[105] *Sesbania* sp., *Sorghum vulgare*,[91] guar (*Cyamopsis tetragonoloba*), cowpea (*Vigna sinensis*), clover (*Melilotus alba*), and (*Sesbania exalta*)[141] have been used for incorporation in soil rotation with cotton. Guar like *Sesbania* is resistant to *P. omnivorum*.[150] Most legumes are susceptible to *P. omnivorum* but they mature well as winter

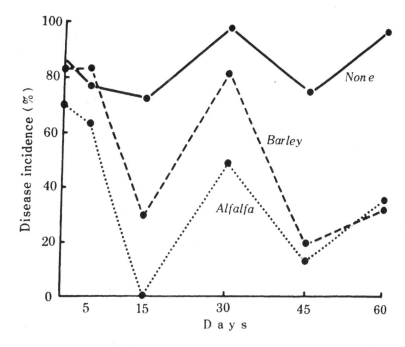

FIGURE 3.   The effect of adding organic amendments to soil on the infection of cotton seedlings with *Macrophomina phaseolina*. (From Ghaffar, A., Zentmyer, G. A., and Erwin, D. C., *Phytopathology*, 59, 1267, 1969. With permission.)

crops when the fungus is inactive, and when incorporated into soil before cotton is planted the legumes reduced the incidence of root rot.[105] One of the theories advanced for marked reduction of Phymatotrichum root rot of cotton is the increase in the population of competitive microorganisms which reduce the activity and pathogenicity of the pathogen. In Arizona the best control was obtained in infested soil which was cropped either to pea or barley in winter, or in soil which received high tonnage of pea plants. Reduction in incidence of Phymatotrichum root rot was attributed to premature germination of overwintering propagules making hyphae vulnerable to competition of microflora.[32]

E. *Rhizoctonia solani* Kühn

Since the report of *R. solani* Kühn (teleomorph = *Thanatophorus cucumeris* (Frank) Donk.) on diseased potato tubers by Julius Kühn in 1858, the fungus has gained the reputation of attacking a tremendous range of plants causing seed decay, damping-off, seedling blight, root rot, crown rot, stem canker, as well as such stem infections as wire stem, soreshin, bud rot, foliage blight, and storage rot in different parts of the world.[17] *R. solani* exists as active mycelium in soil, in the absence of a host, and as dormant sclerotia for up to 6 years.[56]

R. solani is inhibited by *Aspergillus clavatus*,[187] *A. niger*,[24] *Chaetomium cupreum*,[188] the *Corticium* sp.,[123] *Cylindrocarpon destructor, C. olivaceus, Epicoccum nigrum, Fusarium culmorum, F. moniliforme,*[31] *F. oxysporum, F. semitectum, F. udum,*[11] *Gliocladium deliquescens, G. roseum,*[31] *G. virens,*[161] *Papulospora stourei,*[176] *Penicillium cyclopium, P. nigricans, P. vermiculatum,*[21] *P. patulum,*[54] *Rhizoctonia oryzae,*[144] *Trichoderma aureoviride,*[187] *T. hamatum,*[33] *T. harzianum,*[8,31,51,72,78,98] *T. lignorum,*[35,179] *Trichothecium roseum,*[31] the *Streptomyces* sp., *Bacillus subtilis,*[54,162] *Pseudomonas aeruginosa,*[23] *P. fluorescens.*[31] Hyphae of *T. viride* readily entered host cells and digested their contents.[48] The *Trichoderma* spp. attached to *R. solani* by hyphal coils, hooks, or appressoria which produced penetration holes and lysis.[49,50,82] The report of Weindling[178] that hyphae of *R. solani* are vigorously attacked and parasitized by a very common soil inhabiting fungus, *Trichoderma lignorum*, led to extensive studies on biological control using *T. lignorum* as a biological antagonist. Weindling and Fawcett[180] showed that *T. lignorum* introduced as spores into acidified soil (pH 4.0) by application of acid peat moss or aluminum sulphate suppressed damping-off of citrus seedlings caused by *R. solani*. This has now been confirmed by others.[9]

Wheat bran cultures of *Trichoderma harzianum* have been demonstrated to control *R. solani* damping-off of radish under greenhouse conditions.[78] Protection of radish seedlings from damping-off lasted for five successive plantings. *T. harzianum* was effective in preventing the build-up and reducing the population of *R. solani*.[78] Similarly, wheat bran inoculum preparations of *T. harzianum* significantly reduced *R. solani* infection on cotton,[8] strawberry,[49] carnation,[51] and damping-off of bean, tomato, and eggplant seedlings.[72] Use of *T. harzianum* after methyl-bromide treatment reduced the incidence of *R. solani* in strawberry nurseries.[49] Similarly, the application of *T. harzianum* after soil fumigation of groundnut fields showed that reinfestation of *R. solani* was prevented.[52] Benomyl-tolerant biotypes of *T. harzianum* suppressed activity of *R. solani* in the soil more effectively than did the wild strain in suppressing damping-off of cotton and radishes.[130]

Application of conidia of *Trichoderma hamatum* conferred suppression to conducive Fort Collins clay loam and therefore was found effective for controlling *R. solani* in radish seedlings and in beans.[33] Reduction in damping-off was higher when the antagonist was added to acidified soil. Infection of *Phaseolus lunatus* and peas was reduced effectively when *T. viride* was either added to the soil or inoculated into roots of the seedlings prior to inoculation with *R. solani*. The antagonist was, however, not able to inhibit disease development once infection had taken place.[107]

Besides the *Trichoderma* spp., *R. solani* damping-off of lettuce was controlled by *Streptomyces lavendulae* and *Penicillium clavariaeforme* only when organic amendments were simultaneously added to the soil.[185] Nearly complete control of *R. solani* damping-off and seedling blight of peas occurred when sterilized soil was simultaneously infested with cultures of *Penicillium vermiculatum* and *R. solani*. Biological control in nonsterilized soil was, however, not successful since about 60% of peas damped-off in simultaneously infested soil, and approximately 50% of the surviving plants had symptoms of seedling blight.[21] Incidence and severity of cucumber fruit rot incited by *R. solani* was reduced in the field after the application of the *Corticium* sp., and the *Trichoderma* sp. The *Corticium* sp. reduced the saprophytic activity of *R. solani* in soil.[98] Similarly, *Gliocladium virens* reduced the severity of *R. solani* root rot of *Phaseolus vulgaris*,[161] and *Azotobacter chroococcum* prevented infection of potato sprouts planted in soil heavily infested with *R. solani*.[114]

Cotton seed coating with *T. hamatum* and *T. harzianum* reduced *R. solani* disease incidence by up to 83% in the greenhouse and up to 47 to 60% in the field.[53] *Gliocladium virens* previously isolated from parasitized hyphae of *R. solani*, when placed with cotton seed sown in infested soil, suppressed damping-off.[82] Treatment of cotton seed with *Pseudomonas fluorescens*, or its antibiotic pyrolmitrin @200μg/mℓ, at the time of planting was effective in preventing *R. solani* damping-off of cotton seedlings.[83] Similarly, coating Iris bulbs with *T. harzianum* was highly effective in reducing *R. solani* incidence in the greenhouse.[34] Potato tubers treated with a spore suspension of *T. harzianum* and *T. viride* had much less stem and stolon canker and black scurf than untreated tubers.[10] Biological control of *R. solani* black scurf disease of potato was observed when potato seed pieces were dipped in a spore suspension of *T. pseudokoningii* or the *Bacillus* sp.[35] Seed treatment with a suspension of conidia of *T. hamatum* at 10⁶/mℓ controlled seed rot and damping-off of peas and radishes.[75] At lower temperatures and in alkaline soils control was not achieved.[101] *Penicillium oxalicum* used as a seed treatment of peas was as effective as captan for control of *R. solani*.[94] Similarly, pea seeds treated with *Trichoderma pseudokoningii*, *T. longibrachiatum*, *T. hamatum*, and *Penicillium* sp., increased emergence and decreased disease severity of *R.*

solani.[36] Seed coated with the mycelium and sclerotia of the *Corticium* sp. significantly reduced damping-off of beans, soyabeans, and sugarbeet in *R. solani*-infested Nebraska soils.[123,124]

R. solani isolates disappeared from the soil after amendment of the soil by the addition of organic materials or plant residues of corn, wheat, or oat.[142] The organic substrate stimulated the activities of antagonistic organisms resulting in the control of damping-off of broad-leaf tree seedlings.[186] Infection of beans by *R. solani* was reduced by incorporating barley, wheat straw, corn stem, and even pine shavings.[149] Maximum inhibition of *R. solani* was found in soils adjusted to a C/N ratio of 40 to 100 and was less in ratios in high 200 to 400 or lower 5 to 20 ranges.[43,129] Reduction in the activity of *R. solani* has been attributed to the antagonistic effects of other organisms, especially those of bacteria and actinomycetes, as well as an increase in CO_2 concentration and a scarcity of available N in the soil. Papavizas and Davey[128] demonstrated a 3 to 4 fold increase in Streptomycetes antagonistic to *R. solani* in soil amended with young corn or oat straw which reduced *Rhizoctonia* root rot of bean. Similarly, addition of mature oat tissue to soil increased the numbers of bean-rhizosphere microorganisms, mostly streptomycetes and bacteria antagonistic to *R. solani.*[127] Addition of chitin, a by-product of the shellfish industry, increased antagonistic actinomycetes and fungi 3- and 5-fold, accompanied with a decrease in bean root rot and saprophytic activity of *R. solani.*[79]

F. *Sclerotinia minor* Jagger

Sclerotinia minor Jagger, the cause of celery foot rot or sclerotinia rot of sweet potato and lettuce is known to survive for up to 2 years.[119] *Sporodesmium sclerotivorum*, a dematiaceous hyphomycete, was found on sclerotia of *S. minor* in the soil.[166] The fungus parasitized the sclerotia of *S. minor* in natural soil as well as in vitro.[14] The mycoparasite destroyed more than 95% of the sclerotia of *S. minor* within 10 weeks or less, and biological control of lettuce drop was correlated with a reduction in *S. minor* inoculum density. *S. sclerotivorum*, when applied @ 1000 spores/g soil, caused a decline in the survival of sclerotia.[3] The mycoparasite was established and caused the destruction of sclerotia in three successive crops of lettuce in a field experiment.[4] Living of sclerotia of *S. minor* on water agar or in sand were also parasitized by *Teratosperma oligocladum* which destroyed the sclerotia in 10 weeks.[15,166,167] Using baits of sclerotia of *S. minor*, Parfitt et al.[133] isolated *T. oligocladum*, a mycoparasite of fungal sclerotia which may prove useful in the biological control of sclerotial plant pathogens. *Trichoderma viride* and *T. harzianum*, applied as spore suspensions (10^7/mℓ) at transplanting time of lettuce plants previously inoculated with *S. minor*, reduced the number of dead plants and increased the yield.[120]

G. *Sclerotinia sclerotiorum* (Lib.) de By.

Sclerotinia sclerotiorum (Lib.) de By. is plurivorous and produces wilt, root rot, or sclerotinia rot of plants. Survival of sclerotia for a least 2 years has been reported.[184] In dual cultures *S. sclerotiorum* is inhibited by *Aspergillus niger*, *A. flavus*, *A. ustus*,[137] *Aspergillus* sp., *Cephalosporium* sp., *Cylindrocarpon* sp.,[1] *Coniothyrium minitans*[1,29,55,63,85,158,163,175] (Figure 4), *Fusarium lateritium*,[147] *Fusarium* sp.,[1,106,117] *Gliocladium catenulatum*,[84] *G. roseum*,[1] *Mucor* sp.,[1,117] *Pythium* sp.,[1] *Penicillium citrinum*, *P. funiculosum*, *P. pallidum*, *P. steckii*, *Stachybotrys atra*,[137] *Sporotrichum carnis*,[55] *Rhizopus* sp.,[1] *Teratosperma oligocladum*,[15] *Trichoderma viride*,[1,84,167] and *Verticillium* sp.[106] Sclerotia of *S. sclerotiorum*, kept in soil infested with *Penicillium citrinum*, *P. pallidum*, *P. steckii*, *Aspergillus flavus*, *A. niger*, and *A. ustus*, were not able to produce stipes and disintegrated.[137]

Coniothyrium minitans was first isolated and described by Campbell[29] from sclerotia of *S. sclerotiorum* associated with guayule roots in California. The hyperparasite is known from Europe and New Zealand.[163] *C. minitans* invaded sclerotia which became rotted and

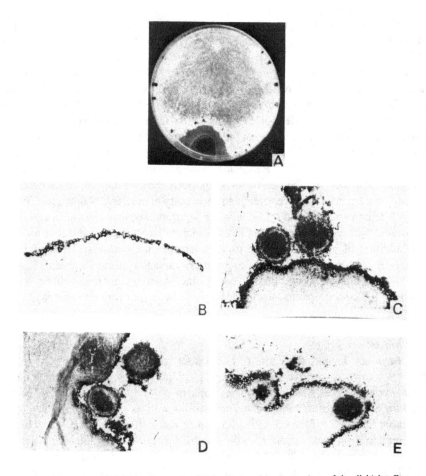

FIGURE 4.   (A) Inhibition of growth of *Sclerotinia sclerotiorum* (top of the dish) by *Coniothyrium minitans* on PDA after 5 days growth at 25°C. (B-E) Photomicrographs (× 150) of microtome sections of agar cultures showing parasitism of *Coniothyrium minitans* on sclerotia of *Sclerotinia sclerotiorum*. (B) Portion of sclerotia of *S. sclerotiorum*. (C) Development of pycnidia of *C. minitans* on top of a sclerotium of *S. sclerotiorum*. (D) Development of the pycnidium inside the sclerotium. (E) As (D) note liberation of pycnospores through the ostiole of the pycnidia of *C. minitans*. (From Ghaffar, A., *Pak. J. Bot.*, 4, 85, 1972. With permission.)

disintegrated easily. In serial microtome sections of *C. minitans*-infected sclerotia, Ghaffar[63] demonstrated pycnidia development both on and inside the sclerotia of *S. sclerotiorum* (Figure 4). Hyphal tips of the parasite penetrate cells of the host killing hyphae and sclerotia. Host cytoplasm disintegrates and the cell wall collapses.[87] Application of inoculum in the form of pycnidial dust to sclerotia placed on compost results in high levels of infection.[163] About 50% of the sclerotia of *S. sclerotiorum* from the rhizosphere soil of sunflower were nonviable and often contaminated by *C. minitans*.[80] *C. minitans* continued to parasitize the pathogen inside the root and inside the base of the sunflower stem.[84] In glass house tests, control of *Sclerotinia* wilt of sunflower and decrease in yield loss was observed when *C. minitans* was introduced in sclerotia-infested soil at seeding time due to control of the primary inoculum of the sclerotia by the hyperparasite.[86] Conidia, pycnidia, and mycelium of *C. minitans* was applied in the field in the autumn, and after 1 month it showed a negligible number of sclerotia[159] and it effectively controlled *S. sclerotiorum* induced disease on bean (*Phaseolus vulgaris*) leaves.[160]

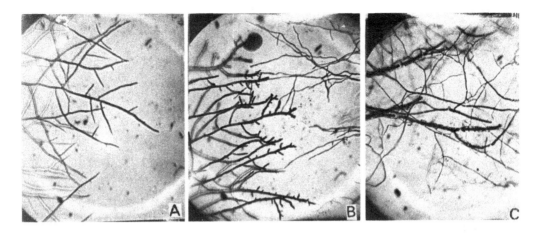

FIGURE 5. *Sclerotium cepivorum* hyphae in agar culture. (A) Healthy. (B) Showing stunted growth when in close proximity to *Trichoderma viride* hyphae. (C) *T. viride* coiling around hyphae of *S. cepivorum*. (From Ghaffar, A., *Mycopath. Mycol. Appl.*, 38, 112, 1969. With permission.)

The parasitism of *Microsphaeropsis centaureae* on sclerotia of *S. sclerotiorum* was found to resemble that of *C. minitans*.[177] *Sporodesmium sclerotivorum*, another mycoparasite, was detected in soil samples from fields which caused a natural decline of sclerotia of *S. sclerotiorum* in field soils.[3] Similarly, introduction of *Fusarium lateritium (Gibbrella baccata)* 7 hr before inoculation provided biological control of *S. sclerotiorum* in lettuce.[147] *Acrostalagmus roseus* parasitized sclerotia,[135] and *T. viride* isolated from decayed sclerotia caused a high level of decay in *S. sclerotiorum*.[90] However, *T. harzianum* from sunflower roots, which caused plasmolysis in hyphae of *S. sclerotiorum*, did not significantly control sclerotinia disease in sunflower.[97] The infection of the sclerotia of *S. sclerotiorum* by *Teratosperma oligocladum* may prove useful in its biological control.[15,167]

H. *Sclerotinia trifoliorum* Erikss.

Sclerotinia trifoliorum Erikss., the cause of crown and stem rot of alfalfa and clover,[6] survives in the form of sclerotia for 7 years.[132] *Mitrula sclerotiorum*,[140] *Trichoderma viride*,[106] and *Coniothyrium minitans*[158] parasitized sclerotia of *S. trifoliorum*. Up to 85 to 99% of the sclerotia were killed by *C. minitans* within 11 weeks, and the mycoparasite remained highly infective to *S. trifoliorum* for at least 14 months. Biological control of *S. trifoliorum* was achieved using pycnidial dust preparations of *C. minitans* which showed up to 98% infection and sclerotial destruction in the field.[163] Sclerotia of *S. trifoliorum* are also infected by *Teratosperma oligocladum* which may prove useful in its biological control.[15,167]

I. *Sclerotium cepivorum* Berk.

Sclerotium cepivorum Berk., the cause of white rot, although restricted to the *Allium* spp., is widespread in many parts of the world.[6] The sclerotia of the fungus survives for up to 4 years.[38] The observations of Scott[143] that mycelium of the fungus cannot grow saprophytically in unsterile soil suggested that antibiosis might be of some significance in the control of white rot disease. In agar culture, *Trichoderma viride* coiled around the hyphae of *S. cepivorum*[60] (Figure 5). *Coniothyrium minitans* parasitized the sclerotia of *S. cepivorum* and produced its pycnidia within them[60] (Figure 6). Microorganisms isolated from soil, viz., *Aleruisma carnis*, *Cladosporium elatum*, *Penicillium expansum*, *P. nigricans*, *P. notatum*, *P. piscarium*, *P. puberulum*, *P. rolfsii*, *P. urticae*, *P. variabile*, *Tilachlidium humicola*, and unidentified isolates of bacteria and actinomycetes produced diffusible antibiotics which inhibited growth of *S. cepivorum* in agar culture.[60]

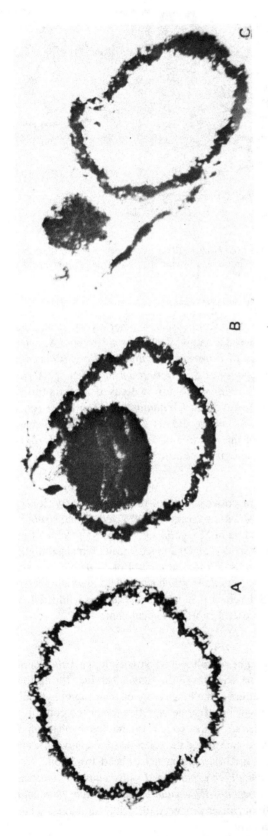

FIGURE 6.   Photomicrographs (× 150) of microtome sections of agar cultures showing parasitism of *Coniothyrium minitans* on the sclerotia of *Sclerotium cepivorum*. (A) A sclerotium of *S. cepivorum*. (B) Development of a pycnidium inside the sclerotium. (C) Development of pycnidium of *C. minitans* on top of a sclerotium of *S. cepivorum*. (From Ghaffar. A.. *Mycopath. Mycol. Appl.*, 38, 112, 1969. With permission.)

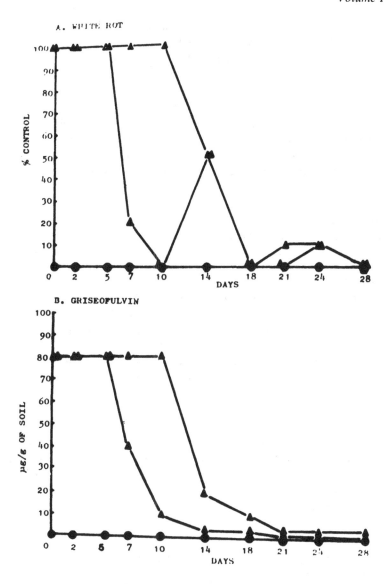

FIGURE 7. Control of white rot of onions and the concentrations of griseofulvin in garden loam alone and those mixed with cultures of *Penicillium nigricans* or *P. nigricans* supplemented with 4% maize meal. (△), garden loam + *P. nigricans*; (▲), garden loam + *P. nigricans* + 4% maize meal; (●), garden loam (control). (From Ghaffer, A., *Mycopath. Mycol. Appl.*, 38, 101, 1969. With permission.)

Biological control of *S. cepivorum* by *P. nigricans*[60,61] and *C. minitans*[7] under greenhouse conditions have been reported. *P. nigricans* cultures mixed with soil (1:4, w/w) were effective in the control of white rot infection up to 1 week after sowing onion seedlings.[61] Further addition of maize meal used as an organic substrate prolonged the effectiveness of *P. nigricans*. Control of white rot was associated with a concentration of 80μg of griseofulvin per gram of soil produced by *P. nigricans*[61] (Figure 7). Similarly, white rot of onion has been successfully controlled in greenhouse experiments by the mycoparasite *C. minitans*.[7] Pycnidial dust of *C. minitans* was applied to soil, and used as a seed dressing to protect onion seedlings planted in *S. cepivorum*-infested soil. *Bacillus subtilis* applied as a seed treatment to onions grown in nonsterile soil reduced the *S. cepivorum* infection on three cultivars in two field trials,[170,171] whereas seed treatment with spores of *P. nigricans* did not provide season-long protection. Absence of white rot in infested fields in Fraser valley of British Columbia in 1977 was attributed to a high level of fungistasis in disease suppressive fields. Lack of germination of the sclerotia of *S. cepivorum* due largely to competing microorganisms, viz., *B. subtilis* and *P. nigricans* which produced antibiotics antagonistic to the growth of *S. cepivorum* in agar culture.[60,170]

Sclerotia of *S. cepivorum* parasitized by *T. harzianum* were completely destroyed.[2] Application of a mycelial and spore suspension of *Trichoderma harzianum* decreased infection of onion in pots and in the field. Sclerotia of *S. cepivorum* were attacked by *Sporodesmium sclerotivorum*[14] and *Teratosperma oligocladum*.[15] Therefore, the use of the fungi as biological control agents also seems promising.

J. *Sclerotium delphinii* Welch

Sclerotium delphinii Welch, the cause of crown rot of Iris, is known to survive in the form of sclerotia in soil for 2 years.[184] In contrast the sclerotia buried below the soil surface quickly rotted and were left as broken hollow shells.[39] Fungi like *Trichoderma hamatum*, *T. harzianum*, *T. koningii*, *T. pseudokoningii*, *T. viride*, *Arthrobotrys oligospora*, the *Fusarium* spp., *Gliocladium roseum*, *Mucor racemosus*, *Pythium echinocarpum*, and the *Stilbum* sp., as well as bacteria were associated with dried sclerotia of *S. delphinii* buried in the field.[39,64] *T. hamatum* was the fungal species encountered most frequently with the sclerotia of *S. delphinii*. On potato dextrose agar (PDA), *T. hamatum* inhibited the growth of *S. delphinii*, eventually overgrew it, and parasitized the sclerotia.[64] Serial microtome sections showed more or less complete replacement of medullary elements by hyphae and chlamydospores of *T. hamatum* (Figure 8).

K. *Sclerotium oryzae* Catt.

Sclerotium oryzae Catt., (teleomorph = *Leptosphaeria salvinii* Catt.), the cause of stem rot of rice, is widespread in rice growing areas of the world.[125] The fungus survives in the form of sclerotia for up to 6 years in soil.[38] In agar culture *Aspergillus flavipes*,[168] *A. fumigatus*, *A. luchuensis*, *A. niger*,[71] *A. rugulosus*, *Penicillium funiculosum*,[168] *P. oxalicum*, *P. piceum*,[71] *P. purpurogenum*, *Pseudoarachniotus roseus*, *Stachybotrys atra*, *Streptomyces albus*, *S. noursei*, *S. rimosus* and the *Bacillus* sp.,[168] have been found to inhibit the growth of *Sclerotium oryzae*. *Aspergillus candidus*, *A. sulphureus*, the *Arachniotus* sp., *Talaromyces flavus*, and *Trichothecium roseum* have also been found to inhibit the growth of *S. oryzae* in agar culture (Shahzad and Ghaffar, unpublished). Decreased viability of the sclerotia of *S. oryzae* at greater soil depth in a paddy field was related to a predominant colonization of sclerotia by bacteria.[169]

L. *Sclerotium rolfsii* Sacc.

Sclerotium rolfsii Sacc. (teleomorph = *Athelia rolfsii* (Curzi) Tu and Kimbrough), a soil-borne pathogen, causes blight and root and stem rot in tropical and subtropical countries on more than 500 species of plants in about 100 dicotyledonous families comprising mostly Compositae and Leguminosae.[13] The sclerotia survive for over 5 years in agar culture.[136] *Trichoderma harzianum*,[5,52,112] *T. viride*, and *Bacillus subtilis*[5,77] were antagonistic to *S. rolfsii* in agar culture. *T. viride* has been isolated from decayed sclerotia of *S. rolfsii*.[42]

Successful biological control of *S. rolfsii*, the cause of southern blight of lupine, tomato, and peanuts in greenhouse and on tomato in the field, has been demonstrated by Wells et al.,[181] with a supplemental food base like rye grass for vigorous growth of the mycoparasite, *T. harzianum*. *T. harzianum* colonizes *S. rolfsii*, the hyphae of which are disrupted and die when *T. harzianum* comes in contact. Backman and Rodriguez Kabana[16] developed a system for the growth and delivery of biocontrol agents to soil. With this system the agent can be

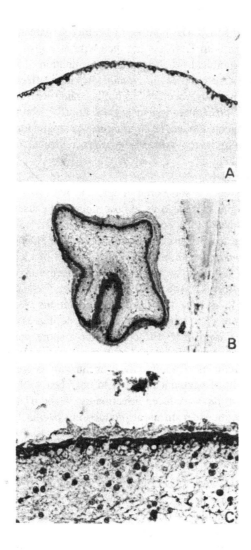

FIGURE 8.   Photomicrograph of microtome sections through sclerotia of *Sclerotium delphinii* showing parasitism of *Triochoderma hamatum*. (A) Portion of sclerotium of *S. delphinii*, control (× 150). (B) Development of conidia and chlamydospores of *T. hamatum* on and inside the sclerotia. (× 25). (C) A portion of infected sclerotium as in B (× 150). (From Coley-Smith, J. R., Ghaffar, A., and Javed, Z., *Soil Biol. Biochem.*, 6, 307, 1974. With permission.)

delivered with conventional farm machinery and can penetrate the foliar canopy. *T. harzianum* was grown on sterile granules for 4 days and applied 70 and 100 days after planting @ 140 kg/ha. A diatomaceous earth granule impregnated with 10% molasses solution (v/ v, pH 5.0, sp.gr. 1300, containing 3g KNO_3 and 3g KH_2PO_4/ℓ) was found suitable for growth and delivery of *T. harzianum* to peanut fields. An average yield increase of 12% with a significant reduction in *S. rolfsii* damage was observed. This is considered as the first economical method for the biological control of a soil-borne pathogen under field conditions.

A wheat bran inoculum preparation of *T. harzianum* significantly reduced *S. rolfsii* infection on bean, cotton, and tomato.[50] Biological control was more efficient at low temperatures of 27°C. Cultures of *T. harzianum* and *Bacillus subtilis* applied to seed rather than soil were more effective in the control of *S. rolfsii* causing collar rot and seedling death of lentil.[5] Seed treatment and soil drenching with *B. subtilis* increased wheat germination and stand in *S. rolfsii*-infested soil.[77] The number of healthy surgarcane sets rose by 20 to 50% when they were inoculated with *T. viride* and then with *R. rolfsii*.[157] Similarly, inoculation with the *Penicillium* sp. inhibited production and germination of sclerotia, providing good control of *S. rolfsii* on tomato.[104] *T. harzianum* provided effective control of *S. rolfsii* infection on groundnut[70] and sugarbeet.[112] Iris bulbs coated with *T. harzianum* were highly effective in reducing *S. rolfsii* infection in a glass house.[34] Reinfestation of *S. rolfsii* in groundnut fields was prevented when *T. harzianum* was applied after soil fumigation. Application of *T. harzianum* to the root zone of tomatoes controlled *S. rolfsii* in infested soil.[52]

M. *Verticillium dahliae* Kleb.

Verticillium dahliae Kleb., a soil-borne fungus, is worldwide in distribution in both temperate and tropical regions. It has a wide host range and causes wilt of many economic crops such as cotton, tomato, tobacco, alfalfa, cucurbit, etc.[88] Microsclerotia have been reported to survive for up to 13 years.[182] Several isolates of *Bacillus subtilis* isolated from American Elm (*Ulmus americana*) inhibited hyphal growth of *V. dahliae* in agar culture.[69] Germination of sclerotia followed by lysis and decay occurs in *V. dahliae* following the addition of certain amendments.[68] In cotton cv. Tashkent, *Trichoderma viride* reduced the disease incidence where *V. dahliae* was used for inoculations.[144] Similarly, application of the *Trichoderma* sp. before ploughing 3-year old lucerne decreased cotton wilt due to *V. dahliae* and increased the yield.[144,165] Effectivenss of a spore preparation of *T. lignorum* increased when the antagonist was introduced after maize but before cotton sowing, or when ploughing in the stubble remains of winter wheat or mustard as green manure.[109] Biological control of *Verticillium* wilt of eggplant in the field has been well demonstrated by Marois et al.[108] where *Talaromyces flavus* reduced infection by 76 and 67%, and increased the yield by 18 and 54% in two fields under different production systems.

III. CONCLUSION

Biological control would seem to be a desirable alternative to chemical control because its use can result in a decrease in the cost of chemical control as well as a decrease in water and soil pollution from pesticides. Although several potential biocontrol agents have been isolated and studied there is a need to improve strains of antagonists or discover new ones. Introduction of a proper strain of antagonist, at a time when the pathogen is more vulnerable, might delay disease development. In most cases a biocontrol agent introduced into the soil would be short-lived and would need a suitable substrate to support growth. The biotic environment of the soil is resistant to newly introduced microorganisms and there is need for a precise ecological knowledge on the behavior of the antagonist and pathogen. There is also a lack of information on the environmental hazards associated with the use of antagonists. There is a need to select suitable, inexpensive food bases and develop a technique for the management of biocontrol agents in the field. The economic feasibility has to be worked out since the production cost of the mycoparasite may generally be lower than cost of implementation by the farmer. Biological control should therefore be used judiciously in the field along with agronomic and cultural practices.

REFERENCES

1. **Abawi, G. S., Grogan, R. G., and Duniway, J. R.,** Microorganisms associated with sclerotia of *Whetzelinia sclerotiorum* in soil and their possible role in the incidence of lettuce drop, *Proc. Am. Phytopathol. Soc.*, 3, 275, 1976.
2. **Abd El Moity, T. H. and Shatla, M. N.,** Biological control of white rot disease of onion *(Sclerotium cepivorum)* by *Trichoderma harzianum, Phytopathol. Z.*, 100, 29, 1981.
3. **Adams, P. B. and Ayers, W. A.,** *Sporodesmium sclerotivorum*: distribution and parasitism in natural biological control of sclerotial fungi, *Phytopathology*, 71, 90, 1981.
4. **Adams, P. B. and Ayers, W. A.,** Biological control of *Sclerotinia* lettuce drop in the field by *Sporodesmium sclerotivorum, Phytopathology*, 72, 485, 1982.
5. **Agrawal, S. C., Khare, M. N., and Agrawal, P. S.,** Biological control of *Sclerotium rolfsii* causing collar rot of lentil, *Indian Phytopathol.*, 30, 176, 1977.
6. **Agrios, G. N.,** *Plant Pathology,* Academic Press, New York, 1978.
7. **Ahmed, A. H. M. and Tribe, H. T.,** Biological control of white rot of onion (*Sclerotium cepivorum*) by *Coniothyrium minitans, Plant Pathol.*, 26, 75, 1977.
8. **Akhtar, C. M.,** Biological control of some plant diseases lacking genetic resistance of the host crop in Pakistan, *Ann. N.Y. Acad. Sci.*, 287, 45, 1977.
9. **Allen, M. C. and Haenseler, C. M.,** Antagonistic action of *Trichoderma* and *Rhizoctonia* and other soil fungi, *Phytopathology*, 25, 244, 1935.
10. **Anonymous,** Control of *Rhizoctonia solani* in potato plants and tubers, *Rep. Long Ashton Res. Stn. Bristol 1981*, 1983, 100.
11. **Arora, D. K. and Dwivedi, R. S.,** Mycoparasitism of *Fusarium* spp. on *Rhizoctonia solani* Kühn., *Plant Soil*, 55, 43, 1980.
12. **Ayanru, W. A. and Green, R. G.,** Alteration of germination patterns of sclerotia of *Macrophomina phaseolina* on soil surface, *Phytopathology*, 64, 595, 1974.
13. **Aycock, R.,** Stem rot and other diseases caused by *Sclerotium rolfsii, N. C. Agric. Exp. Stn. Tech. Bull.*, 174, 1, 1966.
14. **Ayers, W. A. and Adams, P. B.,** Mycoparasitism of sclerotia of *Sclerotinia* and *Sclerotium* species by *Sporidesmium sclerotivorum, Can. J. Microbiol.*, 25, 17, 1978.
15. **Ayers, W. A. and Adams, P. B.,** Mycoparasitism of sclerotial fungi by *Teratosperma oligocladum, Can. J. Microbiol.*, 27, 886, 1981.
16. **Backman, P. A. and Rodriguez Kabana, R.,** A system for the growth and delivery of biological control agents to the soil, *Phytopathology*, 65, 819, 1975.
17. **Baker, K. F.,** Types of *Rhizoctonia* diseases and their occurrence, in *Rhizoctonia solani, Biology and Pathology,* Parameter, J. R., Jr., Ed., University of California Press, Berkeley, 1970, 125.
18. **Baker, K. F. and Cook, R. J.,** *Biological Control of Plant Pathogens,* W. H. Freeman & Co., San Francisco, 1974.
19. **Baker, R.,** Biological control of plant pathogens: Definitions, in *Biological Control in Agriculture. IPM Systems,* Hoy, M. A. and Herzog, D. C., Eds., Academic Press, New York, 1985, 25.
20. **Boogert, P. H., Van Den, J. F., and Jagger, G.,** Accumulation of hyperparasites of *Rhizoctonia solani* by addition of live mycelium to soil, *Neth. J. Plant Pathol.*, 89, 223, 1983.
21. **Boosalis, M. G.,** Effect of soil temperature and green manure amendment of unsterilized soil on parasitsm of *Rhizoctonia solani* by *Penicillium vermiculatum* and *Trichoderma* sp., *Phytopathology*, 46, 473, 1956.
22. **Boosalis, M. G. and Scharen, A. L.,** Methods for microscopic detection of *Aphanomyces eutiches* and *Rhizoctonia solani* and for isolation of *R. solani* associated with plant debris, *Phytopathology*, 49, 192, 1959.
23. **Bora, T.,** Studies on the antagonistic effects of a saprophytic bacterium *Pseudomonas aeruginosa* (Schroeter) Migula on some damping off fungi, *J. Turkish Phytopathol.*, 5, 43, 1976.
24. **Bora, T.,** *In-vitro in vivo* investigations on the effect of some antagonistic fungi against the damping off disease of egg plant, *J. Turkish Phytopathol.*, 6, 17, 1977.
25. **Brathwaite, C. W. D.,** Inhibition of *Sclerotium rolfsii* by *Pseudomonas aeruginosa* and *Bacillus subtilis* and its significance in the biological control of southern blight of pigeon pea (*Cajanus cajan* (L.) Mill., *Int. Congr. Plant Pathol.*, (Abstr.), Munich, West Germany, 1978.
26. **Burdsall, H. H., Jr., Hach, H. C., Boosalis, M. G., and Setliff, E. C.,** *Laetisaria arvalis* (Aphyllophorales, Corticiaceae): a possible biological control agent for *Rhizoctonia solani* and *Pythium* spp., *Mycologia*, 72, 728, 1980.
27. **Butler, G. M.,** Vegetative structures, in *The Fungi.* Vol. 2, Ainsworth, G. C. and Sussman, A. S., Eds., Academic Press, New York, 1966, 83.
28. **Butt, Z. L. and Ghaffar, A.,** Inhibition of fungi, actinomycetes and bacteria by *Stachybotrys atra, Mycopathol. Mycol. Appl.*, 47, 241, 1972.

29. **Campbell, W. A.**, A new species of *Coniothyrium* parasitic on sclerotia, *Mycologia,* 39, 190, 1947.

30. **Chakraborty, U. and Purkayastha, R. P.**, Role of rhizobitoxine in protecting soyabean roots from *Macrophomina phseolina* infection, *Can. J. Microbiol.,* 30, 285, 1984.

31. **Chand, T. and Logan, C.**, Antagonists and parasites of *Rhizoctonia solani* and their efficacy in reducing stem canker of potato under controlled conditions, *Trans. Br. Mycol. Soc.,* 83, 107, 1984.

32. **Chavez, H. B., Bloss, H. E., Boyle, A. M., and Gries, G. A.**, Effect of crop residues in soil on Phymatotrichum root rot of cotton, *Mycopathologia,* 58, 1, 1976.

33. **Chet, I. and Baker, R.**, Isolation and biocontrol potential of *Trichoderma hamatum* from soil naturally suppressive to *Rhizoctonia solani, Phytopathology,* 71, 286, 1981.

34. **Chet, I., Elad, I., Kalfon, A., Hadar, Y., and Katan, J.**, Integrated control of soilborne and bulb borne pathogens in Iris, *Phytoparasitica,* 10, 229, 1983.

35. **Chu, F. F. and Wu, W. S.**, Biological and chemical control of potato black scurf, *Plant Prot. Bull. Taiwan,* 22, 269, 1980.

36. **Chu, F. F. and Wu, W. S.**, Biological and chemical control of *Rhizoctonia solani* by pea seed treatment, *Mem. Coll. Agric. Nat. Taiwan Univ.,* 21, 19, 1981.

37. **Clark, F. E.**, Experiment toward the control of the take all disease of wheat and the Phymatotrichum root rot of cotton, *U.S. Dep. Agric. Tech. Bull.,* 835, 27, 1942.

38. **Coley-Smith, J. R. and Cooke, R. C.**, Survival and germination of fungal sclerotia, *Annu. Rev. Phytopathol.,* 9, 65, 1971.

39. **Coley-Smith, J. R., Ghaffar, A., and Javed, Z. U. R.**, The effect of dry conditions on subsequent leakage and rotting of fungal sclerotia, *Soil Biol. Biochem.,* 6, 307, 1974.

40. **Cook, R. J. and Baker, K. F.**, *The Nature and Practice of Biological Control of Plant Pathogens,* American Phytopathological Society, St. Paul, Minn., 1983, 539.

41. **Cook, G. E., Boosalis, M. G., Dunkle, L. D., and Odvody, G. N.**, Survival of *Macrophomina phaseoli* on corn and sorghum stalk residue, *Plant Dis. Rep.,* 57, 873, 1973.

42. **Curl, E. A. and Hansen, J. D.**, The microflora of natural sclerotia of *Sclerotium rolfsii* and effects upon the pathogen, *Plant Dis. Rep.,* 48, 446, 1964.

43. **Davey, C. B. and Papavizas, G. C.**, Effect of dry mature plant materials and nitrogen on *Rhizoctonia solani* in soil, *Phytopathology,* 50, 522, 1960.

44. **de la Cruz, R. E. and Hubbel, D. H.**, Biological control of the charcoal root rot fungus *Macrophomina phaseolina* on slash pine seedlings by a hyperparasite, *Soil Biol. Biochem.,* 7, 25, 1975.

45. **Dhingra, O. D. and Khare, M. N.**, Biological control of *Rhizoctonia bataticola* on Urid bean, *Phytopathol. Z.,* 76, 23, 1973.

46. **Dhingra, O. D. and Sinclair, J. B.**, Survival of *Macrophomina phaseolina* sclerotia in soil: effects of soil moisture, carbon:nitrogen ratios, carbon sources and nitrogen concentrations, *Phytopathology,* 65, 236, 1975.

47. **Dhingra, O. D. and Sinclair, J. B.**, *An Annotated Bibliography of Macrophomina phaseolina (1905-1976),* Univ. Fed. de Vicosa, Brasil and University of Illinois, 1977, 244.

48. **Durrell, L. W.**, Hyphal invasion by *Trichoderma viride, Mycopathol. Mycol. Appl.,* 35, 138, 1968.

49. **Elad, Y., Chet, I., Zeidan, O., and Henis, Y.**, Control of *Rhizoctonia* root rot in strawberry, *Hort. Abstr.,* 51, 2498, 1980.

50. **Elad, Y., Chet, I., Boyle, P., and Henis, Y.**, Parasitism of *Thrichoderma* sp. on *Rhizoctonia solani* and *Sclerotium rolfsii,* SEM and Fluorescent microscopy, *Phytopathology,* 73, 85, 1983.

51. **Elad, Y., Hadar, Y., Hadar, E., Chet, I., and Henis, Y.**, Biological control of *Rhizoctonia solani* in carnation, *Plant Dis.,* 65, 675, 1981.

52. **Elad, Y., Hadar, Y., Chet, I., and Henis, Y.**, Prevention with *Trichoderma harzianum* Rifai aggr., of reinfestation by *Sclerotium rolfsii* Sace. and *Rhizoctonia solani* Kühn of soil fumigated with methyl bromide and improvement in disease control in tomato and peanuts, *Crop Prot.,* 1, 199, 1982.

53. **Elad, Y., Kalfon, A., and Chet, I.**, Control of *Rhizoctonia solani* in cotton by seed coating with *Trichoderma* spp., spores, *Plant Soil.,* 66, 279, 1982.

54. **ElGoorani, M. A., Farag, S. A., and Shehata, M. R. A.**, The effect of *Bacillus subtilis* and *Penicillium patulum* in in-vitro growth and pathogenecity of *Rhizoctonia solani* and *Phytophthora cryptogea, Phytopathol Z.,* 85, 345, 1976.

55. **Ervio, L.**, Certain parasites of fungal sclerotia, *Maataloustieteellinen Aikak.,* 36, 1, 1964.

56. **Gadd, C. H. and Bertus, L. S.**, *Corticium vagum* B & C, the cause of a disease of *Vigna oligosperma* and other plants in Ceylon, *Ceylon J. Sci.,* 11, 27, 1928.

57. **Garrett, S. D.**, Toward biological control of soilborne plant pathogens, in *Ecology of Soilborne Plant Pathogens,* Baker, K. F. and Snyder, W. C., Eds., University of California Press, Berkeley, 1965, 571.

58. **Ghaffar, A.**, Survival of *Macrophomina phaseoli* (Maubl.) Ashby, the cause of root rot of cotton, *Pak, J. Sci. Res.,* 20, 112, 1968.

59. **Ghaffar, A.,** Interaction of soil fungi with *Macrophomina phaseoli* (Maubl.) Ashby, the cause of root rot of cotton, *Mycopathol. Mycol. Appl.,* 34, 196, 1968.

60. **Ghaffar, A.,** Biological control of white rot of onion. I. Interactions of soil microorganisms with *Sclerotium cepivorum* Berk., *Mycopathol. Mycol. Appl.,* 38, 101, 1969.

61. **Ghaffar, A.,** Biological control of white rot of onion. II. Effectiveness of *Penicillium nigricans* (Bain.) Thom., *Mycopathol. Mycol. Appl.,* 38, 113, 1969.

62. **Ghaffar, A.,** Interaction of actinomycetes with *Macrophomina phaseoli* (Maubl.) Ashby, the cause of root rot of cotton, *Mycopathol. Mycol. Appl.,* 44, 271, 1971.

63. **Ghaffar, A.,** Some observations of the parasitism of *Coniothyrium minitans* on the sclerotia of *Sclerotinia sclerotiorum, Pak. J. Bot.,* 4, 85, 1972.

64. **Ghaffar, A.,** Microorganisms associated with the decay of sclerotia of *Sclerotium delphinii* Welch, 2nd Int. Plant Pathol. Congr., (Abstr.), St. Paul, Minn., 1133, 1973.

65. **Ghaffar, A. and Akhtar, P.,** Survival of *Macrophomina phaseoli* (Maubl.) Ashby on cucurbit roots, *Mycopathol. Mycol. Appl.,* 35, 245, 1968.

66. **Ghaffar, A., Zentmyer, G. A., and Erwin, D. C.,** Effect of organic amendments on severity of *Macrophomina* root rot of cotton, *Phytopathology,* 59, 1267, 1969.

67. **Goel, S. K. and Mehrotra, R. S.,** Biological control of the root rot and collar rot of okra (*Abelmoschus esculentus* (L.) Moench), *Ann. Microbiol.,* 125A, 365, 1974.

68. **Green, R. J., Jr. and Papavizas, G. C.,** The effect of carbon source, C/N ratios and organic amendments on survival of propagules of *Verticillium alboatrum* in soil, *Phytopathology,* 58, 567, 1968.

69. **Gregory, G., Schreiber, L. R., Roberts, N., and Ichida, J.,** Antagonistic activity of *Bacillus subtilis* against the vascular wilt pathogen, *Phytopathology,* 75, 963, 1975.

70. **Grinstein, A., Elad, Y., Katan, J., and Chet, I.,** Control of *Sclerotium rolfsii* by a herbicide and *Trichoderma harzianum, Plant Dis. Rep.,* 63, 823, 1979.

71. **Gupta, A. K., Aggarwal, A., and Mehrotra, R. S.,** *In Vitro* studies on antagonistic microorganisms against *Sclerotium oryzae* Catt., *Geobios,* 12, 3, 1985.

72. **Hadar, Y., Chet, I., and Henis, Y.,** Biological control of *Rhizoctonia solani* damping off with wheat bran cultures of *Trichoderma harzianum, Phytopathology,* 69, 64, 1979.

73. **Hakeem, S. A. and Ghaffar, A.,** Reduction of the number of sclerotia of *Macrophomina phaseolina* in soil by organic amendments, *Phytopath. Z.,* 88, 272, 1977.

74. **Hakeem, S. A and Ghaffar, A.,** Combined effect of organic amendment and soil moisture on the decline in numbers of sclerotia of *Macrophomina phaseolina, Pak. J. Bot.,* 12, 153, 1980.

75. **Harman, G. E., Chet, I., and Baker, R.,** Factors affecting *Trichoderma hamatum* applied to seeds as a biocontrol agent, *Phytopathology,* 71, 569, 1981.

76. **Hawker, L. E.,** Ecological factors and the survival of fungi, *Symp. Soc. Gen. Microbiol.,* 7, 238, 1957.

77. **Hegde, R. K., Kulkarni, S., Sidderamaiah, A. L., and Krishnaprasad, K. S.,** Biological control of *Sclerotium rolfsii,* causal agent of foot rot of wheat, *Curr. Res.,* 9, 67, 1980.

78. **Henis, Y., Ghaffar, A., and Baker, R.,** Integrated control of *Rhizoctonia solani* damping off of radish. Effect of successive planting, PCNB, and *Trichoderma harzianum* on pathogen and disease, *Phytopathology,* 68, 900, 1978.

79. **Henis, Y., Sneh, B., and Katan, J.,** Effect of organic amendments on *Rhizoctonia solani* and accompanying microflora in soil, *Can. J. Microbiol.,* 13, 643, 1963.

80. **Hoes, J. A. and Huang, H. C.,** *Sclerotinia sclerotiorum,* viability and separation of sclerotia from soil, *Phytopathology,* 65, 1431, 1975.

81. **Hornok, L. and Welcz, I.,** *Fusarium heterosporum,* a highly specialized hyperparasite of *Calviceps purpurea, Trans. Brit. Mycol. Soc.,* 80, 377, 1983.

82. **Howell, C. R.,** Effect of *Gliocladium virens* on *Pythium ultimum, Rhizoctonia solani* damping off of cotton seedlings, *Phytopathology,* 72, 496, 1982.

83. **Howell, C. R. and Stipanovic, R. D.,** Control of *Rhizoctonia solani* on cotton seedlings with *Pseudomonas fluorescens* and with an antibiotic produced by the bacterium, *Phytopathology,* 69, 480, 1979.

84. **Huang, H. C.,** Biological control of *Sclerotinia* wilt of sunflower, *Can. Agric.,* 24, 12, 1976.

85. **Huang, H. C.,** Importance of *Coniothyrium minitans* in survival of sclerotia of *Sclerotinia sclerotiorum* in wilted sunflower, *Can. J. Bot.,* 55, 289, 1977.

86. **Huang, H. C.,** Control of sclerotinia wilt of sunflower by hyperparasites, *Can. J. Plant Pathol.,* 2, 26, 1980.

87. **Huang, H. C. and Hoes, J. A.,** Penetration and infection of *Sclerotinia sclerotiorum* by *Coniothyrium minitans, Can. J. Bot.,* 54, 406, 1976.

88. **Isaac, I.,** A comparative study of pathogenic isolates of *Verticillium, Trans. Br. Mycol. Soc.,* 32, 137, 1949.

89. **Jackson, C. R.,** Reduction of *Sclerotium bataticola* infection of peanut kernels by *Aspergillus flavus, Phytopathology,* 55, 934, 1965.

90. **Jones, D. and Watson, D.,** Parasitism and lysis by soil fungi of *Sclerotinia sclerotiorum* (Lib.) deBy., a phytopathogenic fungus, *Nature*, 224, 287, 1969.

91. **Jordan, H. V., Adams, J. E., Hooton, D. R., Porter, D. D., and Blank, L. M.,** Cultural practices as related to incidence of cotton root rot in Texas, *U.S. Dep. Agric. Tech. Bull.*, 948, 42, 1948.

92. **King, C. J., Hope, C., and Eaton, E. D.,** Some microbiological activities affected in manurial control of cotton root rot, *J. Agric. Res.*, 49, 1093, 1934.

93. **King, C. J., Loomis, H. F., and Hope, C.,** Studies on sclerotia and mycelial strands of the cotton root rot fungus, *J. Agric. Res.*, 42, 827, 1931.

94. **Kommendahl, T. and Windels, C. E.,** Evaluation of biological seed treatment for controlling root diseases of pea, *Phytopathology*, 68, 1087, 1978.

95. **Kovoor, A. T. A.,** Some factors affecting the growth of *Rhizoctonia bataticola* in soil, *J. Madras Univ.*, 24, 47, 1954.

96. **Kulkarni, S., Siddaramaiah, A. L., and Basavarajaiah, A. B.,** Antagonistic action of *Streptomyces* sp. on *Rhizoctonia bataticola*, a common soil borne pathogen of forest nurseries, *Indian For.*, 106, 126, 1980.

97. **Lee, Y. A. and Wu, W. S.,** Management of the *Sclerotinia* disease with biological and chemical methods, *Mem. Coll. Agric. Nat. Taiwan Univ.*, 19, 96, 1979.

98. **Lewis, J. A. and Papavizas, G. C.,** Integrated control of *Rhizoctonia* fruit rot of cucumber, *Phytopathology*, 70, 85, 1980.

99. **Likhachev, A. N. and Vasin, V. B.,** Determination of mycoparasitic activity of strains of *Trichoderma lignorum* (Tode) Harz., *Mikol. Fitopatol.*, 9, 248, 1975.

100. **Lily, V. G., Nair, U. K., Padalai, K. M., and Menon, K. P. V.,** Observations on the inhibitory activity of a species of bacterium on some fungi parasitic on the coconut palm, *Indian Coconut J.*, 5, 160, 1952.

101. **Liu, S. and Baker, R.,** Mechanism of biological control in soil suppressive to *Rhizoctonia solani*, *Phytopathology*, 70, 404, 1980.

102. **Logan, C., Cooke, L. R., Copeland, R. B., and Mills, P. R.,** Potato diseases, in *Annu. Rep. Res. Tech. Work 1982*, Department of Agriculture, Belfast, N. Ireland, 1983, 201.

103. **Loveless, A. R.,** Use of honey dew state in the identification of ergot species, *Trans. Br. Mycol. Soc.*, 47, 205, 1964.

104. **Lozano, T. and Pineda Lopez, B.,** Preliminary studies on the biological control of *Sclerotium rolfsii* Sacc. in Cordoba region, *Fitopatol. Columbiana*, 6, 67, 1977.

105. **Lyle, E. W., Dunlap, A. A., Hill, H. O., and Hargrove, B. D.,** Control of cotton root rot of sweet clover in rotation, *Texas Agric. Exp. St. Bull.*, 699, 5, 1948.

106. **Makkonen, R. and Pohjakallio, O.,** On the parasites attacking sclerotia of some fungi pathogenic to higher plants and on the resistance of those sclerotia to their parasites, *Acta. Agric. Scand.*, 10, 105, 1960.

107. **Mall, S.,** *Rhizoctonia* disease of legume crops as affected by *Trichoderma viride*, *Proc. Indian Nat. Sci. Acad. B.*, 41, 559, 1976.

108. **Marois, J. J., Johnston, S. A., Dunn, M. T., and Papavizas, G. C.,** Biological control of *Verticillium* wilt of egg plant in the field, *Plant Dis.*, 66, 1166, 1982.

109. **Marupov, A.,** On the question of methods using *Trichoderma* against *Vericillium* wilt of cotton, *Ref. Zh.*, 11, 55, 1974.

110. **Mass, J. I.,** Factors affecting development of Iris rhizome rot caused by *Botrytis convoluta* Whetzel & Drayton, *Diss. Abstr.*, 29, 1230, 1968.

111. **Mass, J. I.,** Effect of time and temperature of storage on viability of *Botrytis convoluta* conidia and sclerotia, *Plant Dis. Rep.*, 53, 141, 1969.

112. **Mathur, S. B. and Sarbhoy, A. K.,** Biological control of *Sclerotinia* root rot of sugarbeet, *Indian Phytopathol.*, 31, 365, 1979.

113. **Maude, R. B., Bambridge, J. M., and Presly, A. H.,** The persistence of *Botrytis allii* in field soil, *Plant Pathol.*, 31, 247, 1982.

114. **Meshram, S. O. and Jager, G.,** Antagonism of *Azotobacter chroococcum* isolate to *Rhizoctonia solani*, *Neth. J. Plant Pathol.*, 89, 191, 1983.

115. **Mitchell, D. T. and Cooke, R. C.,** Some effects of temperature on germination and longevity of sclerotia of *Claviceps purpurea*, *Trans. Br. Mycol. Soc.*, 51, 721, 1968.

116. **Mitchell, R. B., Hooton, D. R., and Clark, F. E.,** Soil bacteriological studies on the control of *Phymatotrichum* root rot of cotton, *J. Agric. Res.*, 63, 535, 1941.

117. **Merriman, P. R.,** Survival of sclerotia of *Sclerotinia sclerotiorum* in soil, *Soil Biol. Biochem.*, 8, 385, 1976.

118. **Mukhopadhya, D. and Nandi, B.,** Survival of *Macrophomina phaseolina* in presence of antagonistic organisms in jute rhizosphere, *Z. Pflanzenkr. Pflanzenschutz*, 85, 719, 1978.

119. **Nisikado, Y. A. and Hirata, K.,** Studies on the longevity of sclerotia of certain fungi under controlled environmental conditions, *Ber. Ohara Inst. Landwirtsch. Forsch.*, 7, 535, 1937.

120. **Nipoti, P., Sportelli, M. D., and Ercole, N.,** Biological control of collar rot (*Sclerotinia minor*) of glass house lettuce, *Inf. Fitopatol.*, 33, 71, 1983.

121. **Norton, D. C.,** Linear growth of *Sclerotium bataticola* through soil, *Phytopathology*, 43, 633, 1953.
122. **Norton, D. C.,** Antagonism in soil between *Macrophomina phaseoli* and selected soil inhabiting organisms, *Phytopathology*, 44, 522, 1954.
123. **Odvody, G. N., Boosalis, M. G., and Kerr, E. D.,** Biological control of *Rhizoctonia solani* with a soil inhabiting basidiomycete, *Phytopathology*, 70, 655, 1980.
124. **Odvody, G. N., Boosalis, M. G., Lewis, J. A., and Papavizas, G. C.,** Biological control of *Rhizoctonia solani*, *Proc. Am. Phytopathol. Soc.*, 4, 158, 1977.
125. **Ou, S. H.,** *Rice Diseases*, Commonwealth Mycological Institute, Kew, Surrey, England, 1972, 369.
126. **Pal, A. K. and Chaudhary, K. C. B.,** Mycoparasitic activity of *Trichoderma harzianum* on *Rhizoctonia solani*, *Proc. Indian Nat. Acad. Sci.*, 1345, 232, 1975.
127. **Papavizas, G. C.,** Microbial antagonism in bean rhizosphere as affected by oat straw and supplemental nitrogen, *Phytopathology*, 53, 1430, 1963.
128. **Papavizas, G. C. and Davey, C. B.,** *Rhizoctonia* disease of bean as affected by decomposing green plant materials and associated microfloras, *Phytopathology*, 50, 516, 1960.
129. **Papavizas, G. C. and Klag, N. G.,** Isolation and quantitative determination of *Macrophomina phaseolina* from soil, *Phytopathology*, 65, 182, 1975.
130. **Papavizas, G. C., Lewis, J. A., and Abd el Moity,** Evaluation of new biotype of *Trichoderma harzianum* for tolerance to benomyl and enhanced biocontrol capabilities, *Phytopathology*, 72, 126, 1982.
131. **Papavizas, G. C. and Lumsden, R. D.,** Biological control of soilborne fungal propagules, *Annu. Rev. Phytopathol.*, 18, 389, 1980.
132. **Pape, H.,** Beitrage zur Biologie und Bekämpfung des Kleekrebses (*Sclerotinia trifoliorum* Erikks.), *Arb. Biol. Abt. (Anst. Reichsanst.) Biol.*, 22, 159, 1937.
133. **Parfitt, D., Coley-Smith, J. R., and Jeeves T. M.,** *Teratosperma oligocladum*, a mycoparasite of fungal sclerotia, *Plant Pathol.*, 32, 459, 1983.
134. **Park, D.,** Survival of microrganisms in soil, in *Ecology of Soilborne Pathogens. Prelude to Biological Control*, Baker, K. F. and Snyder, W. C., Eds., University of California Press, Berkeley, 1965, 82.
135. **Pohjakallio, O., Salonen, A., Rukola, A., and Ikaheimo, K.,** On a mucous mold fungus, *Acrostalagmus roseus* Bainier, as antagonist to some plant pathogens, *Acta. Agric. Scand.*, 6, 178, 1956.
136. **Povah, A.,** Notes upon reviving old cultures, *Mycologia*, 19, 317, 1927.
137. **Rai, J. N. and Saxena, V. C.,** Sclerotial mycoflora and its role in natural biological control of white rot disease, *Plant Soil*, 43, 509, 1975.
138. **Radha, K.,** The genus *Rhizoctonia* in relation to soil moisture. II. *Rhizoctonia bataticola*, cotton root rot organism, *Indian Coconut J.*, 13, 137, 1960.
139. **Reichert, I. and Hellinger, E.,** On the occurrence, morphology and parasitism of *Sclerotium bataticola*, *Palest. J. Bot.*, 6, 107, 1947.
140. **Roed, H.,** *Mitrula sclerotiorum* Rostr. and its relations to *Sclerotinia trifoliorum* Erikss, *Acta Agric. Scand.*, 4, 78, 1954.
141. **Rogers, C. H.,** Cotton root rot studies with special reference to sclerotia, cover crops, rotations, tillage, seeding rates, soil fungicides and effects on seed quality, *Texas Agric. Exp. St. Bull.*, 614, 45, 1942.
142. **Sanford, G. B.,** Effect of various soil supplements on the virulence and parasitism of *Rhizoctonia solani*, *Sci. Agric.*, 27, 533, 1947.
143. **Scott, M. R.,** Studies of the biology of *Sclerotium cepivorum* Berk. I. Growth of the mycelium in soil, *Ann. Appl. Biol.*, 44, 576, 1956.
144. **Shadmanova, N. A., Kerimov, K. N., and Ishankulova, M.,** Effectiveness of *Trichoderma* against complex infectious backgrounds, *Khlopkovodstvo*, 5, 27, 1981.
145. **Shahjahan, A. K. M., O'Neill, N. R., and Rush, M. C.,** Interaction between *Rhizoctonia oryzae* and other sclerotial fungi causing stem and sheath diseases of rice, *Proc. Am. Phytopathol. Soc.*, 3, 229, 1976.
146. **Singh, R. J. and Mehrotra, R. S.,** Biological control of *Rhizoctonia bataticola* on gram by coating seed with *Bacillus* and *Streptomyces* spp. and their influence on plant growth, *Plant Soil*, 56, 475, 1980.
147. **Sitepu, K. and Wallace, H. R.,** Biological control of *Sclerotinia sclerotiorum* in lettuce by *Fusarium lateritium*, *Aust. J. Exp. Agric. Anim. Husb.*, 24, 272, 1984.
148. **Smith, W. H.,** Germination of *Macrophomina phaseoli* sclerotia as affected by *Pinus lambertiana* root exudates, *Can. J. Microbiol.*, 15, 1387, 1969.
149. **Snyder, M., Schroth, N., and Christou, H. R.,** Effect of plant residues on root rot of beans, *Phytopathology*, 49, 755, 1959.
150. **Streets, R. B.,** Diseases of guar (*Cyamopsis psoraloides*), *Phytopathology*, 38, 918, 1948.
151. **Streets, R. B. and Bloss, H. E.,** Phymatotrichum root rot, *Monogr. Am. Phytopathol. Soc.*, 8, 38, 1973.
152. **Sussman, A. S.,** Longevity and survivability of fungi, in *The Fungi*. Vol. 3, Ainsworth, G. C. and Sussman, A. S., Eds., Academic Press, New York, 1968, 447.
153. **Thirumalachar, M. J.,** *Chainia*, a new genus of the Actinomycetales, *Nature*, 176, 934, 1955.
154. **Thirumalachar, M. J. and O'Brien, M. J.,** Suppression of charcoal rot in potato with a bacterial antagonist, *Plant Dis. Rep.*, 61, 543, 1977.

155. **Thomas, K. M.,** Detailed administration report of the government mycologist, 1937-38, Madras, 1938, 21.

156. **Thomas, K. M.,** Detailed administration report of the government mycologist, 1938-39, Madras, 1939, 30.

157. **Tokeshi, H., Valdebanito, R. M., De Sonza, N. L., and Yokomizo, N. K. S.,** Biological control of *Sclerotium rolfsii* on sugarcane with *Trichoderma* sp. *Summa, Phytopathologica,* 6, 95, 1980.

158. **Tribe, H. T.,** On the parasitism of *Sclerotinia trifoliorum* by *Coniothyrium minitans, Trans. Br. Mycol. Soc.,* 40, 489, 1957.

159. **Trutman, P., Keane, P. J., and Merriman, P. R.,** Reduction in sclerotial inoculum of *Sclerotinia sclerotiorum* with *Coniothyrium minitans, Soil Biol. Biochem.,* 12, 461, 1980.

160. **Trutman, P., Keane, P. J., and Merriman, P. R.,** Biological control of *Sclerotinia sclerotiorum* on aerial parts of the plants by the hyperparasite *Coniothyrium minitans, Trans. Br. Mycol. Soc.,* 78, 521, 1982.

161. **Tu, J. C. and Vaartaja, O.,** The effect of the hyperparasite *(Gliocladium virens)* on *Rhizoctonia solani* and in *Rhizoctonia* root rot of white beans, *Can. J. Bot.,* 59, 22, 1981.

162. **Turchetti, T.,** Prospects of biological control of some diseases of forest plants, *Info. Fitopatol.,* 29, 7, 1979.

163. **Turner, G. J. and Tribe, H. T.,** Preliminary field plot trials on biological control of *Sclerotinia trifoliorum* by *Coniothyrium minitans, Plant Pathol.,* 24, 109, 1975.

164. **Turner, G. J. and Tribe, H. T.,** On *Coniothyrium minitans* and its parasitism of *Sclerotinia* species, *Trans. Br. Mycol. Soc.,* 66, 97, 1976.

165. **Udaidullaeu, K., Yuldashev, A., and Yunosov, M.,** The effectiveness of introduction of *Trichoderma* sp. and lucerne roots under cotton against *Verticillium* wilt, *Tr. Sredneaziat. Nauchno Issled. Inst. Zaschity Rast.,* 11, 79, 1977.

166. **Uecker, F. A., Ayers, W. A., and Adams, P. B.,** A new hyphomycete on sclerotia of *Sclerotinia sclerotiorum, Mycotaxon,* 2, 275, 1978.

167. **Uecker, F. A., Ayers, N. A., and Adams, P. B.,** *Teratosperma oligocladum,* a new hyphomycetous mycoparasite of sclerotia of *Sclerotinia sclerotiorum, S. trifoliorum,* and *S. minor. Mycotaxon,* 10, 421, 1980.

168. **Usmani, S. M. H.,** Biological control of *Sclerotium oryzae* Catt., the cause of stem rot of rice, Ph.D. Thesis, University of Karachi, Pakistan, 1981, 120.

169. **Usmani, S. M. H. and Ghaffar, A.,** Biological control of *(Sclerotium oryzae)* Catt., the cause of stem rot of rice. I. Population and viability of sclerotia in soil, *Pak. J. Bot.,* 6, 157, 1974.

170. **Utkhede, R. S. and Rahe, J. E.,** Biological control of onion white rot, *Soil Biol. Biochem.,* 12, 101, 1980.

171. **Utkhede, R. S. and Rahe, J. E.,** Effect of *Bacillus subtilis* on growth and protection of onion against white rot, *Phytopathol. Z.,* 106, 189, 1983.

172. **Vasudeva, R. S.,** Studies on the root rot disease in the Punjab. I. Symptoms, incidence and cause of the disease, *Indian J. Agric. Sci.,* 5, 496, 1935.

173. **Vasudeva, R. S. and Chakravarthi, B. P.,** The antibiotic action of *Bacillus subtilis* in relation to certain parasitic fungi with special reference to *Alternaria solani* (Ell. & Mart.) Jones & Grout, *Ann. Appl. Biol.,* 41, 612, 1954.

174. **Vasudeva, R. S. and Sikka, M. R.,** Studies on the root rot disease of cotton in the Punjab. X. Effect of certain fungi on the growth of root rot fungi, *Indian J. Agric. Sci.,* 11, 422, 1941.

175. **Voros, J.,** *Coniothyrium minitans* Campbell, a new hyperparasite fungus in Hungary, *Acta. Phytopathol. Acad. Sci. Hung.,* 4, 221, 1969.

176. **Warren, J. R.,** An undescribed species of *Papulospora* parasitic on *Rhizoctonia solani* Kühn, *Mycologia,* 40, 391, 1948.

177. **Watson, A. K. and Miltimore, J. E.,** Parasitism of the sclerotia of *Sclerotinia sclerotiorum* by *Microsphaeropsis centaureae, Can. J. Bot.,* 53, 2458, 1975.

178. **Weindling, R.,** *Trichoderma lignorum* as a parasite of soil fungi, *Phytopathology,* 22, 837, 1932.

179. **Weindling, R.,** Studies on a lethal principle effective in the parasitic action of *Trichoderma lignorum* on *Rhizoctonia solani* and other soil fungi, *Phytopathology,* 24, 1153, 1934.

180. **Weindling, R. and Fawcett, H. S.,** Experiments in the control of *Rhizoctonia* damping off of citrus seedlings. *Hilgardia,* 10, 1, 1936.

181. **Wells, H. D., Bell, D. K., and Jaworski, C. A.,** Efficacy of *Trichoderma harzianum* as a biocontrol for *Sclerotium rolfsii, Phytopathology,* 62, 442, 1972.

182. **Wilhelm, S.,** Longevity of the *Verticillium* wilt fungus in the laboratory and field, *Phytopathology,* 45, 180, 1955.

183. **Willets, H. J.,** The survival of fungal sclerotia under adverse environmental conditions, *Biol. Rev.,* 46, 387, 1971.

184. **Williams, G. H. and Western, J. H.,** The biology of *Sclerotinia trifoliorum* Erikss., and other species of sclerotium forming fungi. II. The survival of sclerotia in soil, *Ann. Appl. Biol.*, 56, 261, 1965.
185. **Wood, R. K. S.,** The control of diseases of lettuce by the use of antagonistic microorganisms. II. The control of *Rhizoctonia solani* Kühn, *Ann. Appl. Biol.*, 38, 217, 1951.
186. **Wright, E.,** Control of damping off of broad leaf seedlings, *Phytopathology*, 31, 857, 1941.
187. **Wu, W. S.,** Antibiotic and mycoparasitic effects of several fungi against seed and soil borne pathogens associated with wheat and oats, *Bot. Bull. Acad. Sin.*, 18, 25, 1977.
188. **Yeh, C. C. and Sinclair, J. B.,** Effect of *Chaetomium cupreum* on seed germination and antagonism to other seed borne fungi of sorghum, *Plant Dis.*, 64, 468, 1980.
189. **Young, P. A.,** Charcoal rot of plants in east Texas, *Texas Agric. Exp. Sta. Bull.*, 712, 1, 1949.
190. **Zak, B.,** Characterization and classification of mycorrhizae of Douglas fir. II. *Pseudotsuga menziesii* and *Rhizopogon vinicolor*, *Can. J. Bot.*, 49, 1079, 1971.

Chapter 11

BIOLOGICAL CONTROL OF DECAY FUNGI IN WOOD

P. C. Mercer

TABLE OF CONTENTS

I. INTRODUCTION

In nature the decay of wood is a process in which material, chiefly from the senescent or dead parts of trees, is recycled through the soil to other trees and plants. In a more artificial environment wooden parts of buildings, fence posts, and other man-made wooden structures may also be decomposed. A large number of microorganisms are associated with decay,[60,99] but those usually credited with a major role are the group of fungi belonging to the basidiomycetes.[41] Thus any attempt to control decay has to address itself to the restriction or cessation of growth of these fungi.

The control strategy will vary depending on whether the wood to be protected is part of a living tree or is a piece of dead timber. Control in the latter case is relatively straightforward using a range of chemicals[20] some of which are more successful than others, e.g., the traditional practice of treating the ends of fence posts and telegraph poles with creosote before placing them in the ground, considerably slows down the decay process, although eventually the wood does succumb.[30] The use of chemicals to control decay in living trees has been much less successful. The traditional approach[11] has been to paint exposed areas of wood with a bitumen preparation in the hope that this will exclude harmful microorganisms until the tree's own protective callus has covered the wound. However, recent work[65,97] has cast strong doubts on the efficacy of this approach and has indicated that in some circumstances decay may be even worse than if wounds were left untreated. There are several possible reasons for this. It is almost impossible to produce a pruning wound free of all propagules of fungi and bacteria. It is also impossible, at present, to obtain a preparation which will remain even visually intact for a substantial period. Mercer et al.[65] found fungal colonization of wounds of common beech only 11 days after treatment with bitumen. The addition of fungicides, especially the type used for treating timber, is also ineffective as they are generally phytotoxic[65] and inhibit the natural defenses of the tree.[96] Mercer et al.[65] found the most effective treatment to be a preparation containing mercuric oxide, which in the most successful tests excluded basidiomycetes for up to 18 months, although colonization by other fungi occurred within 2 months. Further, Mercer[62] found that other wound dressings, including bitumen preparations, could increase the growth of callus, and Mercer et al.[65] found indications of a slowing down or cessation of growth of decay fungi after wound closure. However, among the most common cases an arboriculturist has to deal with is the mature or overmature tree with large overhanging branches growing in a public place or too close to a building. With such a tree any wound resulting from the removal of branches is likely to be large, callus growth which is related to the tree's vigour is likely to be slow, and even with the stimulation of a wound dressing it is unlikely that the wound will cover over in the lifetime of the tree. In this case a wound treatment which prevents the ingress of basidiomycetes for only 18 months is likely to be of limited use in the long term.

Are there any alternatives to chemical control which hold out hope of more long-term success? One to which much attention has been recently paid is the use of biological control.

II. BIOLOGICAL CONTROL IN THE CONTEXT OF WOOD DECAY

For the purposes of this chapter biological control will be defined in the broad terms of Baker and Cooke:[3]

Biological control is the reduction of inoculum density or disease producing activities of a pathogen or parasite in its active or dormant state by one or more organisms accomplished naturally or through manipulation of the environment, host, or antagonist, or by mass introduction of one or more antagonists.

A slight problem with the application of this definition to the control of wood-decay organisms is that they are not generally pathogenic in the strict sense. Some wood-decay fungi can actively enter trees without wounding, e.g., *Heterobasidion annosum* and *Ganoderma lucidum*,[80,112,113] and can therefore be thought of as pathogenic, but it appears that most decay fungi are growing in wood which is either senescent or already dead. However, in so far as substantial decay of the nonliving parts of a tree can result in the complete destruction of the living parts due to windblow or collapse, decay fungi can be looked upon as pathogens in a broad sense. Obviously decay fungi growing in timber structures are purely saprophytic, but as it is likely that microorganisms controlling decay fungi in the living tree will also control the same fungi attacking timber, control of microorganisms in timber has been included for the sake of completeness. However, before exploring the application of biocontrol to decay fungi it is necessary to discuss the process of decay more thoroughly.

III. THE PROCESS OF DECAY

Although Johnson[47] suggested that dry rot in timber was caused by "a visit from a plant" his view does not seem to have gained wide support and the general consensus up till just over a hundred years ago was that decay engendered fungi ("spontaneous generation"). In 1878 Hartig[41] discovered that it was the other way round — that it was the activities of fungi which caused decay. This work gave rise to the classical heartrot concept whereby decay fungi, which were able to gain access to the heartwood of a tree, were able to move at will through it and cause decay. The tree had a purely passive role.[92] No other organisms were thought to be involved — the coincidence of decay and the presence of fruitbodies was so strong. Although other fungi and bacteria were found they were usually dismissed as contaminants. However, in the last 20 years the whole decay process has been much more thoroughly investigated and discussed. Although there is by no means complete agreement, it is seen to be a much more complex process than that suggested by the heartrot concept.

IV. SITUATIONS FOR DECAY

In trees there are three main situations in which decay may occur. In the first situation woody tissue becomes exposed due to wounds. These may occur as the result of artificial pruning,[59] natural pruning,[15,16] branch-breakage due to wind,[63] animals such as squirrels,[64] birds such as woodpeckers[36] and sapsuckers,[91] insects such as bark beetles,[117] and nematodes.[84] In the second situation branches become senescent due to suppression by neighboring trees[10] or die-back in overmature trees.[8] In the third situation potentially healthy and non-stressed trees are attacked by pathogenic fungi which can also cause decay, e.g., *Chondrostereum purpureum* in silver-leaf disease of fruit trees[22] and *Heterobasidion annosum* in butt rot of conifers.[80] Some pathogenic fungi, such as *Nectria galligena*, may not directly cause decay but produce cankers which allow decay by other organisms.

V. REACTION OF THE TREE

No matter how a decay fungus gains entry into a tree there is generally a response from the tree. Depending on the situation this may be adequate to completely restrict the fungus to the point of entry, or at the other extreme it may be completely inadequate and the tree may be killed and decayed.

The response of tree bark to wounding and/or the entry of decay fungi has been researched and discussed less than the subsequent responses of woody tissues. In 1984, however, Biggs et al.[4] reviewed the subject, describing the work of Mullick[68] in which he outlined two nonspecific processes in wounded bark. In the first process a superficial injury triggers phellogen regeneration probably within a few hours of injury. Biggs et al.[4] suggested this

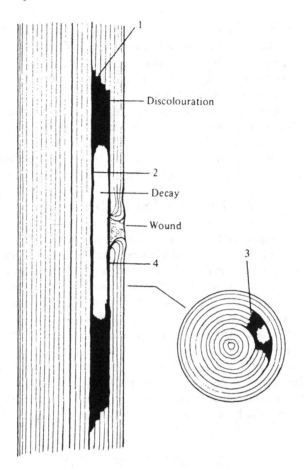

FIGURE 1.   Diagram of a longitudinal and a transverse section through a wound, showing the compartment produced by wall types 1 to 4 (see text). (After Shigo, A. L. and Marx, H. G., *U.S. Dept. Agric. For. Ser. Info. Bull.*, 405, 73, 1977.)

phellogen was similar to the phellogen observed in host-pathogen interactions reported in the literature. In the second process deeper injuries to the bark trigger regeneration of the vascular cambium. These two processes may or may not be able to restrict fungal growth and damage to the bark. Tippet et al.[111] observed the restriction of areas of dead bark by periderm (derived from phellogen) around drill wounds in the roots of Balsam fir and Western hemlock.

If wounds are deeper, allowing decay fungi to penetrate below the bark, they initiate what is popularly known as compartmentalization.[96] In this theory the area around a wound is regarded as a compartment with four different types of walls (Figure 1) which actively restrict the growth of invading decay fungi. The first wall is formed by the plugging of vessels above and below the point of wounding; the second wall is formed by existing annual rings; the third wall is formed by medullary rays; and the fourth wall by a barrier zone which is formed between the wood present at the time of wounding and the wood formed subsequently. The first through third walls together can be looked on as the "reaction zone" described earlier by Shain.[90] The walls are not all equally effective against fungal advance; their impermeability increases from the first wall through the fourth. This gives rise to the most common compartment shape — a wedge (Figure 2) usually elongated vertically (Figure 3). The walls appear to be formed by a combination of resistance due to anatomical structures and antifungal chemicals. Pearce and Rutherford[72] noted suberization of the fourth wall in

FIGURE 2.   Wound of lime showing restriction (↑) of decay (compartmentalization). (From Mercer, P. C., *Arboricul. J.*, 6, 131, 1982. With permission.)

response to fungal invasion of the wounds of oak. They also noted some suberization of ray parenchyma cells suggesting their possible involvement in the third wall. However, Boddy and Rayner[9] have questioned the whole idea of active compartmentalization suggesting that the restriction of invading decay fungi is purely accidental, resulting from differences in water content of the various tissues and also from extant tree anatomy. They point out that most fungi have difficulty growing in tissues with a high water content, and that as long as the tree is able to maintain its sapwood at a sufficiently high water content fungi will be excluded. Production of gums, tyloses, and suberized layers would help to seal off infected areas and retain a high moisture content in the remaining tissues. This alternative theory would, however, place trees in a position different from all other higher plants where active defense reactions such as induced resistance occur.[45] The theory also tends to dismiss a role for the high content of polyphenols found in wounded tissue. Schuck[86] found that wood of *Picea abies* reacts to injuries with an alteration of the terpene fraction in the resin. It is unlikely that this is pure coincidence. However, neither the theories of Shigo and Marx[96] or Boddy and Rayner[9] at present answer all the questions, and whichever (or neither) is correct it is still clear that decay fungi entering a tree wound tend to be restricted in their growth. The processes of containment may be adequate to restrict microorganisms to a small area around their point of entry or they may not, in which case extensive decay of any tissue present at the time of entry can occur.[96] Generally the larger the breach in a tree's surface the more extensive the subsequent decay.[63]

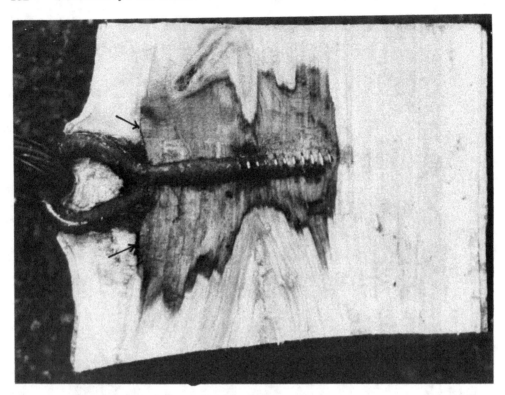

FIGURE 3. Section of London plane containing a cable-securing screw. Note elongation of stain above and below the screw and restriction of stain to the wood present at the time of wounding by the screw. Compartmentalization wall 4 (↑).

VI. INTERACTION OF DECAY FUNGI WITH OTHER MICROORGANISMS

Apart from wound size, growth of decay fungi may be governed by a number of factors, e.g., the season at the time of entry, the microclimate at the point of entry, the vigor and age of the tree, and any interaction with other microorganisms. Although other microorganisms such as bacteria have the capacity to decay wood and bark,[27] basidiomycetes (and some xylariaceous ascomycetes) are generally the chief agents of wood decay. Trees may be attacked by basidiomycetes acting alone. Wikström[118] observed *Phellinus tremulae* causing heartrot of *Populus tremula* without the involvement of other organisms. Boddy and Rayner[8] found basidiomycetes to be the predominant colonizers of attached branches of English oak, some species acting as pioneers. However, it is more common for basidiomycetes to be found in association with other microorganisms which may assist or inhibit the decay process.[99,101] Successions of microorganisms may occur, either resulting from the availability of nutrients[9] or the effects of antimicrobial substances produced by the host or other microorganisms.[99] One such succession was shown by Mercer[60] following the wound of a beech:

Injury → Specialized basidiomycetes; bacteria; (pioneer non-decay fungi) → Discoloration of wood → Non-decay fungi; bacteria; basidiomycetes → Reduction in wood discoloration → Basidiomycetes; bacteria → Decay

However, occasionally *Trichoderma viride* was present in which case basidiomycetes were generally absent. Shortle and Cowling[99] found similar successions in wounds of Sweetgum

FIGURE 4.   Section of incubated beech wood showing fungal individuals demarcated by zone lines
(↑). (From Mercer, P. C., *Arboricul. J.*, 6, 131, 1982. With permission.)

and Yellow poplar. However, it is probable that more than one type of succession can occur
with different tree species, and even in the same tree species but in different situations.
There may also be different successions in different parts of the tree. Decayed wood is
frequently divided up into a number of regions (Figure 4) which may contain either different
species of fungus or genetically incompatible mycelium of the same species.[7,76]

VII. DECAY IN TIMBER

Much of the basic substrate available for decay fungi in trees will also be available in
timber, and the decay process in the latter, therefore, has many similarities with that described

above with the main difference being a lack of any response from the "host". This difference has a marked effect on the relative resistance of sapwood and heartwood in trees and timber. Thus, in trees the living sapwood is able to resist decay more readily than the relatively inert heartwood, even though the latter often contains high concentrations of fungitoxic polyphenols. In timber, however, the opposite is the case and sapwood, lacking polyphenols, is more readily attacked than heartwood. However, the anatomy of timber is almost identical to that of the standing tree and decay fungi, as in the tree, can travel more readily along the vascular system ("grain") than at right angles to it.

Timber is liable to decay in various situations but the common denominator is water content.[20,28] Thus timber structures which have become wet due to rain, such as window and door frames (especially where grain is exposed at joints) are liable to wet rot. Flooring and skirting timbers in contact with damp masonry may develop dry rot. Timber posts embedded in the ground are very liable to rot, especially at ground level — above ground the wood is usually too dry for fungal growth and below ground it is too wet and has insufficient air.

VIII. THE APPLICATION OF BIOLOGICAL CONTROL TO DECAY

Because the process of decay can be affected by the wood/decay fungus combination as well as by the microclimate, the microecology, and other factors, the application of biocontrol has to be considered differently in each situation.

IX. BIOLOGICAL CONTROL OF DECAY-FUNGI IN TIMBER

Natural biological control of decay fungi in timber can be effected by keeping the environment sufficiently dry for the moisture content in the timber to be so low that fungal growth is impossible. This is relatively easily achieved in internal timber in buildings. Where this is not easy, as in window frames, door frames, fencing, etc., timber can be used which is naturally resistant to decay fungi either because of its density or inclusion of high levels of naturally occurring fungitoxic substances.[20] Heartwood, as already noted, is more resistant than sapwood and the most resistant timber tends to come from tropical hardwood species. However, such timber is expensive, and relatively susceptible but cheaper conifer softwood is much more commonly used.[28] Such wood can be protected relatively easily by chemicals, and there has not been much impetus for the use of microorganisms for decay control. It is difficult to envisage active control by microorganisms in structures such as window frames which are normally expected to be dry and which only become decayed when water permeates through cracks. Any biological control agent would have to maintain itself during the time the timber remained dry only coming into action when wetting occurred. On the other hand, Bruce and King[12] noted that *Scytalidium* and *Trichoderma* spp., which had been growing in blocks of *Pinus* and *Tilia* for 3 months, left behind residues which were fungitoxic to *Lentinus lepideus* even after the blocks had been sterilized and leached with water. Treatment of timber with such fungi might improve its resistance when the level of natural fungicides is low.

In many countries of the world, fence posts and telegraph poles, being embedded in the ground, are constantly being wetted and failure of these structures due to decay is common, even with chemical treatment. *Poria carbonica* is a fungus commonly associated with the decay of such poles, and Ricard and Bollen[79] obtained its control, when growing in wood chips, with a *Scytalidium* sp. also isolated from a telegraph pole. Bruce and King[12] and Klingström and Johansson[51] noted fungitoxicity due to *Scytalidium* even after the substrate had been sterilized by autoclaving. Ricard[77] claimed control of *P. carbonica* by *Scytalidium* in the poles themselves, but only one pole was examined in detail. *Lentinus lepideus* is another fungus associated with decay in telegraph poles, and Bruce et al.[13] obtained control in laboratory tests using *Trichoderma* spp. isolated from creosoted wood, while Ricard[78] successfully established growth of a mixture of *T. viride* and *T. polysporum* at the base of creosoted telegraph poles also to control *L. lepideus*. Control of another *Lentinus* sp.(*L. edodes*) by *Trichoderma, Gliocladium*, and other members of the *Hypocrea* was demonstrated by Komatsu and Inada.[52] Generally, however, although tests have been performed on the antagonism of microorganisms towards decay fungi on logs and wood blocks[44,66] the tests have not primarily been aimed at the control of decay in timber as such (although applicable), but in either the living or recently felled tree.

X. BIOLOGICAL CONTROL OF DECAY-FUNGI IN STUMPS

Some decay fungi, such as *Heterobasidion annosum* and *Phellinus weirii*, infect trees via root grafts from the roots of neighboring stumps which are themselves infected. In the case of *H. annosum* the stumps become infected by spores of *H. annosum* during the thinning process. However, if the stumps are colonized by competing fungi before *H. annosum* is able to infect them then subsequent infection by *H. annosum* is greatly reduced. Several fungi have been tried for this purpose, and the use of one of these, *Phlebiopsis (Peniphora) gigantea*, was one of the first examples of the biological control of tree decay. Rishbeth[81] obtained a measure of control in pine plantations in England after application of *P. gigantea* to the stumps. However, it was not until 11 years later[83] that a practical method was developed, and this is now used routinely in many forests following thinning. Results with spruce have not been as consistent as those on pine. Although Kallio[49] obtained control by stump inoculation in Finland, Seaby,[87] in Ireland, and Holdenrieder,[42] in Germany, found poor and negligible control, respectively. Greig[37] discovered that it was possible to inoculate *P. gigantea* successfully onto pine stumps by adding its spores to the lubricating oil of the chainsaw used in the felling. However, this also inoculated the basal end of the felled tree which was then also liable to decay.

Trichoderma spp. have also been used for control of *H. annosum*. However, results to date have been mixed and there is as yet no commercial exploitation of this control. Kallio[49] compared *T. viride* unfavorably with *P. gigantea* in spruce stumps. Seaby[87] obtained good control in both spruce and pine stumps when *T. viride* was inoculated in the spring, but poor control when inoculated in the winter when spores of *H. annosum* germinated more readily than did those of *T. viride*. Lundeborg and Unestam[57] and Holdenrieder[42] found control by *T. harzianum* both in vivo and in vitro. However, Seaby[88] always found *T. viride* superior to *T. harzianum*. Lundeborg and Unestam[57] also found control with *T. polysporum*. Siertoa[102] showed protection of wood blocks with filtrates of *T. viride*, but the degree of protection depended on the substrate on which the *T. viride* had been produced. Gibbs[35] found variation in control between *Trichoderma* isolates from acidic and alkaline soils. Inoculation via chainsaw lubricating oil was also attempted with *T. viride*. Early problems with suspension of the spores were overcome by the use of a mixture of chainsaw oil, water, polyethylene glycol, and lecithin.[89] Spores remained viable in the mixture for at least seven months. The method is still being evaluated, but, if successful, it could prove to be a useful alternative to *P. gigantea* since *Trichoderma* spp. are not decay fungi and there would therefore be no risk to felled timber.

Other potential control organisms have been less well researched. Klingström and Johansson[51] observed the elimination or inhibition of *H. annosum* from infected trees following inoculation with *Scytalidium album*. Aufsess[2] also observed control on slides, plates, and wood samples, but Holdenrieder[42] could show no effect with *S. lignicola* in spruce stumps. Pratt[74] observed inhibition of *H. annosum* by *Cryptosporiopsis abietina* in autoclaved Sitka spruce heartwood, but although Stillwell[105] also observed inhibiting effects of this fungus, Pratt frequently observed it growing in close proximity to rot columns caused by *H. annosum* in Sitka spruce trees.

Phellinus weirii, which causes a root rot, also spreads to healthy trees from infected stumps and/or roots. Nelson and Thies,[69,70] in a bid to accelerate the natural replacement of the pathogen, were able to successfully inoculate *Trichoderma viride* into infected stumps by drilling holes in the stumps and inserting pellets or dowels containing the fungus. After one year *T. viride* was isolated more frequently from those stumps with advanced decay than from sound stumps or stumps which were only stained. However, the authors concluded that a greater degree of colonization by *T. viride* would be required before it could be used as a practical biocontrol agent. Unlike Seaby's experience[87] with *Trichoderma* and *H. annosum*, there was a greater success with autumn and winter inoculations than with summer inoculations.

XI. BIOLOGICAL CONTROL OF PATHOGENIC DECAY FUNGI IN TREES

The majority of decay fungi entering a tree are saprophytic or at the most mildly pathogenic. One exception to this is *Chondrostereum purpureum*, the causal agent of silver-leaf disease of fruit trees. The fungus requires fresh wounds (usually pruning) for entry and can then move quickly through the tree colonizing individual branches and causing partial detachment of the leaf cuticle giving rise to a silvery appearance. Sometimes, as yet unexplained, remission occurs and the tree may recover. More usually the branch is killed and bears the numerous characteristic fruit bodies of the pathogen. If sufficient branches are affected the entire tree may be killed. Grosclaude[38] was the first to try to control *C. purpureum* biologically. He claimed complete protection against the infection of wounds of plum trees by application of the spores of *Trichoderma viride*. Dubos and Ricard[29] also found the treatment effective in peach trees. Grosclaude et al.[39] devised a method of inoculating the spores via pruning shears, as did Jones et al.[48] Corke,[22] however, although he did not get the complete protection in plum trees that Grosclaude had achieved, did effect a degree of recovery in infected pear trees by drilling holes in the trunks and inserting wood dowels colonized by *T. viride*. Symptoms were reduced by 61% which was significantly greater than the reduction due to natural remission. Woodgate-Jones,[119] examining the effects of temperature inoculated *T. viride* into infected plum trees in January, April, July, and October but found no significant differences in the percentage of recovery.

A group of fungi which do not directly cause decay but may eventually lead to it are those causing cankers, e.g., *Nectria*, *Seiridium*, and *Eutypa* spp. *Nectria galligena* causes a canker of apple trees. It enters either through pruning cuts[17] or via leaf-scars following leaf-fall in the autumn.[107] Research on control has been aimed at both these points of entry. Thus Swinburne,[107] Swinburne and Brown,[108] and Swinburne et al.,[110] in Ireland, found that *Bacillus subtilis*, a bacterium isolated from leaf-scars, was highly antagonistic to *N. galligena* in vitro, and if applied to leaf-scars immediately following leaf-fall it reduced subsequent infection by *N. galligena*. However, although results were good following artificial defoliation of trees they were not better than those obtained by conventional fungicidal sprays when compared under orchard conditions. Corke and Hunter,[23] in England, used a range of microorganisms on freshly pruned shoots and also obtained control of *N. galligena* with *B. subtilis* and with a *Penicillium* sp., although there were some reservations about the latter's suitability as it could also cause fruit rot. Inoculation with *Trichoderma viride*, on the other hand, generally aggravated cankering. However, Magro et al.,[58] in N. Italy, were successful in controlling *Seiridium cardinale*, agent of cypress canker, using *Trichoderma viride*, and they showed that it was almost invariably a part of the saprophytic flora of cypress shoots. Carter[18] and Carter and Price,[19] in Australia, showed inhibition of the growth of *Eutypa armeniacae*, the cause of a canker of apricot trees, by *Fusarium lateritium* in vitro, and they also obtained a significant level of protection in trees by the inoculation of the same fungus onto pruning wounds.

XII. BIOLOGICAL CONTROL OF DECAY FUNGI IN WOUNDS

Although pathogenic decay fungi and canker fungi can enter via pruning wounds, the majority of decay fungi found in wounds are saprophytic or only mildly pathogenic. Most research on biological control had centered on the former fungi and it is only relatively recently that attention has been turned towards the control of fungi causing decay in wounds. Pottle et al.,[73] in the U.S., showed that pruning wounds of Red maple could be protected from basidiomycetes for over 2 years by inoculating the wounds with a suspension (in glycerol) of *Trichoderma harzianum* isolated from Red maple wood (Figure 5). There were, however, signs of a breakdown in effectiveness by 31 months. There were suggestions that the inoculation of *T. harzianum* with a bacterium which had been isolated along with *T. harzianum*, was more effective than *T. harzianum* alone. Mercer and Kirk[66,67] used a preliminary test on malt agar in Petri dishes to screen 514 fungal and 22 bacterial isolates from wood on a range of basidiomycetes. The most commonly occurring genus with controlling properties was *Trichoderma* although there was considerable variation even within the same species (Table 1).

Other fungi showing a degree of inhibition were *Cryptosporiopsis fasiculata* and *Fusarium lateritium*. Of the bacteria three isolates of *Bacillus* spp. showed a moderate measure of control. Further in vitro tests on wood strips indicated one of the *T. viride* isolates to be the most effective, and it together with isolates of *C. fasiculata*, *F. lateritium*, and a *Bacillus* sp. were tested further by inoculation into trunk wounds of *Fagus sylvatica*. The trials, which spanned four years and relied on natural infection of the wounds by basidiomycetes, showed similar trends to the laboratory tests in that the most effective microorganism was the *T. viride* isolate. This reduced levels of basidiomycetes in inoculated wounds to 15% of that in uninoculated wounds (Figure 6). Inoculation of *T. viride* in glycerol improved its establishment over inoculation in water. Lower numbers of basidiomycetes were also isolated from wounds inoculated with *C. fasiculata* and *F. lateritium* than from uninoculated wounds, but the differences could not be shown to be significant. Field tests with the *Bacillus* isolate appeared totally unsuccessful as no *Bacillus* could be reisolated only $8^{1}/_{2}$ months after inoculation and the trial was abandoned. However, no decay fungi were isolated from inoculated or uninoculated wounds and it is possible that if the trial had continued differences might have been found due to fungitoxic residues.[56] Preston et al.[75] found six isolates of the *Bacillus* spp. from sound aged wood to be active in controlling a range of basidiomycetes.

XIII. ALTERATION OF THE ENVIRONMENT TO ENCOURAGE BIOLOGICAL CONTROL

As well as simply applying inoculum of an antagonistic microorganism to a wound or stump, or injecting it into a tree, it may be possible to alter the microenvironment either to encourage an artificial inoculant or to build up numbers of naturally occurring microorganisms which have controlling properties. Depending on the situation this can be done in a number of ways. Chemicals may be added with an inoculant, e.g., the use of glycerol to improve the establishment of *Trichoderma viride*.[67,73] Or the chemicals may be added separately, e.g., inoculation of *T. viride* to creosote-treated telegraph poles.[78] The addition of only a chemical may alter the substrate in such a way that it is more attractive to a biocontrol agent. Thus Bliss[6] discovered that the successful control of *Armillaria mellea* root rot in citrus orchards in the U.S. was at least partially achieved by the build-up of large amounts of *T. viride* subsequent to fumigation. Similarly, Rishbeth[82] noted that part of the effect of creosote

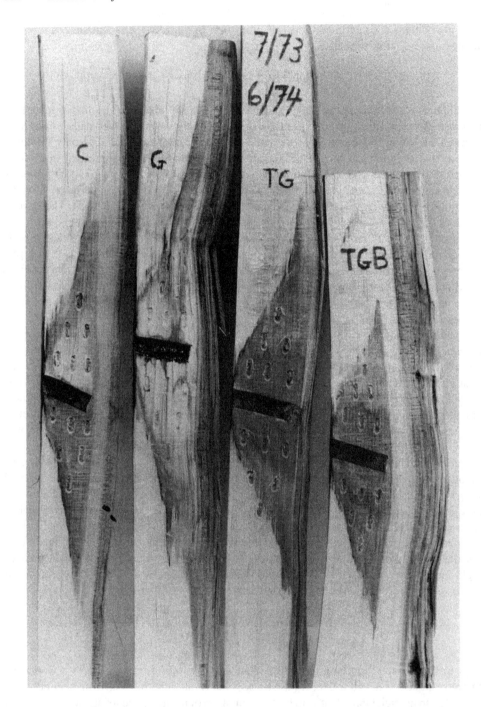

FIGURE 5.   Section of Red maple drilled and treated as follows: (C) control; (G) glycerol; (TG) *Trichoderma harzianum* inoculated in glycerol; and (TGB) as TG but with bacterial culture originally isolated along with *T. harzianum*. Note decayed wood in sections C and G. The small holes indicate sampling points. (From Shigo, A. L., *Northeastern For. Exp. Stn. Gen. Tech. Rep.*, U.S. Dept. Agric., NE 82, 167, 1983.)

or ammonium sulphamate added to pine stumps to control infection by *Heterobasidion annosum* could be explained by the encouragement of microorganisms such as *T. viride*. Seaby[87] noted improved colonization by *T. viride* of pine and spruce stumps when the inoculum was made up in 1% ammonium sulphamate.

Table 1
EFFECT OF ISOLATES OF *TRICHODERMA*
SPP. ON THE GROWTH[a] OF
***CHONDROSTEREUM PURPUREUM* ON MALT**
AGAR

Days after inoculation	Isolate number[b]					
	1	2	3	4	5	6
2	−0.02	0.06	0.00	0.16	0.00	0.00
3	−0.22	0.25	0.10	—	0.05	0.58
4	—	—	—	—	0.13	1.55

Note: S.E., ± 0.052.

[a] Difference(cm) from *C. purpureum* growing without *Trichoderma* isolates. Measurements ceased when cultures met.
[b] 1 to 3, three isolates of *T. viride*; 4, *T. harzianum*; 5, *T. polysporum*; 6, *T. koningii*.

After Mercer, P. C. and Kirk, S. A., *Ann. Appl. Biol.*, 104, 221, 1984.

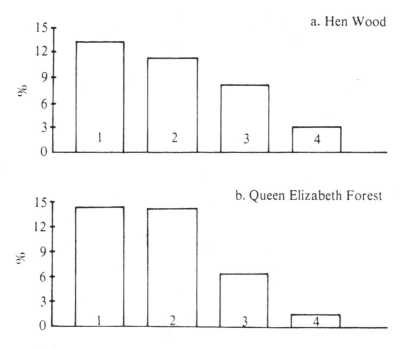

FIGURE 6.   Percentage of trunk wounds of beech yielding decay fungi four years after wounding. There were two sites — Hen Wood and Queen Elizabeth Forest. (1) uninoculated; (2) inoculated with *Fusarium lateritium*; (3) inoculated with *Cryptosporiopsis fasiculata*; and (4) inoculated with *Trichoderma viride*. Treatment 4 was significantly different from 1 at the 5% level at Hen Wood and the 1% level at Queen Elizabeth Forest. No other comparisons were significantly different. (From Mercer, P. C. and Kirk, S. A., *Ann. Appl. Biol.*, 104, 221, 1984. With permission.)

The effects of some chemicals in producing biological control may have been partly fortuitous, but it may also be possible to devise a method whereby a biocontrol agent and chemical are specifically selected to act in tandem. Thus Papavizas et al.[71] induced a level of tolerance to benomyl in cultures of *Trichoderma harzianum* which had been irradiated by UV light. Such cultures could then be inoculated either in benomyl or to substrates already treated with the chemical. This particular example would not be directly applicable to the control of basidiomycetes as most of these are highly tolerant of benomyl. However, the principle remains and inducing resistance in a fungus such as *T. viride* to a fungicide such as triadimenol could prove very useful in the treatment of pruning wounds where chemical treatment is only effective for a short time.[34] Chemicals and biocontrol agents may also be added at different times of the year as part of a management program. For example, Swinburne and Brown,[108] found that *Nectria* cankers in apple trees were only effectively controlled by *Bacillus subtilis* if there had been an earlier spray with the fungicide dithianon.

The microenvironment may also be altered by nonchemical means. Both Shortle and Shigo,[100] and Mercer and Kirk[67] noted that wrapping wounds in black polyvinyl chloride (PVC) reduced the number of basidiomycetes compared with unwrapped wounds. The amount of callus in wrapped wounds was also greater, and since the closure of wounds by callus growth appears to restrict subsequent decay[62] this would be beneficial in wounds where closure could be expected in 2 to 3 years. A similar situation was observed by Mercer et al.[65] where a chemically inert, flexible wound covering (Lac Balsam®) controlled basidiomycete colonization for over a year and also increased callus growth.

Another method of altering the microenvironment was recommended by Leach.[54] He observed that the removal of a strip of bark round the base of a tree infected with *Armillaria mellea* reduced the carbohydrate reserves in the roots and encouraged colonization by saprophytic soil fungi at the expense of *A. mellea*.

XIV. ENHANCEMENT OF BIOLOGICAL CONTROL BY THE HOST

As well as encouraging the growth of biocontrol agents it is also possible to improve the natural resistance of the tree. For example, irrespective of the theories of active or passive restriction of decay fungi, it is clear that there is considerable variation in restriction both between species and within species (Figure 7). Copony and Barnes[21] and French and Manion[33] noted clonal variation in the incidence of *Hypoxylon* canker on trembling aspen. Shigo et al.[98] suggested that the restriction of decay fungi in hybrid poplar was at least partially under genetic control. However, Johansson and Unestam,[46] in a good review of the search for host resistance to *Heterobasidion annosum*, concluded that although there was evidence for a genetic component in host resistance to the fungus it was still not clear, in spite of much research, how breeding for resistance can be accomplished as trees coming from so-called resistant clones still had a very wide range of response.

As noted above the closure of wounds by callus tissue appears an effective method of restricting the growth of decay fungi.[65] Some trees are more effective at closure than others, and it is possible that lines with quick growing callus could be bred. Growth of callus is also related to general tree growth and vigor.[62] Younger trees callus faster than older ones and this is a good reason for pruning while the trees are young. Wound size will also tend to be smaller. The improvement in general vigor by the addition of fertilizer has also been shown by Mercer[62] to improve callus growth in lime trees. However, improvement of wound closure is not necessarily correlated with restriction in the growth of decay fungi by mechanisms other than closure. Thus Shigo et al.[98] found no correlation in hybrid poplar, and Hallaksela[40] found no differences in infection by *Heterobasidion annosum* of a fertilized and an unfertilized stand of Norway spruce, but did find larger numbers of *Graphium* spp. in the fertilized stand. Blanchette and Sharon,[5] on the other hand, did show an improvement

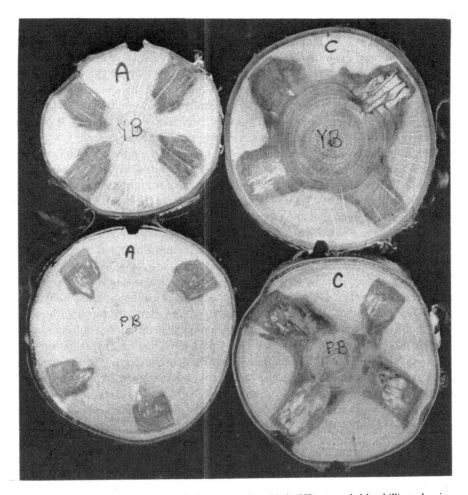

FIGURE 7.   Sections from paper birch (PB) and yellow birch (YB), wounded by drilling, showing variations in internal response. Trees YB(A) and PB(A) were clear to the pith before wounding indicating efficient walling-off of branch stubs. Damage from drill wounds was also effectively restricted. Trees YB(C) and PB(C) had large columns of central discolored wood before wounding suggesting poor walling-off of branch stubs. Wounds resulted in larger areas of damage and decay than in YB(A) or PB(A). (From Shigo, A. L., *Northeastern For. Exp. Stn. Gen. Tech. Rep.*, U.S. Dept. Agric., NE 82, 167, 1983.)

both in wound closure and restriction of decay fungi by the addition of an inoculum of *Agrobacterium tumefaciens* to wounds of Yellow birch.

Although fungi entering a tree via either senescent branches or stubs left by pruning wounds can move relatively easily along the branch or stub, several authors have noted a naturally resistant area of tissue where the branch joins the trunk.[1,9,63,94] This area will often restrict decay fungi to either the branch or stub. Often this area is demarcated by an obvious collar of tissue. Avoiding cutting into this area when pruning will allow the tree's natural control of decay to be fully realized.[93,95] Flush pruning will reduce the tree's defenses and will also create a considerably larger wound which will take much longer to callus over.

XV. MODES OF ACTION OF BIOCONTROL AGENTS

As the microorganisms and the conditions for the initiation of wood decay are varied it would be expected that any biological control methods would also be likely to operate in different ways and with different control microorganisms. Surprisingly, however, in spite of the amount of research on biological control, only a relatively small number of groups of microorganisms have been found which show any significant control under field conditions. In the research described above the most commonly employed group of fungi were those of the genus *Trichoderma*, e.g., Mercer and Kirk[66] found that out of a group of over 500 test fungi the largest group showing active control in vitro were the *Trichoderma* spp., although there was considerable variation between both species and isolates, a point also noted by Wells and Bell.[116] The antagonistic nature of the *Trichoderma* genus has long been known. Saccardo[85] noted *Trichoderma* spp. as being parasitic on the fruit bodies of basidiomycetes. Weindling[115] noted the control of *Rhizoctonia solani* by *T. lignorum* and suggested that the *Trichoderma* spp. might be suitable candidates for biological control agents. He observed that the hyphae of *T. harzianum* coiled around the hyphae of *R. solani*, behavior also noted by Ferrera Cerrato[31] and by Dennis and Webster in a detailed examination of the mode of action of the *Trichoderma* isolates.[24-26] They also discovered both volatile and nonvolatile inhibitors of a range of fungi in vitro. Mercer and Kirk[66] and Bruce et al.[13] also noted inhibition of *Chondrostereum purpureum* and *Lentinus edodes*, respectively, by volatiles from *Trichoderma* spp. In spite of this range of control methods other possibilities have also been put forward. Hulme and Shields[43] suggested that control may be due to the removal by *Trichoderma* of the nonstructural carbohydrates necessary for rapid fungal colonization. Smith et al.[104] suggested a more complex process in the control of basidiomycetes in wounds of Red maple by *T. harzianum*. They postulated that the first colonizers of the wounds were fungi such as the *Phialophora* spp. which could grow in the phenol-rich conditions associated with wounding and which could detoxify the tissue sufficiently for basidiomycetes (which are sensitive to phenols) to then be able to colonize. *T. harzianum* also grew in the phenol-rich conditions but did not detoxify and basidiomycetes were therefore not able to grow. Control of *Heterobasidion annosum* in conifer stumps may occur in a similar manner. The lack of sensitivity of *Trichoderma* spp. to phenols could also explain the successful inoculation of *T. viride* into the phenol-rich conditions of creosote-treated telegraph poles.[78] It is possible that isolates of *Trichoderma* can also induce high levels of phenols. Mercer and Kirk[67] noted significantly more discoloration and therefore more phenols in beech wounds inoculated with *T. viride* than in uninoculated wounds. The discolored areas were largely sterile. However, the method of control by *T. viride* of *Chondrostereum purpureum* in silver-leaf disease is still not clear. Translocation of *T. viride* propagules does not appear to occur and activity must therefore be at a distance, possibly by translocated metabolites. On the other hand, fruit from trees treated with *T. viride* was shown by Corke[22] to have none of the known *Trichoderma* antibiotics.

Scytalidium spp., another group of fungi commonly used for biological control, have also been shown to produce antibiotics active against basidiomycetes.[12,79] Klingström and Johansson[51] purified a fraction from a culture of *S. album* which inhibited growth of *Heterobasidion annosum* and stimulated the production of phenol oxidase. Unlike *Trichoderma* spp., volatile inhibitors do not appear to be produced.[13] Aufsess[2] noted both a chemical action and a coiling of the hyphae of *Scytalidium* around the hyphae of *H. annosum*.

Although *Cryptosporiopsis* spp. showed antagonistic effects towards decay fungi in vitro the effects in vivo were less obvious.[66,67,104] Nevertheless, *Cryptosporiopsis* spp. have been shown to have a range of at least four antibiotics active against fungi.[32,106] *Fusarium lateritium*, although successful in controlling apricot canker,[18] was even less successful than *Cryptosporiopsis fasiculata* in controlling decay fungi in beech wounds.[67] Nevertheless, *Fusarium* spp., like *Cryptosporiopsis* spp., have been shown to produce antibiotics active against fungi. Kapoor and Hoffman[50] noted strong antagonism of cereal eyespot and sharp eyespot pathogens by *F. avenaceum*. Inhibition was related to antibiotic production rather than nutrient depletion. Burmeister et al.[14] also noted the production of an antibiotic produced by isolates of *F. roseum* which caused swelling of hyphae of *Penicillium digitatum*.

The only group of bacteria commonly found in trials on the biological control of decay fungi are *Bacillus* spp. Here too antibiotic production has been implicated in their controlling activity. Landy et al.[53] isolated bacillomycin, an antifungal antibiotic, from *B. subtilis*. Wakayama et al.[114] isolated a peptide (mycocerein) from the culture filtrate of *B. cereus*, and this showed antifungal activity when tested against a range of fungi, though basidiomycetes were not tested. Loeffler et al.[56] observed antifungal effects against *Rhizoctonia solani* by an antibiotic (fengymicin) produced by *B. subtilis*, while another antibiotic from *B. subtilis* (bacilysin) inhibited yeast and bacteria but not filamentous fungi.

Improved biological control can sometimes be achieved by the use of additives, either with an inoculant or to a substrate, to encourage naturally occurring microorganisms, but it is not always clear how this works. For example, possible reasons for the improved control by *Trichoderma viride* of *Heterobasidion annosum* by using an inoculant in an ammonium sulphamate solution have been suggested by Seaby:[87] (1) the supply of nitrogen may be limiting to *T. viride*, (2) antagonistic bacteria may be inhibited, or (3) host cells may be killed enabling *T. viride* to colonize faster. It is also possible that production of phenols may be increased which could favor *T. viride*.

The addition of a PVC wrap to wounds was shown by Shortle and Shigo[100] and Mercer and Kirk[67] to reduce colonization by basidiomycetes. Although natural colonization by *T. viride* was encouraged by the addition of PVC it did not prove possible to correlate this with basidiomycete reduction. Natural colonization by *T. viride* generally proved inferior in controlling basidiomycetes to artificial inoculants, possibly because of strain selection or the swamping effect of large numbers of propagules applied immediately following wounding. It is possible that the effect of PVC and of the more long-lasting wound paints is to keep the moisture content of the wood sufficiently high to discourage growth of basidiomycetes. Die-back of host tissue may also be reduced and if an active host response is important this may be more readily realized. Improvement in active host response is suggested by the reduction of decay fungi by the addition of *Agrobacterium tumefaciens*[5] but at present there is no clear idea how this is accomplished.

XVI. TECHNIQUES AND PROBLEMS IN BIOLOGICAL CONTROL OF DECAY

Unlike the control of diseases of crop plants, the control of decay has to be very long-term — a matter of years if not decades. This makes research on possible biocontrol agents potentially difficult. It is clearly not practical to do long-term field trials with a large number of microorganisms; some preliminary screening is necessary. However, there is a problem of the translation of laboratory results to the field. Corke and Hunter,[23] Lundeborg and Unestam,[57] Pratt,[74] and Hallaksela[40] all found good laboratory control of decay fungi was not necessarily translated into control under field conditions, and Stillwell[105] noted good control of basidiomycetes in peeled birch logs but not in unpeeled ones. Mercer and Kirk[66,67] used three laboratory tests to screen fungi and bacteria. The initial screen was on malt agar, followed by wood strips on malt agar, and finally wood strips on nutrient-free beds of glass beads. Candidates from the final test were then used in field trials. Those microorganisms that performed well in the field had also performed well in laboratory tests, but the converse was not necessarily true. For fungal biocontrol agents the best correlation with field trials was with the final wood strip test. On the other hand, although a *Bacillus* sp. isolate performed relatively well in this test, it totally failed to maintain itself in field trials, a point also noted by Swinburne et al.[109] with *B. subtilis* in apple tree bark. As a preliminary screening is necessary it is best that conditions should be as natural as possible. The high nutrient status of malt agar in Petri dish tests can seriously distort antagonistic effects. Testing on wood should be carried out at some stage, and several methods have been described, e.g., Leben[55] devised a test on surface-sterilized wood discs which provides a relatively simple and effective method for large scale screening.

There still remains the possibility of microorganisms not performing well in laboratory tests but doing so in field trials. Mercer and Kirk[67] tested two such fungi — *Alternaria alternata* and *Aureobasidium pullulans*, but results were inconclusive because of insufficient colonization by decay fungi. This highlights another problem — the artificial inoculation of decay fungi as against natural infection. All the field trials by Mercer et al[65] relied on natural infection and in the main part of the work there was sufficient colonization to show significant effects of biocontrol agents. Pottle et al.[73] also showed significant effects with natural infection. It would be preferable if this could be done in all field trials but there is the obvious problem of not knowning for several years whether or not it has been successful.

Another factor raised by Seaby[87] is the performing of biocontrol experiments at room temperature in the laboratory and then extrapolating results into the field. He noted large differences between spring and winter inoculations of *Trichoderma viride* to control *Heterobasidion annosum*. There were also suggestions in the work of Pottle et al,[73] that control of basidiomycetes in Red maple wounds by *T. harzianum* was more effective when inoculations were made in the summer than in the winter. On the other hand, Woodgate-Jones[119] found no differences in inoculation times with *T. viride* against *Chondrostereum purpureum* and Mercer and Kirk (unpublished) showed no differences in beech wounds with inoculations of *T. viride* throughout the year. It is possible, however, that factors other than temperature, such as host metabolism and sap flow, which vary with time could also affect the balance of control.

The amount of inoculum of a biocontrol agent may also be important, especially if control is by a swamping effect. However, Smith et al.[103] showed no effect of the spore load of *T. harzianum* in wounds of Red maple.

Much of the work on the biological control of disease like the work on fungicidal control, has a large empirical element. Some groups such as *Trichoderma* have been shown to have above average controlling properties and for that reason some workers have concentrated on this group. Mercer and Kirk[66] attempted to increase the number of biocontrol agents by a wide-scale screening but, in spite of this, found *Trichoderma* isolates to be the most successful in controlling decay fungi in wounds. Other groups of microorganisms may yet be discovered to have good controlling properties, but at present there are only about five. With fungicidal control too, there is often reliance on only a small number of chemically related substances, e.g., in the control of cereal mildew at present only two groups of fungicides are being and this has led to resistance problems. Whether the widespread use of a small number of biocontrol agents would also result in resistance is purely a matter of conjecture and would probably only be resolved by a better understanding of the mechanisms of biological control.

XVII. CONCLUSIONS OF PRESENT WORK AND THE FUTURE OF BIOLOGICAL CONTROL OF DECAY

Biological control has been shown to be an effective alternative to chemical control for some decay situations, e.g., the control of *Heterobasidion annosum*. In other situations, such as the control of decay fungi in wounds, biological control is superior to chemical control, while in the control of *Chondrostereum purpureum* in silver-leaf disease, biological control is the only effective treatment. Decay can also be limited by good practices (biological control in a wider sense), e.g., in timber, the use of naturally resistant wood and the maintenance of dry conditions, and in trees, correct pruning while the tree is young and also the maintenance of vigor.

Future work could include a search for more possible candidates for biological control agents. Combinations of microorganisms may also be effective. It would also be beneficial, however, to obtain a better understanding both of the mechanism of biological control and of the host response. This would enable better use to be made of natural host resistance and of naturally occurring organisms, and also of maximizing the establishment and controlling activities of artificially applied inoculum. Novel combinations of chemical and biological control may also be devised.

REFERENCES

1. **von Aufsess, H.,** The formation of a protective zone at the base of branches of broadleaved and coniferous trees and its effectiveness in preventing fungi from penetrating into the heartwood of living trees, *Forstwiss. Centralbl.,* 94, 140, 1975.
2. **von Aufsess, H.,** On the effect of different antagonists on the mycelial growth of some wood destroying fungi, *Mater. Org.,* 11, 183, 1976.
3. **Baker, K. F. and Cooke, R. J.,** *Biological Control of Plant Pathogens,* W. H. Freeman & Co., San Francisco, 1974, 433.
4. **Biggs, A. R., Merrill, W., and Davis, D. D.,** Discussion: responses of bark tissues to injury and infection, *Can. J. For. Res.,* 14, 351, 1984.
5. **Blanchette, R. A. and Sharon, E. M.,** *Agrobacterium tumefaciens,* a promoter of wound healing in *Betula alleghaniensis, Can. J. For. Res.,* 5, 722, 1975.
6. **Bliss, D. E.,** The destruction of *Armillaria mellea* in citrus soils, *Phytopathology,* 41, 665, 1951.
7. **Boddy, L. and Rayner, A. D. M.,** Population structure, intermycelial interaction and infection biology of *Stereum gausapatum, Trans. Br. Mycol. Soc.,* 78, 337, 1982.
8. **Boddy, L. and Rayner, A. D. M.,** Ecological roles of basidiomycetes forming decay communities in attached oak branches, *New Phytol.,* 93, 77, 1983.
9. **Boddy, L. and Rayner, A. D. M.,** Origins of decay in living deciduous trees: The role of moisture content and a re-appraisal of the expanded concept of tree decay, *New Phytol.,* 94, 623, 1983.
10. **Boddy, L. and Thompson, W.,** Decomposition of suppressed oak trees in even-aged plantations. I. Stand characteristics and decay of aerial parts, *New Phytologist,* 93, 261, 1983.
11. **Brown, G. E.,** *The Pruning of Trees, Shrubs and Conifers,* Faber & Faber, London, 1972, 351.
12. **Bruce, A. and King, B.,** Biological control of wood decay by *Lentinus lepideus* Fr. produced by *Scytalidium* and *Trichoderma* residues, *Mater. Org.,* 18, 17, 1983.
13. **Bruce, A., Austin, W. J., and King, B.,** Control of growth of *Lentinus lepideus* by volatiles from *Trichoderma, Trans. Br. Mycol. Soc.,* 82, 423, 1984.
14. **Burmeister, H. R., Ellis, J. J., and Vesonder, R. F.,** Survey for *Fusaria* that produce an antibiotic that causes conidia of *Penicillium digitatum* to swell, *Mycopathologia,* 74, 29, 1981.
15. **von Butin, H. and Kowalski, T.,** Die natürliche Astreinigung und ihre biologischen Vorausetzungen I. Die Pilzflora der Buche (*Fagus sylvatica* L.), *Eur. J. For. Pathol.,* 13, 322, 1983.
16. **von Butin, H. and Kowalski, T.,** Die natürliche Astreinigung und ihre biologischen Vorausetzungen II. Die Pilzflora der Stieleiche (*Quercus robur* L.), *Eur. J. For. Pathol.,* 13, 428, 1983.
17. **Byrde, R. J. W., Evans, S. G., and Rennison, R. W.,** The control of apple canker in two Somerset orchards by a copper spray programme, *Plant Pathol.,* 14, 143, 1965.
18. **Carter, M. V.,** Biological control of *Eutypa armeniacae, Aust. J. Exp. Agric. Anim. Husb.,* 11, 687, 1971.
19. **Carter, M. V. and Price, T. V.,** Biological control of *Eutypa armeniacae.* III. A comparison of chemical, biological and integrated control, *Aust. J. Agric. Res.,* 26, 537, 1975.
20. **Cartwright, K. St. G. and Findlay, W. P. K.,** *Decay of Timber and its Prevention.* Her Majesty's Stationery Office, London, 1946, 294.
21. **Copony, J. A. and Barnes, B. U.,** Clonal variation in the incidence of *Hypoxylon* canker in trembling aspen, *Can. J. Bot.,* 52, 1475, 1974.
22. **Corke, A. T. K.,** *Rep. Long Ashton Res. Stn. 1979,* Bristol, 1980, 190.
23. **Corke, A. T. K. and Hunter, T.,** Biocontrol of *Nectria galligena* infection of pruning wounds on apple shoots, *J. Hort. Sci.,* 54, 47, 1979.
24. **Dennis, C. T. and Webster, J.,** Antagonistic properties of species groups of *Trichoderma.* I. Production of non-volatile antibiotics, *Trans. Br. Mycol. Soc.,* 57, 25, 1971.
25. **Dennis, C. T. and Webster, J.,** Antagonistic properties of species groups of *Trichoderma.* II. Production of volatile antibiotics, *Trans. Br. Mycol. Soc.,* 57, 41, 1971.
26. **Dennis, C. T. and Webster, J.,** Antagonistic properties of species groups of *Trichoderma.* III. Hyphal interaction, *Trans. Br. Mycol. Soc.,* 57, 363, 1971.

27. **Deschamps, A. M.,** Nutritional capacities of bark and wood decaying bacteria with particular emphasis on condensed tannin degrading strains, *Eur. J. For. Pathol.,* 12, 252, 1982.

28. **Dickinson, D. J.,** *Decomposer Basidiomycetes,* 4th Br. Mycol. Symp., Cambridge University Press, London, 1982, 179.

29. **Dubos, B. and Ricard, J. L.,** Curative treatment of peach trees against silver leaf disease (*Stereum purpureum*) with *Trichoderma viride* preparations, *Plant Dis. Rep.,* 58, 147, 1974.

30. **Duncan, C. G. and Deverall, F. J.,** Degradation of wood preservatives by fungi, *Appl. Microbiol.,* 12, 57, 1964.

31. **Ferrera Cerrato, R.,** Hyperparasitism of *Trichoderma viride* (Hyphomycetes) on phytopathogenic and saprophytic fungi, *Rev. Latinoam. Microbiol.,* 18, 77, 1976.

32. **Fisher, P. J., Anson, A. E., and Petrini, O.,** Novel antibiotic activity of an endophytic *Cryptosporiopsis* sp. isolated from *Vaccinium myrtillus, Trans. Br. Mycol. Soc.,* 83, 145, 1984.

33. **French, J. R. and Manion, P. D.,** Variability of host and pathogen in *Hypoxylon* canker of aspen, *Can. J. Bot.,* 53, 2740, 1975.

34. **Gendle, P., Clifford, D. R., Mercer, P. C., and Kirk, S. A.,** Movement, persistence and performance of fungitoxicants applied as pruning wound treatment on apple trees, *Ann. Appl. Biol.,* 102, 281, 1983.

35. **Gibbs, J. N.,** A study of the epiphytic growth habit of *Fomes annosus, Ann. Bot.,* 31, 755, 1967.

36. **Gibbs, J. N.,** An oak canker caused by a gall midge, *Forestry,* 55, 69, 1982.

37. **Greig, B. J. W.,** Inoculation of pine stumps with *Peniophora gigantea* by chainsaw felling, *Eur. J. For. Pathol.,* 6, 286, 1976.

38. **Grosclaude, C.,** Premier éssais de protection biologique de blessure de taille vis à vis du *Stereum purpureum* Pers., *Ann. Phytopathol.,* 2, 507, 1970.

39. **Grosclaude, C., Ricard, J. L., and Dubos, B.,** Inoculation of *Trichoderma viride* spores via pruning shears for biological control of *Stereum purpureum* on plum tree wounds, *Plant Dis. Rep.,* 57, 25, 1973.

40. **Hallaksela, A.-M.,** Bacteria and their effect on the microflora in wounds of living Norway spruce (*Picea abies*), *Comm. Int. Forestalis Fenniae,* 121, 25, 1984.

41. **Hartig, R.,** *Die Zersetzungerscheinungen des Holzes der Nadelholzbaume und der Eiche in forstlicher, botanischer und chemischer Richtung,* Springer Verlag, Berlin, 1878.

42. **von Holdenrieder, O.,** Untersuchungen zur biologischen Bekämpfung von *Heterobasidion annosum* an Fichte (*Picea abies*) mit antagonistischen Pilzen II. Interaktionstests auf Holz, *Eur. J. For. Pathol.,* 14, 137, 1984.

43. **Hulme, M. A. and Shields, J. K.,** Biological control of decay fungi in wood by competition for non-structural carbohydrates, *Nature,* 227, 300, 1970.

44. **Hulme, M. A. and Shields, J. K.,** Effect of primary fungal infection upon secondary colonisation of birch bolts, *Mater. Org.,* 7, 177, 1972.

45. **Hwang, B. K. and Heitefuss, R.,** Induced resistance of spring barley to *Erysiphe graminis* f. sp. *hordei, Phytopathol. Z.,* 103, 41, 1982.

46. **Johansson, M. and Unestam, T.,** The search for resistance to *Heterobasidion* root rot in Norway spruce — old and new approaches in studies of infection biology, *Eur. J. For. Pathol.,* 12, 346, 1982.

47. **Johnson,** *Trans. Soc. Encouragement Arts, Manufacturers, Commerce,* 21, 286, 1803.

48. **Jones, K. G., Morgan, N. G., and Corke, A. T. K.,** *Rep. Long Ashton Res. Stn. 1974,* Bristol, 1975, 107.

49. **Kallio, T.,** Protection of spruce stumps against *Fomes annosus* (Fr.) Cooke by some wood inhabiting fungi, *Acta For. Fenn.,* 117, 20, 1971.

50. **Kapoor, I. J. and Hoffman, G. M.,** Antagonistic effects of soil microbes on *Rhizoctonia solani* and *Rhizoctonia*-like fungi (*Ceratobasidium* sp.) associated with foot rot of cereals, *Z. Pflanzenkr. Pflanzenschutz,* 91, 186, 1984.

51. **Klingström, A. E. and Johansson, S. M.,** Antagonism of *Scytalidium* isolates against decay fungi, *Phytopathology,* 63, 473, 1973.

52. **Komatsu, M. and Inada, S.,** *T. viride* as an antagonist of the wood inhabiting Hymenomycetes IX. Antifungal action of *Trichoderma, Gliocladium* and other species of the *Hypocrea* to *Lentinus edodes* (Berk) Sing., *Rep. Tottori Mycol. Inst.,* 7, 19, 1964.

53. **Landy, M., Warren, G. H., Roseman, S. B., and Colis, L. G.,** Bacillomycin. An antibiotic from *Bacillus subtilis* active against pathogenic fungi, *Proc. Soc. Exp. Biol.,* 67, 539, 1948.

54. **Leach, R.,** Biological control and ecology of *Armillaria* (Vahl.) Fr., *Trans. Br. Mycol. Soc.,* 23, 320, 1939.

55. **Leben, C.,** Biological control of decay fungi — A wood disk evaluation method, *For. Sci.,* 24, 560, 1978.

56. **Loeffler, W., Tschen, J. S.-M., Vanittanakom, N., Kugler, M., Knorpp, E., Hsieh, T. F. and Wu, T.-G.,** Antifungal effects of Bacilysin and Fengymycin from *Bacillus subtilus,* F-29-3. A comparison with activities of other *Bacillus* antibiotics, *J. Phytopathol.,* 115, 204, 1986.

57. **Lundeborg, A. and Unestam, T.,** Antagonism against *Fomes annosus.* Comparison between different test methods *in vitro* and *in vivo, Mycopathologia,* 70, 107, 1981.

58. **Magro, P., Di Lenna, P., and Marciano, P.,** *Trichoderma viride* on cypress shoots and antagonistic action against *Seiridium cardinale, Eur. J. For. Pathol.,* 14, 165, 1984.
59. **Mercer, P. C.,** Attitudes to pruning wounds, *Arboric. J.,* 3, 457, 1979.
60. **Mercer, P. C.,** *Decomposer Basidiomycetes,* 4th Br. Mycol. Soc. Symp., Cambridge University Press, London, 1982, 143.
61. **Mercer, P. C.,** Tree wounds and their treatment, *Arboric. J.,* 6, 131, 1982.
62. **Mercer, P. C.,** Callus growth and the effect of wound dressings, *Ann. Appl. Biol.,* 103, 527, 1983.
63. **Mercer, P. C.,** An investigation of conditions occurring within pruning wounds, *Eur. J. For. Pathol.,* 14, 1, 1984.
64. **Mercer, P. C.,** The effect on beech of bark-stripping by grey squirrels, *Forestry,* 57, 199, 1984.
65. **Mercer, P. C., Kirk, S. A., Gendle, P., and Clifford, D. R.,** Chemical treatments for the control of decay in pruning wounds, *Ann. Appl. Biol.,* 102, 435, 1983.
66. **Mercer, P. C. and Kirk, S. A.,** Biological treatments for the control of decay in tree wounds. I. Laboratory tests, *Ann. Appl. Biol.,* 104, 211, 1984.
67. **Mercer, P. C. and Kirk, S. A.,** Biological treatments for the control of decay in tree wounds II. Field tests, *Ann. Appl. Biol.,* 104, 221, 1984.
68. **Mullick, D. B.,** The nonspecific nature of defense in bark and wood during wounding, insect and pathogen attack, *Recent Adv. Phytochem.,* 11, 395, 1977.
69. **Nelson, E. E. and Thies, W. G.,** Colonisation of *Phellinus weirii*-infested stumps by *Trichoderma viride* I. Effect of isolate and inoculum base, *Eur. J. For. Pathol.,* 15, 425, 1985.
70. **Nelson, E. E. and Thies, W. G.,** Colonisation of *Phellinus weirii*-infested stumps by *Trichoderma viride* II. Effects of season of inoculation and stage of wood decay, *Eur. J. For. Pathol.,* 16, 56, 1986.
71. **Papavizas, G. C., Lewis, J. A., and Abd-El Moity, T. H.,** Evaluation of new biotypes of *Trichoderma harzianum* for tolerance to benomyl and enhanced biocontrol capabilities, *Phytopathology,* 72, 126, 1982.
72. **Pearce, R. B. and Rutherford, J.,** A wound-associated suberized barrier to the spread of decay in the sapwood of oak (*Quercus robur* L.), *Physiol. Plant Pathol.,* 19, 359, 1981.
73. **Pottle, H. W., Shigo, A. L., and Blanchard, R. O.,** Biological control of wound hymenomycetes by *Trichoderma harzianum, Plant Dis. Rep.,* 61, 687, 1977.
74. **Pratt, J. E.,** *Fomes annosus* butt rot of Sitka spruce IV. Observations on the distribution of *Cryptosporiopsis abietina* (Petrak) in the stems of rotted trees, *Forestry,* 55, 183, 1982.
75. **Preston, A. F., Erbisch, F. H., Kramm, K. R., and Lund, A. E.,** Developments in the use of biological control for wood preservation, *Proc. Am. Wood Preservers' Asso. 1982,* 78, 53, 1982.
76. **Rayner, A. D. M. and Todd, N. K.,** *Decomposer Basidiomycetes,* 4th Br. Mycol. Symp., Cambridge University Press, London, 1982, 109.
77. **Ricard, J. L.,** Biological control of decay in Douglas fir poles — seven years' perspective, *Eur. J. For. Pathol.,* 5, 175, 1975.
78. **Ricard, J.,** Biological control of decay in standing creosote-treated poles, *J. Inst. Wood Sci.,* 7, 6, 1976.
79. **Ricard, J. L. and Bollen, W. B.,** Inhibition of *Poria carbonica* Overh. by *Scytalidium* sp., an imperfect fungus isolated from Douglas fir poles, *Can. J. Bot.,* 46, 643, 1968.
80. **Rishbeth, J.,** Observations on the biology of *Fomes annosus* with particular reference to East Anglian pine plantations III. Natural and experimental infection of pines and some factors affecting severity of the disease, *Ann. Bot.,* 58, 221, 1951.
81. **Rishbeth, J.,** Control of *Fomes annosus* Fr., *Forestry,* 25, 41, 1952.
82. **Rishbeth, J.,** *Fomes annosus* on stumps, *Trans. Br. Mycol. Soc.,* 40, 167, 1957.
83. **Rishbeth, J.,** Stump protection against (fomes annosus) III. Inoculation with *Peniophora gigantea, Ann. Appl. Biol.,* 52, 63, 1963.
84. **Ruehle, J. L.,** Nematodes, the overlooked enemies of tree roots, *Proc. Int. Shade Tree Conf.,* 40, 60, 1964.
85. **Saccardo, P. A.,** *Sylloge fungorum,* 2, 469, 1883.
86. **Schuck, H. J.,** The chemical composition of the monoterpene fraction in wounded wood of *Picea abies* and its significance for the resistance against wound infecting fungi, *Eur. J. For. Pathol.,* 12, 175, 1982.
87. **Seaby, D. A.,** The possibility of using *Trichoderma viride* for the control of *Heterobasidion annosum* on conifer stumps, Proc. Semin. Biological Control, Royal Irish Academy, Dublin, 1977, 57.
88. **Seaby, D. A.,** Annu. Rep. Res. Tech. Work 1983, Department of Agriculture, Belfast, N. Ireland, 1984, 204.
89. **Seaby, D. A.,** Annu. Rep. Res. Tech. Work 1984, Department of Agriculture, Belfast, N. Ireland, 1985, 210.
90. **Shain, L.,** Resistance of sapwood in stems of Loblolly pine to infection by *Fomes annosus, Phytopathology,* 57, 1034, 1967.
91. **Shigo, A. L.,** Ring shake associated with sapsucker injury, in *U.S. Forest Service Research Paper, Northeastern,* No. 8, Upper Darby, Pa., 1963.

92. **Shigo, A. L.,** Tree decay an expanded concept, in *U.S. Dept. Agric. For. Ser. Inf. Bull.*, 419, 73, 1979.
93. **Shigo, A. L.,** Branches, *J. Arboric.*, 6, 300, 1980.
94. **Shigo, A. L.,** Tree defects — a photo guide, in *Northeastern For. Exp. Stn. Gen. Tech. Rep.*, U.S. Dept. Agric., NE 82, 167, 1983.
95. **Shigo, A. L.,** How tree branches are attached to trunks, *Can. J. Bot.*, 63, 1391, 1985.
96. **Shigo, A. L. and Marx, H. G.,** Compartmentalization of decay in trees, *U.S. Dept. Agric. For. Ser. Inf. Bull.*, 405, 73, 1977.
97. **Shigo, A. L. and Wilson, C. L.,** Are tree wound dressings beneficial?, *Arborist's News*, 36, 85, 1971.
98. **Shigo, A. L., Shortle, W. C., and Garrett, P. W.,** Genetic control suggested in compartmentalization of discolored wood associated with tree wounds, *For. Sci.*, 23, 179, 1977.
99. **Shortle, W. C. and Cowling, E. B.,** Development of discoloration, decay and microorganisms following wounding of Sweetgum and Yellow poplar trees, *Phytopathology*, 68, 609, 1978.
100. **Shortle, W. C. and Shigo, A. L.,** Effect of plastic wrap on wound closure and internal compartmentalization of diseased and decayed wood in Red maple, *Plant Dis. Rep.*, 62, 999, 1978.
101. **Shortle, W. C., Menge, J. A., and Cowling, E. B.,** Interaction of bacteria, decay fungi and live sapwood in discoloration and decay of trees, *Eur. J. For. Pathol.*, 8, 293, 1978.
102. **Sierota, Z. H.,** Inhibiting effect of *Trichoderma viride* Pers. ex Fr. filtrates on *Fomes annosus* (Fr.) Cke. in relation to some carbon sources, *Eur. J. For. Pathol.*, 7, 164, 1977.
103. **Smith, K. T., Blanchard, R. O., and Shortle, W. C.,** Effect of spore load of *Trichoderma harzianum* on wood-invading fungi and volume of discolored wood associated with wounds in *Acer rubrum*, *Plant Dis. Resp.*, 63, 1070, 1979.
104. **Smith, K. T., Blanchard, R. O., and Shortle, W. C.,** Postulated mechanism of biological control of decay fungi in Red maple wounds treated with *Trichoderma harzianum*, *Phytopathology*, 71, 496, 1981.
105. **Stillwell, M. A.,** A growth inhibitor produced by *Cryptosporiopsis* sp. an imperfect fungus isolated from yellow birch *Betula alleghaniensis* Britt., *Can. J. Bot.*, 44, 259, 1965.
106. **Stillwell, M. A., Wood, F. A., and Strunz, G. M.,** A broad-spectrum antibiotic produced by a *Cryptosporiopsis* sp. against a range of basidiomycetes and other fungi, *Can. J. Microbiol.*, 15, 501, 1965.
107. **Swinburne, T. R.,** Microflora of apple leaf scars in relation to infection by *Nectria galligena*, *Trans. Br. Mycol. Soc.*, 60, 389, 1973.
108. **Swinburne, T. R. and Brown, A. E.,** A comparison of the use of *Bacillus subtilis* with conventional fungicides for the control of apple canker (*Nectria galligena*), *Ann. Appl. Biol.*, 82, 365, 1976.
109. **Swinburne, T. R., Barr, J. G., and Brown, A. E.,** Production of antibiotics by *Bacillus subtilis* and their effect on fungal colonists of apple leaf scars, *Trans. Br. Mycol. Soc.*, 65, 211, 1975.
110. **Swinburne, T. R., Cartwright, J., Flack, N. J., and Brown, A. E.,** A comparison of phenylmercuric nitrate, mbc generators and *Bacillus subtilis* for the control of autumn infections by *Nectria galligena* in apple orchards, *Rec. Agric. Res.*, (Department of Agriculture for Northern Ireland), 25, 57, 1977.
111. **Tippett, J. T., Bogle, A. L., and Shigo, A. L.,** Response of balsam fir and hemlock roots to injuries, *Eur. J. For. Pathol.*, 12, 357, 1982.
112. **Toole, E. R.,** Root rot caused by *Polyporus lucidus*, *Plant Dis. Rept.*, 50, 945, 1966.
113. **Wagener, W. W. and Cave, M. S.,** Pine killing by the root fungus *Fomes annosus* in California, *J. For.*, 44, 47, 1946.
114. **Wakayama, S., Ishikawa, F., and Oishi, K.,** Mycocerein, a novel antifungal peptide antibiotic produced by *Bacillus cereus*, *Antimicrob. Agents Chemother.*, 26, 939, 1984.
115. **Weindling, R.,** *Trichoderma lignorum* as a parasite of other soil fungi, *Phytopathology*, 22, 837, 1932.
116. **Wells, H. D. and Bell, D. K.,** Variable antagonistic reaction *in vitro* of *Trichoderma harzianum* against several pathogens, *Phytopathology*, 69, 1048, 1979.
117. **Whitney, H. S. and Cobb, F. W.,** Non-staining fungi associated with bark beetles on Ponderosa pine, *Can. J. Bot.*, 50, 1943, 1972.
118. **Wilström, C.,** The decay pattern of *Phellinus tremulae* (Bond.) Bond. et Borisov in *Populus tremula* L., *Eur. J. For. Pathol.*, 6, 291, 1976.
119. **Woodgate-Jones, P.,** *Rep. Long Ashton Res. Stn. 1979*, Bristol, 1980, 129.

INDEX

Verticillium chlamydosporium, 121
Verticillium dahliae, 154
  control of
    amoebae and, 8
    host-pathogen-environment interactions, 168
    production and use of agents against, 23
    soil treatments, 41
    weed control with, 144
Verticillium lamellicola, 122
Verticillium lecanii, 46
  activity of, 62, 63
  mass production and use, 25
Verticillium leptobactrum, 122
Verticillium psalliotae, 62
Verticillium wilt, 7, 8
Vesicular arbuscular mycorrhiza, 8
Vicia sativa, 158
Vigna sinensis, 158
Vinclozolin, 27
Viridin, 74
Virulence, 6, 8
Viruses, 3—4
  insect, see Baculoviruses
  leaf surface chemicals, 3
  nematode control, 115—116
  weed control, 143
Virus-like molecule, and virulence, 6

W

Water/moisture, 7
  bacterial control agents and, 89
  sclerotial disease control, 156
  tree tissue content, 181, 184
  weed control, 147
Weak pathogens, 44
Weeds
  fungal pathogens, 45—46
  fungal pathogens, classical strategy, 131—138
    blackberry rust, 136—137
    diseases, 132
    effectiveness, assessment of, 132
    host specificity, 132—133
    pamakani leaf spot, 137—138
    selection paf agent, 132
    skeleton weed rust, 133—136
  fungal pathogens, development of methods, 138—

139
  integrated control, 139—142
  present status, 142—146
  strategies for control, 131
  suitability of, for biocontrol, 130—131
Western hemlock, 180
Wood, 25, see also Trees
  application methods, 26, 42
  pathogens, production and use of control agents,
    22, 23
Wood decay fungi, see also Trees
  biological control, 178—179
    application, 184
    conclusions and future prospects, 194
    host factors enhancing control, 190—191
    mechanisms of action of agents, 76, 191—193
    pathogenic fungi in trees, 186—187
    in stumps, 185—186
    techniques and problems, 193—194
    in timber, 184—185
    in wounds, 187
  decay process, 179
  environmental alterations, 187—188
  interactions with other microorganisms, 182—183
  situations for decay, 179
  timber, decay in, 183—184
  tree reaction, 179—181
Wounding
  biological barriers, 2
  decay of wood and, 179
  fungal control agents, 42—43
    treatment, 190
    Trichoderma, 76, 78
  pathogens, 26

X

Xanthium, 144
Xanthium strumarium, 61
Xanthomonads, 84
Xanthomonas campestris pv. *vesicatoria*, 5
Xylan, 58

Y

Yeasts, 61
Yellow poplar, 183

Printed in the United States
by Baker & Taylor Publisher Services